Field and
Laboratory Methods
for
General Ecology

Third Edition

Field and
Laboratory Methods
for
General Ecology

Third Edition

James E. Brower
Brookhaven National Laboratory

Jerrold H. Zar
Northern Illinois University

Carl N. von Ende
Northern Illinois University

WCB Wm. C. Brown Publishers

Book Team
Editor *Kevin Kane*
Developmental Editor *Marge Manders*
Production Coordinator *Kay Driscoll*

 Wm. C. Brown Publishers

President *G. Franklin Lewis*
Vice President, Publisher *George Wm. Bergquist*
Vice President, Publisher *Thomas E. Doran*
Vice President, Operations and Production *Beverly Kolz*
National Sales Manager *Virginia S. Moffat*
Advertising Manager *Ann M. Knepper*
Marketing Manager *Craig S. Marty*
Editor in Chief *Edward G. Jaffe*
Production Editorial Manager *Colleen A. Yonda*
Production Editorial Manager *Julie A. Kennedy*
Publishing Services Manager *Karen J. Slaght*
Manager of Visuals and Design *Faye M. Schilling*

Some of the laboratory experiments included in this text may be
hazardous if materials are handled improperly or if procedures are
conducted incorrectly. Safety precautions are necessary when you
are working with chemicals, glass test tubes, hot water baths,
sharp instruments, and the like, or for any procedures that
generally require caution. Your school may have set regulations
regarding safety procedures that your instructor will explain to you.
Should you have any problems with materials or procedures,
please ask your instructor for help.

Cover image © 1989 Woody Logsdan

Cover design by John R. Rokusek

Library of Congress Catalog Card Number: 89-61102

ISBN 0-697-05145-5

Printed in the United States of America by Wm. C. Brown Publishers,
2460 Kerper Boulevard, Dubuque, IA 52001

10   9   8   7

contents

preface

The subject of ecology is that of organisms, and groups of organisms, interacting with their biotic and abiotic environment. This book is a compendium of field and laboratory procedures used by ecologists to describe and analyze plant and animal aggregations and their environments. That the title refers to *general* ecology emphasizes that the orientation is not from a plant, animal, or physical factor point of view exclusively, nor does it center on specific geographic areas. Rather, the book stresses *interactions* among biotic entities, and between them and abiotic factors.

This book is designed for use in an introductory course in ecological methods, or in conjunction with a beginning ecology text. Ecological terms are defined and concepts developed as they appear. No previous ecological knowledge is assumed. However, an elementary biological background is helpful; and a few references to elementary chemistry will appear. Quantitative procedures are included where appropriate; but only a knowledge of introductory algebra is needed to deal with them. The coverage reflects the breadth and vigor of contemporary ecological investigation, as well as the classical work that led to them. If a portion of the book considers chemical analysis or mathematical manipulation, it is because ecological study has need of them; but the thrust of the material is nonetheless biological.

Each section consists of one or more methods for the study of a given ecological topic. Rather than presenting "cookbook" exercises and blanks to fill with data, we show the principles and practice of each procedure, explaining how to collect, record, and analyze data, with remarks on possible sources of error. Suggested exercises at the end of sections can guide instructors and students to formulate appropriate field and laboratory studies. Examples of data collection and analysis are given throughout, to enable students to handle their own observations.

We recommend that you study sections 1A and 1C (ecological sampling and report writing) first and refer to them frequently during subsequent sections. Section 1B on statistical analysis presents elementary but ecologically important considerations. (It may, however, be deemed unsuited to some introductory ecology courses. This discussion, along with references to statistical testing elsewhere in the book, may be omitted without detracting greatly from the rest of the material.)

The sequence of study of units 2 through 6 is not prescribed; it may follow the direction of the instructor or textbook. For example, studies in unit 4, 5, and 6 (on populations, communities, and production) may logically follow or accompany experiences in unit 3 (on biotic sampling); and parts of unit 2 (on habitat analysis) may be studied in conjunction with portions of units 3 through 6. Nowhere do the discussions depend on a specific flora or fauna, thus providing considerable flexibility to both instructor and student in applying the procedures presented.

In an effort to offer our readers the most effective and balanced coverage of ecological theory and technique, some new features have been implemented in the third edition. Unit four has an added discussion of predation, which was written by Carl von Ende, a specialist in population ecology. The discussion of population growth in this section was also significantly rewritten for greater clarity and comprehension. In response to the considerable popularity of the computer programs, three new programs have been added in the areas of spatial distribution, species diversity, and community similarity. The reference sections throughout the book have also been updated to offer current information on ecological studies.

Drafts and previous editions of this work have been used by general ecology students for several years, and their valuable remarks and criticisms have helped us develop the final form of this book. We are also indebted to the computer-services facilities at Northern Illinois University, with which some of the book's original tables and graphs were produced. Forty-three of the most demanding line drawings were expertly prepared by Jane K. Glaser and Carol Garner.

J. E. Brower
J. H. Zar
C. N. von Ende

introduction

Just as in many other areas of science, quantitative techniques in ecological methods have become more sophisticated in the last few decades. Ecological investigations today seldom end with descriptive surveys; they are more problem oriented. Older concepts and theories are being evaluated and improved in ways that previously were impossible or impractical.

This unit is a foundation on which ecological studies may be designed, the data analyzed, and the results reported. An ecological study, like any scientific study, involves the following processes:

1. Perceiving the problem
2. Defining the entity or entities to be studied
3. Designing an experiment or study
4. Selecting a sampling (i.e., data-collection) procedure
5. Obtaining representative samples
6. Observing and measuring the samples to obtain data
7. Objectively analyzing the data
8. Interpreting and drawing conclusions from the data
9. Reporting the findings

Defining the problem, designing the experiment, and drawing conclusions involve mental processes based on perception, experience, reasoning, and creativity. Although statistical considerations are helpful in designing studies and drawing conclusions from data, perception of the problem is one aspect of science that is almost solely dependent on the intuitive and creative processes of the human mind. Without this process steps 2 through 9 would have little value. Although exercises have been suggested throughout this book, we have seldom tried to define specific problems, to give specific experimental designs, or to interpret possible results. These processes are left to the student and instructor. This unit is designed as a guide to be used in conjunction with any of the other units in the book.

unit 1

collecting, analyzing, and reporting ecological data

1. Introduction

Ecologists generally wish to collect quantitative information about a habitat, community, or population, for quantitative data may allow objective and illuminating presentation, summary, and interpretation of ecological phenomena. However, it usually is impossible or impractical to monitor the entire habitat or to obtain measurements of all the organisms in a given area. Biologists rarely can collect all of the data about which they wish to draw conclusions. For example, it may be desired to draw conclusions about the body weights of all mice in a particular habitat. The only way to make statements about the weights of all mice with 100 percent confidence would be to weigh every mouse, probably an impossible task. Instead, only some of the total number of mice are weighed, and we can then infer from this portion of the total the weights of all the mice. The entire set of data of interest (i.e., the weights of all of the mice) is called a *statistical population,* and the actually-measured portion, or subset, of the population is a *statistical sample.*

Established sampling procedures exist for obtaining information about organisms and their environment. In this section we shall deal with the general principles of sampling underlying the specific techniques of sampling habitats and biological populations given in units 2 and 3. The theoretical bases for ecological sampling procedures may be found in such texts as Grieg-Smith (1983), Pielou (1977), Poole (1974), Seber (1982), and Southwood (1978).

A statistical population is that entire set of data about which one wishes to draw conclusions. This is not to be confused with a *biological population,* which is the aggregation of individual organisms of a single species inhabiting a given area. A statistical population, then, is an entire set of measurements from a habitat, a community, a biological population, or a portion of a biological population. Though a statistical sample is a portion of a larger set of data (the statistical population), a *physical sample* is a portion, or subset, of a collection of one or more material objects, either biotic or abiotic. As an example of physical sampling, we can take a 1-liter sample of pond water (meaning we collected a portion of the entire volume of water in the pond), or a sample of vegetation from a forest (i.e., a small portion of all the forest vegetation), or a sample of 100 mice from an entire biological population of that species. A statistical sample, on the other hand, refers to a collection of data such as measurements of the temperature or phosphate content of pond water, the biomass of vegetation, or the tail lengths of mice.

When collecting samples in an ecological study, one *must* know what natural entity is being sampled. A particular study may require a precise definition of the strata, zones, microhabitats, and/or times being sampled. Also, one may wish to study only a certain taxon or a particular collection of taxa. For example, if we obtain a collection

1a

ecological sampling

of pond animals with a fine-mesh plankton net, we have not sampled all the pond fauna. Rather, we must be aware of the particular kinds of animals the particular sampling procedure can collect. Sweeping an insect net through the herbaceous vegetation of a forest would not yield a sample of all animals in that forest, but only a sample of those forms inhabiting a particular portion of the ecological community (i.e., the herb stratum, rather than the soil, shrub, or tree stratum), and only those not escaping capture by the net. Also, a sample of an ecological population seldom contains all the stages of the life cycle, which is important to realize when making inferences about a population or community. No single sampling device or technique can provide data on an entire habitat, community, or biological population. This is why we must always define the ecological entity actually sampled by a given procedure.

2. Selecting samples

After defining the ecological entity to be sampled and choosing the sampling technique (detailed in unit 3), one can then do the actual sampling. However, assurance of a truly representative sample of the defined population, community, or habitat is usually a difficult problem in ecology. Normally, samples should be taken at random. Random sampling implies that each measurement in the population has an equal opportunity of being selected as part of the sample, and that the occurrence of one measurement in a sample in no way influences the inclusion of another. Sampling procedures are *biased* if some members of the population are more likely to be recorded than others, or if the recording of some affects the recording of others. If the sample is taken at random from a statistical population, legitimate conclusions may be drawn (with known chance of error) about that population, even though only a small portion of it has been measured.

A table of random numbers (table 1A.1) often helps obtain random samples. In table 1A.1, each integer from

Table 1A.1. *Random Numbers*

72965	92280	85318	98478	05200	26558	04697	63195	41679	24133
25182	09959	91375	97794	50193	25930	47938	95633	22271	15628
78812	39100	81576	84683	47466	04204	86339	31919	83404	48293
87264	75327	92529	25409	52589	20914	58768	46171	32657	89750
21571	57796	67813	88705	52576	51712	12407	00644	81748	04204
98532	11191	63198	79306	04193	00859	83906	30625	67175	37774
38981	76006	33931	22225	00014	37716	67499	90402	08962	88602
11305	19964	22932	62300	64508	32996	05699	06536	22619	89725
96753	89989	67869	65743	65353	55722	91650	77833	05353	05950
28316	27206	32507	96140	83430	75357	57822	75247	93486	20481
24390	09214	19493	94975	71393	54675	51712	00581	11187	73464
23995	32726	41075	32118	63946	62464	60599	81670	73097	78553
41920	60706	55864	70343	61238	06810	53263	07815	56588	29384
78281	15410	26154	70445	27828	38282	29051	13433	84405	82969
92910	17017	92704	25210	63833	04909	02571	58402	62649	86771
29265	89779	95437	51929	75534	70858	54623	99661	87146	16775
60422	65242	57037	95091	25582	76743	95890	09033	08368	62677
42748	43783	94238	97764	64110	68935	21057	14994	94235	53722
39611	11320	52913	20490	84147	59510	45967	93742	71756	09298
74011	92403	54878	91689	20402	20287	05402	16617	86101	28192
49056	17282	52320	73306	91759	85329	88229	62615	25802	28655
06572	13935	69948	12322	84900	85760	67583	36717	75897	39169
32726	45220	41600	61236	55701	08181	26259	49841	88968	83197
13800	03061	28494	09432	95359	92550	11251	76533	51923	34450
09838	95794	39792	06406	81584	49541	20520	91941	43448	91692
86499	23583	61444	72616	78692	50822	10283	23499	17883	21908
19618	23145	32406	91793	50163	72615	61939	18183	20368	51482
04145	26409	44737	98157	14158	94981	66518	84956	65372	00578
44083	35657	49215	93131	41815	34454	46347	02783	27988	86461
13883	40605	76333	56473	27866	16074	00939	05149	14090	70080
08697	34971	19204	70701	56065	23839	45794	62036	07594	36604
86447	56887	61107	63246	88350	51579	95387	03708	16441	64848
37914	39110	60363	95348	96498	17447	18058	36020	57301	50492
08771	12569	06379	51277	88233	45879	89353	82759	16691	20680
65529	84747	61160	19575	98709	23055	37992	82397	62884	63738
53783	03060	00563	21869	41559	85468	37401	81331	62733	10999
40881	01466	66439	92600	95878	43878	76006	93166	20603	76173
81424	81842	17993	63784	39351	41580	89006	47888	92753	45323
47362	92940	89774	05283	49461	21521	72572	37403	90574	22562
79898	44180	49706	58783	47012	90892	89032	56904	56473	38246
98433	36491	48288	53653	77220	82969	70063	58551	20025	83414
79849	94549	69691	11789	43233	46831	08737	25992	11296	69195
26004	14598	80743	25043	45287	35345	46914	71487	10345	48236
46218	40835	82386	91946	14266	77484	02759	92164	77842	21600
49618	10730	47690	44746	09566	36769	39108	47001	62935	10227
66259	25266	88651	56018	68181	45119	91387	37257	83610	53138
65170	81485	14727	22898	63815	17317	68293	06449	91890	49994
82679	72969	04512	11079	95969	87389	46263	96780	78124	04120
37900	90316	47434	60701	89649	51773	26139	39231	72264	17654
27111	31679	71539	61375	58691	20215	91170	44290	91396	90173

This table was prepared using an International Business Machines Corporation (1968:77) algorithm. Larger tables of random numbers are found in Dixon and Massey (1983:446–450), Rohlf and Sokal (1981:72–75), Snedecor and Cochran (1980:463–466), Steel and Torrie (1980:572–575), and Zar (1984:653–656).

0 to 9 has an equal and independent chance of occurring at any location in the table, each two-digit number from 00 to 99 has a random chance of occurring anywhere in the table, and so on. Each time this table is used, it should be entered at random; that is, do not always begin at the same point in the table. Once entered, numbers in the table may be read in any predecided direction—horizontally, vertically, or diagonally. If members of a population of objects (e.g., mice or trees) could be numbered, then a random sample of n objects from that population could be designated by considering n different numbers from the random number table. This is equivalent to placing each member of the population in a hat and drawing n of them by chance. However, this method generally is impractical since numbering the individuals in the population would mean obtaining all of its members; if this could be done there would be little need for sampling.

Random numbers may be used to select random map coordinates or numbered sampling sites. Sampling sites can be numbered easily by arbitrarily selecting a point within the habitat and marking off four compass directions (N, E, S, W) from this point to define four quadrants. A randomly-selected number could represent the number of meters, or tens of meters, along one axis of a quadrant, and a second random number could do the same along the other axis for that quadrant. Thus, each pair of random numbers would establish a specific point in the quadrant at which to collect a physical sample. A quadrant could be selected at random by picking a random number from 1 to 4 and this process repeated until a sufficient number of random points had been selected.

3. Sampling replication

A single measurement generally is insufficient to draw conclusions about an ecological characteristic. This is because a single datum is not adequate to judge how reliably that characteristic had been estimated. Repeated measurements may vary greatly; hence a single value could have an uncomfortably high probability of being far from the average value. Therefore, a series of repeated, or *replicated,* measurements should be taken. From this collection of replicates (i.e., the statistical sample) we can estimate the mean of the statistical population and determine how much error exists in making this estimate (see sections 1B.2.1 and 1B.2.4).

How many replicate data are needed to obtain a reliable estimate of some aspect of a statistical population (i.e., of a characteristic of an ecological population, community, or habitat)? There is no set answer, but a number of procedures can aid in determining whether enough measurements have been collected. Two common methods—the species-sample curve and the performance curve—are discussed here. A procedure using statistical considerations is discussed in section 1B.2.5.

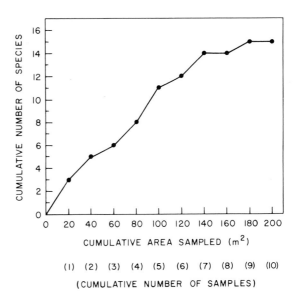

Figure 1A.1. A species-area curve for the data in table 1A.2, plotting cumulative number of species against area sampled. If the cumulative number of species is plotted against the cumulative number of ecological samples (indicated in parentheses), this would be a species-sample curve.

Table 1A.2. *Data for generating the species-area curve of figure 1A.1. Each ecological sample is from a 20-m²$ area.*

Sample number	Cumulative area sampled (m²)	Number of species	Number of new species	Cumulative number of new species
1	20	3	3	3
2	40	4	2	5
3	60	5	1	6
4	80	3	2	8
5	100	4	3	11
6	120	4	1	12
7	140	4	2	14
8	160	3	0	14
9	180	5	1	15
10	200	4	0	15

In a *species-sample curve,* the cumulative number of species is plotted against the cumulative number of physical samples, where each sample might be a plot, transect interval, point-quarter point, insect-net effort, seine haul, etc. (see unit 3). If the cumulative number of species is plotted against the cumulative size of the area sampled, this is called a *species-area curve.*

Figure 1A.1 is a presentation of the data in table 1A.2. Here each datum is a species enumeration for a 20-m² area. One finds three species in the first sample. Since the second sample has four species, but two are species found in sample 1 and two are species newly found in sample 2,

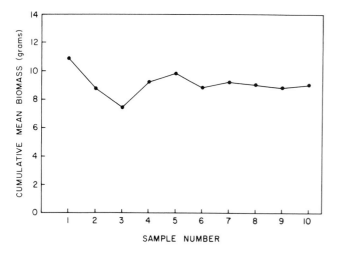

Figure 1A.2. A performance curve for the data in table 1A.2, plotting cumulative mean biomass against cumulative number of samples.

Table 1A.3. *Biomass data for generating the performance curve plotted in figure 1A.2.*

Sample number	Biomass (g)	Cumulative mean biomass (per sample) (g)
1	10.9	10.9
2	6.7	8.8
3	4.9	7.5
4	14.7	9.3
5	12.3	9.9
6	3.9	8.9
7	11.7	9.3
8	7.7	9.1
9	7.3	8.9
10	10.9	9.1

then there are 3 + 2, or 5, species found in a total of 40 m² of sampling. The number of samples is considered sufficient after the curve levels off (see figure 1A.1). However, if the curve levels off after only a very few samples, then the area in each sample is too large. The species-sample curve is an aid in evaluating both the number of replicates and the size of the physical sample. Physical samples that are too small may require a very large number of replicates. On the other hand, if the physical samples are too large then fewer samples may be taken than necessary to allow for a satisfactory estimate of statistical error. The species-area curve is also useful for comparing the diversity of different communities and may be used in conjunction with sections 5A and 5B.

A *performance curve* examines the mean value of a set of measurements for an ecological variable. For example, the mean density or biomass for a given species (or for all species) may be plotted as a function of the cumulative number of samples or the cumulative area sampled (figure 1A.2). It is analogous to a species-area curve, but it plots a cumulative mean of some variable rather than the cumulative number of species. For a small number of ecological samples, such a mean fluctuates widely from sample to sample, but as the number of replicates increases the fluctuation of the mean decreases (see figure 1A.2). The number of replicates may be considered sufficiently large when such fluctuations are so slight that the cumulative mean has become insensitive to variations in the data. For example, the data of table 1A.3 represent ten measurements of biomass as determined from ten physical samples.

4. Subsampling

Occasionally, ecological samples are taken in the field and only portions of them, or *subsamples,* are later examined in the laboratory. The principles of subsampling are like those of sampling: The subsample must be randomly taken from the sample. This may require (as in a chemical analysis) shaking, mixing, or blending the sample before taking the subsample. In this way subsample characteristics reflect the characteristics of the entire sample.

5. Experimental design

Closely associated with the concept of sampling is that of *experimental design*—the planning of field or laboratory studies. Experimental design deals with the questions to be asked in a study, the selection of variables to be studied, and the choice of a sampling program. The design is constructed, prior to the data collection, with specific procedures of sampling and data analysis in mind (see section 1B and units 2 and 3). There are many complex designs by which data may be collected and analyzed. A few of the simplest and most common will be discussed here and in section 1B.

The most commonly used experimental design in ecological work is the two-sample comparison. Here, one selects two situations in which all conditions but one are equal (or nearly equal). For example, one may measure the population density of caddisfly larvae in a stream to conclude whether there is a difference between the densities in two different current velocities. One then selects two sites with similar habitat characteristics (dissolved oxygen, stream substrate, depth, etc.) but with different current velocities. On examining the collected data, you may conclude that the population density of caddisfly larvae is different at the two current conditions. However, you cannot automatically conclude a direct cause-and-effect relationship and assert that the difference in population size was due to the current *per se* (e.g., faster current may result in more food availability or better protection from predators.)

6. Selected references

Andrewartha, H. G. 1971. Introduction to the study of animal populations. University of Chicago Press, Chicago.

Barbour, M. G., J. H. Burk, and W. D. Pitts. 1980. Terrestrial plant ecology. Benjamin/Cummings Publishing Co., Menlo Park, Calif.

Bormann, F. H. 1953. The statistical efficiency of sample plot size and shape in forest ecology. Ecology 34:474–487.

Connor, E. F., and E. D. McCoy. 1979. The statistics and biology of the species-area relationship. Amer. Natur. 113:791–833.

Dixon, W. J., and F. J. Massey, Jr. 1983. Introduction to statistical analysis. 3d ed. McGraw-Hill Book Co., New York.

Green, R. H. 1979. Sampling Design and Statistical Methods for Environmental Biologists. John C. Wiley & Sons, New York.

Greig-Smith, P. 1983. Quantitative plant ecology. 3d ed. University of California Press, Berkeley, Calif.

International Business Machines Corporation. 1968. System/360 scientific subroutine package (360A-CM-03X). Version III. Programmer's manual. White Plains, N.Y.

Kershaw, K. A., and J. H. H. Looney. 1985. Quantitative and dynamic plant ecology. 3d ed. Edward Arnold, London.

Pielou, E. C. 1977. Mathematical ecology. John C. Wiley & Sons, New York.

Poole, R. W. 1974. An introduction to quantitative ecology. McGraw-Hill Book Co., New York.

Prince, S. D. 1986. Data analysis, 345–375. In P. D. Moore and S. B. Chapman (eds.), Methods in plant ecology. 2d ed. Blackwell Scientific Publications, Oxford, England.

Rohlf, F. J., and R. R. Sokal. 1981. Statistical tables. W. H. Freeman and Co., San Francisco.

Seber, G. A. F. 1982. The estimation of animal abundance and related parameters. 2d ed. Macmillan Publishing Co., New York.

Snedecor, G. W., and W. G. Cochran. 1980. Statistical methods. Iowa State University Press, Ames, Iowa.

Sokal, R. R., and F. J. Rohlf. 1981. Biometry: The principles and practice of statistics in biological research. 2d ed. W. H. Freeman and Co., San Francisco.

Southwood, T. R. E. 1978. Ecological methods. Methuen and Co., London.

Steel, R. G. D., and J. H. Torrie. 1980. Principles and procedures of statistics: A biometrical approach. 2d ed. McGraw-Hill Book Co., New York.

Zar, J. H. 1984. Biostatistical analysis. Prentice-Hall, Englewood Cliffs, N.J.

1b

data analysis

1. Introduction

Procedures called *statistical methods* allow an ecologist to engage in three very important activities: (1) quantitatively describing and summarizing characteristics of sets of data; (2) drawing conclusions about large sets of data (from habitats, communities, or biological populations), having data from only small portions (samples) of them; and (3) objectively assessing differences and relationships between sets of data. This section presents some basic statistical concepts and methods used by ecologists. More thorough coverage is found in biostatistics texts such as Sokal and Rohlf (1981) and Zar (1984). Other statistical references frequently helpful to biological investigators are Dixon and Massey (1983), Snedecor and Cochran (1980), and Steel and Torrie (1980).

Basic to the consideration of statistical procedures are the concepts of a statistical population and a statistical sample, as introduced in section 1A.1. Recall from that discussion that the statistical population is the entire set of data about which we wish to draw conclusions, and a statistical sample is a portion, or subset, of the statistical population. If a statistical sample is taken at random from a statistical population (see section 1A.2), then conclusions may be drawn about the population, with a known chance of error, even though only a small portion of it was measured.

2. Descriptive statistics

A measure that describes or characterizes an entire population of data is called a *parameter.* However, as it is generally not possible to collect all data in a population, we can not calculate parameters directly but must estimate them by computing *statistics,* descriptive measures derived from the data in samples taken from the population.

2.1. Averages A very useful measure of the central tendency, or average, of a population is the *mean;* and a population mean (symbolized by statisticians by the Greek mu, μ) may be estimated by the mean of a random sample from the population. If X represents a datum (e.g., a grasshopper weight, a tree height, a water temperature, or the number of phytoplankton in a milliliter of water), the mean of a sample of such data is computed as:

$$\bar{X} = \Sigma X/n, \tag{1}$$

where $\bar{X}$ (pronounced "X bar") is the conventional symbol for the sample mean, ΣX indicates the summation of all values of X in the sample, and n is the number of data in the sample. This calculation is demonstrated below for the following data—tree heights—in meters: 10.1, 11.4, 11.7, 12.1, 13.3 m.

$$\Sigma X = 10.1 + 11.4 + 11.7 + 12.1 + 13.3 \text{ m} = 58.6 \text{ m}$$
$$n = 5$$
$$\bar{X} = \Sigma X/n = 58.6 \text{ m}/5 = 11.72 \text{ m}.$$

As long as our sample of tree heights was obtained at random from the entire population of tree heights, we may assert that the sample mean of 11.72 m is a good estimate of the population mean (i.e., a good estimate of the mean tree height of the entire population). Just how precise an estimate we have calculated can be expressed as shown in section 2.4 below. As a general rule, the mean should not be considered more accurate than one decimal place beyond the accuracy of the original data. (In the present example, the data are accurate to 0.1 m; therefore, our mean is rounded to the nearest 0.01 m.)

Another average sometimes encountered is the *median,* which is simply the middle measurement in a ranked listing of data. In the above example, 11.7 m is the median, for there are as many heights less than 11.7 m as there are heights greater than this value. (If there are an even number of data, there will be two middle measurements, and the median is the mean of the two; e.g., the median of 2, 3, 5, 6, 8, and 11 is 5.5.) There is no widely accepted symbol for the median of a population or of a sample.

2.2. Measures of Variability Calculating a mean or other average gives only a partial description of a set of data. Observe that each of these two samples of data have the same mean (namely 11): 1, 6, 11, 16, 21 and 10, 11, 11, 11, 12. So, to help describe these samples we also need a measure of how variable, or how dispersed, the data are.

One measure of data dispersion is the *range,* simply the difference, between the largest and smallest data in the collection. However, the sample range nearly always is a biased estimate of the population range in that it tends to underestimate it. Also, consider that the following two sets of data have the same range: 5, 19, 20, 20, 20, 21, 35 and

5, 10, 15, 20, 25, 30, 35. By using only two data (the smallest and the largest), the range does an incomplete job of describing variability.

Of special importance and utility in statistical analysis are those measures of data dispersion based on the deviation of data from their mean. We define a quantity termed the *sum of squared deviations from the mean,* referred to simply as *sum of squares* (abbreviated SS), as:

$$SS = \Sigma(X - \bar{X})^2. \qquad (2)$$

Thus for the tree height data presented above,

$$SS = (10.1 - 11.72)^2 + (11.4 - 11.72)^2$$
$$+ (11.7 - 11.72)^2 + (12.1 - 11.72)^2$$
$$+ (13.3 - 11.72)^2$$
$$= 5.37 \text{ m}^2.$$

(Note that the units of measurement of SS are the squares of the original units.) The computation of equation 2 can become tedious for large numbers of data, but fortunately many calculators have the ability to compute the quantities ΣX and ΣX^2 very simply (and often simultaneously), and equation 3 (often called the "machine formula" for SS) is mathematically equivalent to equation 2:

$$SS = \Sigma X^2 - (\Sigma X)^2/n. \qquad (3)$$

Using equation 3 for the tree height data:

$$\Sigma X = 10.1 + 11.4 + 11.7 + 12.1 + 13.3 = 58.6 \text{ m}$$
$$\Sigma X^2 = (10.1)^2 + (11.4)^2 + (11.7)^2 + (12.1)^2$$
$$+ (13.3)^2$$
$$= 692.16 \text{ m}^2$$
$$n = 5$$
$$SS = 692.16 \text{ m}^2 - (58.6 \text{ m})^2/5$$
$$= 692.16 \text{ m}^2 - 686.79 \text{ m}^2$$
$$= 5.37 \text{ m}^2,$$

the same result as with equation 2 above.

The sample *variance* is:

$$s^2 = SS/DF, \qquad (4)$$

where DF is a quantity called *degrees of freedom* (often symbolized by the Greek nu, ν), defined as:

$$DF = n - 1 \qquad (5)$$

when computing sample variance. Thus for our tree height data,

$$DF = 5 - 1 = 4$$
$$s^2 = 5.37 \text{ m}^2/4 = 1.34 \text{ m}^2$$

and the sample variance is a good estimate of the population variance, usually symbolized by the Greek sigma squared, σ^2 (i.e., a good estimate of the variance we would compute if we had all the data in the population of interest).

The sample *standard deviation* (abbreviated s, or SD) is:

$$s = \sqrt{s^2}, \qquad (6)$$

which generally is reported as a measure of variability in preference to the variance because it has the same units as the original data. It is an estimate of σ, the standard deviation in the entire population of data. For the tree heights:

$$s = \sqrt{1.34 \text{ m}^2} = 1.16 \text{ m}.$$

The sample mean and standard deviation are very useful in describing the population of data from which our sample came. Many electronic calculators can compute s without need for the user to deal with any intermediate results (e.g., SS or s^2). However, some of these machines do not calculate s in the same fashion as described here, thus giving a value of s that is not suitable for many purposes.

Sometimes, the standard deviation is expressed relative to the mean, yielding a unitless statistic termed the *coefficient of variation:*

$$\text{coefficient of variation} = s/\bar{X}, \qquad (7)$$

which, for our sample data, is:

$$\text{coefficient of variation} = 1.16/11.72$$
$$= 0.10, \text{ or } 10\%.$$

2.3. Accuracy and Precision While the nonscientist may consider accuracy and precision to be synonyms, it is important for us to distinguish between these two terms. *Accuracy* is the closeness of a measured value to the true value. Thus, for example, an accurate measurement of the length of a fish would be one that is very near the true fish length. If a measure is consistently high, or consistently low, it is said to be *biased.*

Precision refers to the closeness of repeated measurements to each other. If one has repeated measurements of the same quantity, then the standard deviation or the coefficient of variation may be used as a measure of precision (these measures being lowest when the precision is greatest). Sample 1 in figure 1B.1 is an example of having low precision (i.e., there is much variability among the data; $s = 4.1$ m and coefficient of variation $= 0.19$); however the mean of the data ($\bar{X} = 21.5$ m) is an accurate estimate of the population mean of 22 meters. The sample 2 data in figure 1B.1 exhibit more precision ($s = 0.5$ m, coefficient of variation $= 0.02$), but the estimate of the mean population tree height ($\bar{X} = 27.8$ m) is inaccurate, for it is consistently high. Bias of this sort is typically due to error in technique (such as using an uncalibrated instrument) or using nonrandom sampling.

Precision also refers to the closeness of a computed estimate to the actual value being estimated. For example,

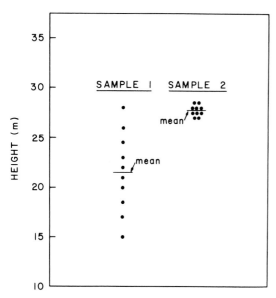

Figure 1B.1. Two hypothetical statistical samples, each of which is composed of data that are ten repeated measurements of the height of a single tree, the true height of which is 22 meters. Sample 1 has low precision but high accuracy; Sample 2 has high precision but low accuracy (see section 2.3).

if a sample mean is very close to the mean of the sampled population, then the former is a precise estimate of the latter. If an estimate is consistently low or consistently high it is said to be *biased*. A confidence interval (section 2.4 below) is a good way to express the precision of an estimate.

2.4. Confidence in Estimating Population Means

When we estimate a population mean by calculating a sample mean, we may wonder how precise the estimate of the mean is. This is answered by considering that repeated samples from the same population will each have a somewhat different mean. The variability among these possible sample means is:

$$s_{\bar{X}} = \sqrt{s^2/n} = s/\sqrt{n}, \qquad (8)$$

a very important statistic known as the *standard deviation of the mean*, the *standard error of the mean*, or simply the *standard error* (SE). For our tree height data,

$$s_{\bar{X}} = \sqrt{1.34 \text{ m}^2/5} = 0.52 \text{ m}$$

or, equivalently,

$$s_{\bar{X}} = 1.16 \text{ m}/\sqrt{5} = 0.52 \text{ m}.$$

Using the standard error, one can express a *confidence interval*, an interval that, with a stated level of confidence, may be said to include the population mean, μ:

$$(1 - \alpha) \text{ confidence interval for } \mu = \bar{X} \pm t s_{\bar{X}}. \qquad (9)$$

Here, $\bar{X}$ is the sample mean and $s_{\bar{X}}$ is the standard error. The value of t is obtained from a statistical distribution

Table 1B.1. *Critical Values of Student's* t.

DF	$\alpha = 0.10$	$\alpha = 0.05$	$\alpha = 0.02$	$\alpha = 0.01$
1	6.31	12.71	31.82	63.66
2	2.92	4.31	6.96	9.92
3	2.35	3.18	4.54	5.84
4	2.13	2.78	3.75	4.60
5	2.01	2.57	3.36	4.03
6	1.94	2.45	3.14	3.71
7	1.89	2.36	3.00	3.50
8	1.86	2.31	2.90	3.36
9	1.83	2.26	2.82	3.25
10	1.81	2.23	2.76	3.17
11	1.80	2.20	2.72	3.11
12	1.78	2.18	2.68	3.06
13	1.77	2.16	2.65	3.01
14	1.76	2.14	2.62	3.00
15	1.75	2.13	2.60	2.95
16	1.75	2.12	2.58	2.92
17	1.74	2.11	2.57	2.90
18	1.73	2.10	2.55	2.88
19	1.73	2.09	2.54	2.86
20	1.72	2.09	2.53	2.85
22	1.72	2.07	2.51	2.82
24	1.71	2.06	2.49	2.80
26	1.71	2.06	2.48	2.78
28	1.70	2.05	2.47	2.76
30	1.70	2.04	2.46	2.75
35	1.69	2.03	2.44	2.72
40	1.68	2.02	2.42	2.70
45	1.68	2.01	2.41	2.69
50	1.68	2.01	2.40	2.68
60	1.67	2.00	2.39	2.66
70	1.67	1.99	2.38	2.65
80	1.66	1.99	2.37	2.64
90	1.66	1.99	2.37	2.63
100	1.66	1.98	2.36	2.63
120	1.66	1.98	2.36	2.62
150	1.66	1.98	2.35	2.61
200	1.65	1.97	2.35	2.61
300	1.65	1.97	2.34	2.59
500	1.65	1.96	2.33	2.59
∞	1.65	1.96	2.33	2.58

The above values were computed as described by Zar (1984:485). More extensive tables of Student's *t* are found in Rohlf and Sokal (1981) and Zar (1984:484–485).

known as *Student's* t, a portion of which is given in table 1B.1. In this table, DF is the degrees of freedom $(n - 1)$, and α is the "significance level" for *t*. A significance level of 5% is most frequently used in biological research, for this convention allows for a reasonable balance between the kinds of errors inherent in statistical testing (although significance levels of 1% and 10% are occasionally employed). Using $\alpha = 0.05$ allows us to compute a 95% confidence interval (and confidence levels of 1% and 10% lead to confidence intervals of 99% and 90%, respectively). For the tree height data:

$$95\% \text{ confidence interval for } \mu = 11.72 \text{ m}$$
$$\pm (2.78)(0.52 \text{ m})$$
$$= 11.72 \text{ m} \pm 1.45 \text{ m}.$$

Thus we may say that the mean of the entire population of tree heights from which our sample came is 11.72 ± 1.45 m (i.e., the mean is between a *lower confidence limit* of 10.27 m and an *upper confidence limit* of 13.17 m). This assertion may be made with 95% confidence, meaning that there is a 5% chance that the statement is wrong (i.e., that the population mean actually is either less than 10.27 m or greater than 13.17 m). This also implies that if one calculated the means of 100 random samples from this population, 95 of them would lie between 10.27 m and 13.17 m.

Note in table 1B.1 that small significance levels, α, have associated with them large values of t. Therefore if $\alpha = 0.01$ were used rather than $\alpha = 0.05$ to compute a 99% confidence interval, instead of a 95% confidence interval, the interval would be larger, namely,

$$99\% \text{ confidence interval for } \mu = 11.72 \text{ m}$$
$$\pm (4.60)(0.52 \text{ m})$$
$$= 11.72 \text{ m} \pm 2.39 \text{ m},$$

with confidence limits of 9.33 m and 14.11 m; and we would assert that there is only a 1% chance that the true population mean is outside this interval (i.e., either below 9.33 m or above 14.11 m).

2.5. Selecting Statistical Sample Size   As can be seen from equation 8, the magnitude of the standard error, $s_{\bar{X}}$, and thus the width of the confidence interval, is inversely related to n, the number of data in our sample. That is, a larger n in general results in a more precise estimate of a population mean. Therefore, we can determine how many data should comprise the statistical sample in order to estimate the population mean with a specified precision.

For the above example, it was calculated that for a sample of five data the population mean, μ, could be estimated to within 1.45 m. Let us say that we want to derive an estimate of μ, precise to within about 1.0 m; this would mean that we desire that:

$$ts_{\bar{X}} = 1.0 \text{ m}.$$

We know that an n greater than 5 is needed, so let us guess that $n = 10$ is required; with this guess:

$$DF = 10 - 1 = 9$$
$$t = 2.26$$
$$s_{\bar{X}} = \sqrt{1.34 \text{ m}^2/10} = 0.37 \text{ m}$$
$$ts_{\bar{X}} = (2.26)(0.37 \text{ m}) = 0.84 \text{ m}.$$

As 0.84 m is less than the desired 1.0 m, $n = 10$ is a sample size larger than that necessary for $ts_{\bar{X}} = 1.0$ m. Let us try $n = 7$:

$$DF = 7 - 1 = 6$$
$$t = 2.45$$
$$s_{\bar{X}} = \sqrt{1.34 \text{ m}^2/7} = 0.44 \text{ m}$$
$$ts_{\bar{X}} = (2.45)(0.44 \text{ m}) = 1.08 \text{ m}.$$

Therefore, a sample size, n, of 7 is predicted to be necessary for the desired precision in estimating the mean of our population of tree heights. (Incidentally, mathematicians call this directed trial-and-error procedure *iteration*.)

3. Comparing statistical populations

One of the most common of biostatistical procedures is drawing conclusions about the similarity or difference between the means of sampled populations of data. For example, an ecologist might wonder whether on the average the plant biomass is the same in two different geographical areas (or in two different seasons, or under two different experimental regimes). The question refers to two statistical populations, and a completely confident answer would require the impractical and probably impossible measurement of the biomass of all plant material in each area. Therefore, one takes a sample from each of the two populations and then infers from the two sample means and variability whether those populations have the same or different means. The sampling of populations in a way that enables the drawing of objective conclusions about them is called *experimental design* (see section 1A.5).

Note that the statistical procedures of sections 3.1 and 3.2 are based on some underling assumptions, indicated in section 3.3.

3.1. Two-Sample Testing   Statistical analysis for a two-sample experimental design is commonly done by a type of "t-testing," where the statistic t is calculated as:

$$t = \frac{|\bar{X}_1 - \bar{X}_2|}{s_{\bar{X}_1 - \bar{X}_2}}. \tag{10}$$

In this computation, $\bar{X}_1$ and $\bar{X}_2$ are the means of samples 1 and 2, respectively; $|\bar{X}_1 - \bar{X}_2|$ tells us to use the absolute value of the difference between the means (i.e., if $\bar{X}_1 - \bar{X}_2$ is negative, drop the negative sign to make the difference between the two means positive). The quantity $s_{\bar{X}_1 - \bar{X}_2}$ is called the standard error of the difference between the means, and it is computed as:

$$s_{\bar{X}_1 - \bar{X}_2} = \sqrt{(s_p^2/n_1) + (s_p^2/n_2)}. \tag{11}$$

The quantities n_1 and n_2 are the two sample sizes (the number of data, biomass measurements in this case) for statistical samples 1 and 2, respectively, and s_p^2 is computed as:

$$s_p^2 = \frac{SS_1 + SS_2}{DF_1 + DF_2}, \tag{12}$$

where SS_1 and SS_2 are the sums of squares (equation 2 or 3), and DF_1 and DF_2 are the degrees of freedom (equation 5) for samples 1 and 2, respectively; s_p^2 is called the *pooled variance* and is a measure of the variability of data within the two samples.

Let us consider hypothetical plant biomass data where seven measurements from one geographical location are: 438, 421, 430, 413, 409, 428, and 419 grams dry weight per square meter, and six biomass data from a second location are: 442, 451, 428, 446, 459, and 437 grams dry weight/m². Based on these thirteen sample data we wish to conclude whether or not the mean plant biomass in one geographic location is the same as the mean in the second location.

By performing the necessary calculations, we arrive at:

$$n_1 = 7$$
$$n_2 = 6$$
$$\bar{X}_1 = 422.6 \text{ g/m}^2 \text{ (by equation 1)}$$
$$\bar{X}_2 = 443.8 \text{ g/m}^2 \text{ (by equation 1)}$$
$$SS_1 = 613.71 \text{ (g/m}^2)^2$$
$$\text{(by either equation 2 or 3)}$$
$$SS_2 = 586.83 \text{ (g/m}^2)^2$$
$$\text{(by either equation 2 or 3)}$$
$$DF_1 = 6 \text{ (by equation 5)}$$
$$DF_2 = 5 \text{ (by equation 5)}$$
$$s_p^2 = (613.71 + 586.83)/(6 + 5)$$
$$= 109.14 \text{ (g/m}^2)^2 \text{ (by equation 12)}$$
$$s_{\bar{X}_1 - \bar{X}_2} = \sqrt{109.14/7 + 109.14/6}$$
$$= \sqrt{33.78 \text{ (g/m}^2)^2} = 5.8 \text{ g/m}^2$$
$$\text{(by equation 11)}$$
$$t = \frac{|422.6 \text{ g/m}^2 - 443.8 \text{ g/m}^2|}{5.8 \text{ g/m}^2} = 3.66$$
$$\text{(by equation 10).}$$

Small t values indicate high probability that the two population means are the same; by contrast, large t values imply lower probability. In more formal terms, the statistician's *null hypothesis (H_0)* is that the means of the two populations are the same (i.e., H_0: $\mu_1 = \mu_2$), and the *alternate hypothesis (H_A)* is that the two population means are not the same (i.e., H_A: $\mu_1 \neq \mu_2$). If the computed t value (from equation 10) is at least as large as the appropriate value of t from table 1B.1 (the so-called *critical value* of t), then the null hypothesis is considered probably not true and is rejected; and the alternate hypothesis is considered to be true. The appropriate critical value of t is that for which the pooled degrees of freedom, DF, are $DF_1 + DF_2$ (i.e., $n_1 - 1 + n_2 - 1 = n_1 + n_2 - 2$), and where α is the desired *significance level* (most commonly 5%). For the biomass data above, the critical value of t

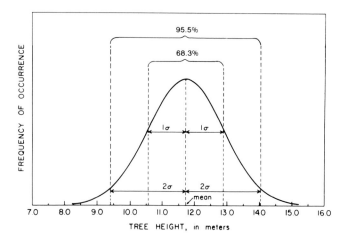

Figure 1B.2. A normal distribution. The data are a hypothetical population of trees heights *(X)*, with a mean, μ, of 11.72 m, and a standard deviation, σ, of 1.16 m. The mean ± 1 standard deviation includes 68.3% of the area under any normal curve; the mean ± 2 standard deviations encompasses 95.5%; and $\mu \pm 3\sigma$ includes 99.7%.

(from table 1B.1) is 2.20. Since the computed t value (3.66) is larger than the tabled t, we may reject the null hypothesis of equality of means and conclude that the two population means are not the same.

3.2. Multisample Testing   Instances abound where differences among more than two means should be tested. If, for example, we collected plant biomass data from three rather than two areas, then the null hypothesis would be H_0: $\mu_1 = \mu_2 = \mu_3$ and the alternate hypothesis would be H_A: $\mu_1 = \mu_2 \neq \mu_3$, or H_A: $\mu_1 \neq \mu_2 = \mu_3$, or H_A: $\mu_1 \neq \mu_2 \neq \mu_3$. To select objectively from among these four possible conclusions, one should not use t-testing. Instead, use procedures known as *analyses of variance*, which employ a test statistic called *F*, rather than the t statistic (see Sokal and Rohlf, 1981: 208–222; Zar, 1984: 162–176), and multiple comparisons (see Sokal and Rohlf, 1981: 242–262; Zar, 1984: 185–198).

3.3. Theoretical Assumptions   Some theoretical foundations must be satisfied for population means to be compared validly by t-testing, or by the analysis of variance or multiple comparison procedures referred to above. All populations must have equal variances, and each population must be composed of data that conform to the "normal distribution," a specific kind of symmetrical, "bell-shaped" distribution of measurements (figure 1B.2). Many kinds of biological data (lengths, weights, heights, rates) conform reasonably well to these conditions. Furthermore, moderate departures from these assumptions have only slight affect on the validity of most statistical testing. However, some kinds of ecological data are predicted from experience or theory to be distinctly non-normal. As a rule they should not be subjected to the types of statistical testing referred to in sections 3.1 and 3.2. These include:

1. data that are proportions or percentages (e.g., percent sky cover or percent water in soil or in plant or animal tissue);
2. data that are counts, such as densities (e.g., numbers of plants in sampling plots, numbers of plankton per milliliter of water, or number of fish per net haul); and
3. data that are measured on a nonlinear scale (e.g., pH or extinction coefficient).

There are two major approaches to the statistical handling of such data. One is to transform the data—to change the measurements into another form (logarithms or square roots are common, depending on the characteristics of the data), after which the transformed values often may be handled validly by the above-mentioned statistical procedures (see Sokal and Rohlf, 1981: 417–428; Zar, 1984: 236–243).

A second and rather simple approach to analyzing data unsuitable for the previously discussed statistical tests is to employ nonparametric methods of testing.

3.4. Nonparametric Testing   Many statistical testing procedures do not depend on such assumptions as normality or equality of variances. These are called *nonparametric,* or *distribution-free* methods. One of the most commonly used is the *Mann-Whitney test* (also called the Wilcoxon-Mann-Whitney test), by which one can test for differences between two populations of data by examining a sample of data from each population. Nonparametric methods like the Mann-Whitney test can be used in instances where *t*-testing is inappropriate, as well as in those where *t*-testing is valid. In the latter case, the relative simplicity and speed of the nonparametric procedures may compensate for their slight disadvantage (consult a statistical reference).

Consider the following hypothetical data: five bottom grabs from a pond result in 1, 14, 15, 11, and 8 worms per 0.1 m²; seven bottom hauls from a second pond revealed 9, 13, 7, 10, 6, 7, and 11 worms per 0.1 m². The Mann-Whitney test allows us to ask whether both areas have the same worm density; the null hypothesis is "H_0: The two populations of worms have the same density," and the alternate hypothesis is "H_A: The two worm populations have different densities." (As this is a nonparametric test, we do not speak of parameters, such as population means, in the hypotheses.)

The first step in the Mann-Whitney procedure is to order the n_1 data in sample 1, order the n_2 data in sample 2, and then assign ranks to all twelve data ($n_1 + n_2 = 12$),

as follows (where the data are numbers of worms per 0.1 m²):

sample 1		sample 2	
data	ranks	data	ranks
8	4	6	1
11	8	7	2.5
11	8	7	2.5
14	11	9	5
15	12	10	6
		11	8
		13	10

As 6 worms/0.1 m² is the smallest datum, it is assigned rank 1; the largest datum (15 worms/0.1 m²) receives rank 12. If two or more data are identical ("tied"), each is assigned a rank that is the mean of their ranks had they not been tied. For example, the second and third smallest data are each 7 worms/0.1 m²; therefore, each is assigned the rank of $(2 + 3)/2 = 2.5$; and the seventh, eighth, and ninth data each have the value of 11 worms/0.1 m², so each is given the rank of $(7 + 8 + 9)/3 = 8$.

Two Mann-Whitney statistics are then calculated:

$$U = (n_1)(n_2) + [(n_1)(n_1 + 1)/2] - R_1 \quad (13)$$

and

$$U' = (n_1)(n_2) - U, \quad (14)$$

where n_1 and n_2 are the sizes of sample 1 and 2, respectively, and R_1 is the sum of the ranks for sample 1. For our worm samples:

$$R_1 = 4 + 8 + 8 + 11 + 12 = 43$$
$$U = (5)(7) + [(5)(6)/2] - 43 = 7$$
$$U' = (5)(7) - 7 = 28.$$

Then one consults table 1B.2, which contains critical values of the Mann-Whitney statistic. If either U or U' is greater than or equal to the tabled value, then H_0 is rejected and H_A is declared true. For our example, the two sample sizes are 5 and 7, so the critical value for testing at the 0.05 significance level is 30. Since neither U nor U' is as large as 30, the null hypothesis, H_0, is not rejected, and it is not concluded that the two samples of worms came from populations having different densities.

Note that table 1B.2 allows us to perform the Mann-Whitney test only for situations where neither n_1 nor n_2 exceeds 20. If sample sizes are greater than this limit, the larger critical value table of Zar (1984: 550–562) may be consulted, or the following procedure (called the *normal approximation*) may be used. First compute:

$$t = \frac{|U - (n_1)(n_2)/2|}{\sqrt{(n_1)(n_2)(n_1 + n_2 + 1)/12}} \quad (15)$$

and compare it to the critical value of Student's t (table 1B.1) for infinity degrees of freedom (DF = ∞). If the

Table 1B.2. *Critical Values of the Mann-Whitney Test Statistic.*

$\alpha = 0.10$

n_1	$n_2=2$	3	4	5	6	7	8	9	10	11	12	13	14	15	16	17	18	19	20
2				10	12	14	15	17	19	21	22	24	25	27	29	31	32	34	36
3		9	12	14	16	19	21	23	26	28	31	33	35	38	40	42	45	47	49
4		12	15	18	21	24	27	30	33	36	39	42	45	48	50	53	56	59	62
5	10	14	18	21	25	29	32	36	39	43	47	50	54	57	61	65	68	72	75
6	12	16	21	25	29	34	38	42	46	50	55	59	63	67	71	76	80	84	88
7	14	19	24	29	34	38	43	48	53	58	63	67	72	77	82	86	91	96	101
8	15	21	27	32	38	43	49	54	60	65	70	76	81	87	92	97	103	108	113
9	17	23	30	36	42	48	54	60	66	72	78	84	90	96	102	108	114	120	126
10	19	26	33	39	46	53	60	66	73	79	86	93	99	106	112	119	125	132	138
11	21	28	36	43	50	58	65	72	79	87	94	101	108	115	122	130	137	144	151
12	22	31	39	47	55	63	70	78	86	94	102	109	117	125	132	140	148	156	163
13	24	33	42	50	59	67	76	84	93	101	109	118	126	134	143	151	159	167	176
14	25	35	45	54	63	72	81	90	99	108	117	126	135	144	153	161	170	179	188
15	27	38	48	57	67	77	87	96	106	115	125	134	144	153	163	172	182	191	200
16	29	40	50	61	71	82	92	102	112	122	132	143	153	163	173	183	193	203	213
17	31	42	53	65	76	86	97	108	119	130	140	151	161	172	183	193	204	214	225
18	32	45	56	68	80	91	103	114	125	137	148	159	170	182	193	204	215	226	237
19	34	47	59	72	84	96	108	120	132	144	156	167	179	191	203	214	226	238	250
20	36	49	62	75	88	101	113	126	138	151	163	176	188	200	213	225	237	250	262

$\alpha = 0.05$

n_1	$n_2=2$	3	4	5	6	7	8	9	10	11	12	13	14	15	16	17	18	19	20
2							16	18	20	22	23	25	27	29	31	32	34	36	38
3				15	17	20	22	25	27	30	32	35	37	40	42	45	47	50	52
4			16	19	22	25	28	32	35	38	41	44	47	50	53	57	60	63	66
5		15	19	23	27	30	34	38	42	46	49	53	57	61	65	68	72	76	80
6		17	22	27	31	36	40	44	49	53	58	62	67	71	75	80	84	89	93
7		20	25	30	36	41	46	51	56	61	66	71	76	81	86	91	96	101	106
8	16	22	28	34	40	46	51	57	63	69	74	80	86	91	97	102	108	113	119
9	18	25	32	38	44	51	57	64	70	76	82	89	95	101	107	114	120	126	132
10	20	27	35	42	49	56	63	70	77	84	91	97	104	111	118	125	132	138	145
11	22	30	38	46	53	61	69	76	84	91	99	106	114	121	129	136	143	151	158
12	23	32	41	49	58	66	74	82	91	99	107	115	123	131	139	147	155	163	171
13	25	35	44	53	62	71	80	89	97	106	115	124	132	141	149	158	167	175	184
14	27	37	47	57	67	76	86	95	104	114	123	132	141	151	160	169	178	188	197
15	29	40	50	61	71	81	91	101	111	121	131	141	151	161	170	180	190	200	210
16	31	42	53	65	75	86	97	107	118	129	139	149	169	179	181	191	202	212	222
17	32	45	57	68	80	91	102	114	125	136	147	158	169	180	191	202	213	224	235
18	34	47	60	72	84	96	108	120	132	143	155	167	178	190	202	213	225	236	248
19	36	50	63	76	89	101	114	126	138	151	163	175	188	200	212	224	236	248	261
20	38	52	66	80	93	106	119	132	145	158	171	184	197	210	222	235	248	261	273

$\alpha = 0.01$

n_1	$n_2=2$	3	4	5	6	7	8	9	10	11	12	13	14	15	16	17	18	19	20
2																			38
3								27	30	33	35	38	41	43	46	49	52	54	57
4					24	28	31	35	38	42	45	49	52	55	59	62	66	69	72
5				25	29	34	38	42	46	50	54	58	63	67	71	75	79	83	87
6			24	29	34	39	44	49	54	59	63	68	73	78	83	87	92	97	102
7			28	34	39	45	50	56	61	67	72	78	83	89	94	100	105	111	116
8			31	38	44	50	57	63	69	75	81	87	94	100	106	112	118	124	130
9		27	35	42	49	56	63	70	77	83	90	97	104	111	117	124	131	138	144
10		30	38	46	54	61	69	77	84	92	99	106	114	121	129	136	143	151	158
11		33	42	50	59	67	75	83	92	100	108	116	124	132	140	148	156	164	172

Table 1B.2—*(Continued)*

										$\alpha = 0.01$									
n_1	$n_2 = 2$	3	4	5	6	7	8	9	10	11	12	13	14	15	16	17	18	19	20
12		35	45	54	63	72	81	90	99	108	117	125	134	143	151	160	169	177	186
13		38	49	58	68	78	87	97	106	116	125	135	144	153	163	172	181	190	200
14		41	52	63	73	83	94	104	114	124	134	144	154	164	174	184	194	203	213
15		43	55	67	78	89	100	111	121	132	143	153	164	174	185	195	206	216	227
16		46	59	71	83	94	106	117	129	140	151	163	174	185	196	207	218	230	241
17		49	62	75	87	100	112	124	136	148	160	172	184	195	207	219	231	242	254
18		52	66	79	92	105	118	131	143	156	169	181	194	206	218	231	243	255	268
19		54	69	83	97	111	124	138	151	164	177	190	203	216	230	242	255	268	281
20	38	57	72	87	102	116	130	144	158	172	186	200	213	227	241	254	268	281	295

The values in the above table are derived, with permission of the publisher, from the extensive tables of Milton (1964, J. Amer. Statist. Assoc. 59:925–934). See Zar (1984:550–562) for some sample sizes and significance levels not included above.

calculated t is equal or greater than this tabled t, then the null hypothesis of population equality is rejected. If there are very many tied ranks, then equation 15 will be more accurate if the denominator is modified somewhat (see Zar, 1984: 143).

If more than two populations of data are to be compared nonparametrically, then the Mann-Whitney test is generally inappropriate. Instead, use nonparametric analogs to analysis of variance (e.g., the Kruskal-Wallis test) (see Sokal and Rohlf, 1981: 429–437; Zar, 1984: 176–179) and nonparametric multiple comparisons (see Zar, 1984: 199–202).

4. Goodness of fit

Ecological data often take the form of a distribution of frequencies of occurrences, in which case we may wish to ask whether the observed distribution is significantly different from some hypothesized distribution. For example, let us assume that we have determined that the bottom of a section of stream is 50% sand, 30% gravel, and 20% silt. We have further observed that for a certain species of fish, 8 individuals were in the sand areas, 18 were in the gravel portions, and 4 were in the vicinity of silt. A typical null hypothesis would be that the fish have no preference among the three substrates—that individuals of this species distribute themselves in the stream without respect to substrate type. The alternate hypothesis is that the fish are not distributed independent of substrate, but that they do show substrate preference.

If the null hypothesis were true, and the distribution of fish is independent of substrate type, then 50% of the total number of fish (15 of the total sample of 30 fish) would have been expected to be in sand areas, 30% (9) in the gravel, and 20% (6) in silt. (The expected or hypothesized frequencies need not be integers; for example, if the total number of fish were 32, instead of 30, the three expected

frequencies would have been 16, 9.6, and 6.4, respectively.) To test whether the observed frequencies deviate significantly from the frequencies expected by the null hypothesis, we may employ a chi-square (χ^2) test, called a "goodness of fit" procedure because it assesses how well an observed distribution of frequencies conforms to a hypothetical one. The test statistic is computed as:

$$\chi^2 = \sum \frac{(f - F)^2}{F}, \tag{16}$$

where f is an observed frequency and F is its associated expected or hypothesized frequency. For the fish example,

$$\chi^2 = \frac{(8 - 15)^2}{15} + \frac{(18 - 9)^2}{9} + \frac{(4 - 6)^2}{6}$$
$$= 3.2667 + 9.0000 + 0.6667$$
$$= 12.933.$$

Note that the chi-square calculation uses frequencies only; it *never* uses percentages or proportions.

The larger the disparity between observed and expected frequencies, the larger the resultant chi-square, and the lower the probability that the null hypothesis actually is true. The appropriate critical value of χ^2 may be obtained from table 1B.3, where the degrees of freedom, DF, are the number of frequency categories minus 1, and the significance level, α, typically is 0.05. If the computed χ^2 is at least as large as the critical value, then the null hypothesis is rejected. In our fish example, there are three categories (sand, gravel, and silt), so:

$$\text{degrees of freedom} = DF = 3 - 1 = 2,$$

and the appropriate critical value from table 1B.3 is $\chi^2 = 5.991$. As the computed chi-square of 12.933 is greater than 5.991, we may reject the null hypothesis and conclude that the fish frequencies of occurrence were dependent on substrate type.

Table 1B.3. *Critical Values of Chi-Square.*

DF	$\alpha = 0.10$	$\alpha = 0.05$	$\alpha = 0.025$	$\alpha = 0.01$
1	2.706	3.841	5.024	6.635
2	4.605	5.991	7.378	9.210
3	6.251	7.815	9.348	11.345
4	7.779	9.488	11.143	13.277
5	9.236	11.070	12.833	15.086
6	10.645	12.592	14.449	16.812
7	12.017	14.067	16.013	18.475
8	13.362	15.507	17.535	20.090
9	14.684	16.919	19.023	21.666
10	15.987	18.307	20.483	23.209
11	17.275	19.675	21.920	24.725
12	18.549	21.026	23.337	26.217
13	19.812	22.362	24.736	27.688
14	21.064	23.685	26.119	29.141
15	22.307	24.996	27.488	30.578
16	23.542	26.296	28.845	32.000
17	24.769	27.587	30.191	33.409
18	25.989	28.869	31.526	34.805
19	27.204	30.144	32.852	36.191
20	28.412	31.410	34.170	37.566
21	29.615	32.671	35.479	38.932
22	30.813	33.924	36.781	40.289
23	32.007	35.172	38.076	41.638
24	33.196	36.415	39.364	42.980
25	34.382	37.652	40.646	44.314
26	35.563	38.885	41.923	45.642
27	36.741	40.113	43.195	46.963
28	37.916	41.337	44.461	48.278
29	33.711	39.087	42.557	45.722
30	40.256	43.773	46.979	50.892
31	41.422	44.985	48.232	52.191
32	42.585	46.194	49.480	53.486
33	43.745	47.400	50.725	54.776
34	44.903	48.602	51.966	56.061
35	46.059	49.802	53.203	57.302
36	47.212	50.998	54.437	58.619
37	48.363	52.192	55.668	59.893
38	49.513	53.384	56.896	61.162
39	50.660	54.572	58.120	62.428
40	51.805	55.758	59.342	63.691

The above values were computed as described by Zar (1984:482). More extensive tables of chi-square are found in Rohlf and Sokal (1981:98–99) and Zar (1984:479–481). Values of chi-square for degrees of freedom (v) greater than 40 may be approximated very accurately (Zar, 1984:482), as follows:

$$\chi^2 \alpha, v = v(1 - 2/9v + c\sqrt{2/9v})^3,$$

If frequency data fall into three or more categories, use the above chi-square procedure, but if there are only two categories, then use a slightly modified chi-square calculation:

$$\chi^2 = \sum \frac{(|f - F| - 0.5)^2}{F}, \qquad (17)$$

where $|f - F|$ means to take the absolute value of $f - F$ (if the quantity is negative, drop the negative sign and make it positive). The subtraction of 0.5 is called the *Yates correction for continuity.*

For example, you might wish to test the null hypothesis that males and females occur with the same frequency in a particular biological population. If a random sample of 26 animals from that population consisted of 10 males and 16 females, then the f values would be 10 and 16, the F values would be 13 and 13 (under the null hypothesis of equal sex frequencies), and,

$$\chi^2 = \frac{(|10 - 13| - 0.5)^2}{13} + \frac{(|16 - 13| - 0.5)^2}{13}$$
$$= 0.481 + 0.481$$
$$= 0.962.$$

As there are two categories of data (male and female) the degrees of freedom are $2 - 1 = 1$, and the critical value of χ^2 (from table 1B.3) is 3.841. As 0.962 is less than 3.841, the null hypothesis is considered true. Hence, males and females are considered to occur with the same frequency in the sampled population.

5. Contingency tables

Another use of the chi-square statistic is in analyzing *contingency tables,* as in tabulations of data from more than one statistical sample, to examine differences in the distribution of frequencies in the several sampled populations.

For example, the numbers of birds of a particular species might be tabulated according to the region of a forest in which they are observed to feed. Here we tabulate f,

Where the appropriate value of c is:

$\alpha =$	0.10	0.05	0.025	0.01
$c =$	1.28155	1.64485	1.95996	2.32635

For example, the critical value of χ^2 for the 5% significance level and 99 degrees of freedom is computed as

$$\chi^2_{0.05,99} = 99[1 - 2/(9)(99)$$
$$+ (1.64485)\sqrt{2/(9)(99)}]^3$$
$$= 99(1 - 0.00224 + 0.07793)^3$$
$$= 99(1.24469)$$
$$= 123.22$$

the frequency observed, in each season in each forest region:

	in trees	in shrubs	on ground	total
in spring	30	20	9	59
in autumn	13	22	26	61
total	43	42	35	120 (= n)

The null hypothesis would be that birds feed in these three forest regions in the same proportion in the spring as in the autumn. If the null hypothesis were true, then (probability considerations tell us) the frequencies of birds expected in the six positions in the contingency table are calculated as

$$F = RC/n, \qquad (18)$$

where R and C are, respectively, the row and column frequencies for the table position, and n is the total number of data in all six positions. (The statistician calls these six positions the six "cells" of the contingency table.) For example, the expected frequency for birds feeding in trees in spring is $F = (59)(43)/120 = 21.14$, the F for shrubs in autumn is $F = (61)(42)/120 = 21.35$, and so on, as follows:

	in trees	in shrubs	on ground	total
in spring	21.14	20.65	17.21	59.00
in autumn	21.86	21.35	17.79	61.00
total	43.00	42.00	35.00	120.00

Note that the row and column totals (and n) are the same for both observed and expected frequencies; this will always be the case, thus providing a good arithmetic check.

The chi-square to be computed is just as in equation 16:

$$\chi^2 = \Sigma\Sigma \frac{(f - F)^2}{F}, \qquad (19)$$

except that we write it with two summation signs (Σ) to indicate that we sum values of $(f - F)^2/F$ across all rows as well as across all columns.[1] For the above contingency table:

$$\chi^2 = \frac{(30 - 21.14)^2}{21.14} + \frac{(20 - 20.65)^2}{20.65} + \frac{(9 - 17.21)^2}{17.21}$$
$$+ \frac{(13 - 21.86)^2}{21.86} + \frac{(22 - 21.35)^2}{21.35}$$
$$+ \frac{(26 - 17.79)^2}{17.79}$$
$$= 3.713 + 0.020 + 3.917 + 3.591 + 0.020$$
$$+ 3.789$$
$$= 15.050.$$

The critical value of χ^2 is read from table 1B.3 with degrees of freedom for contingency table being

$$DF = (\text{no. of rows} - 1)(\text{no. of columns} - 1). \quad (20)$$

As this table of forest-bird data has 2 rows and 3 columns,

$$DF = (2 - 1)(3 - 1) = 2$$

and the critical value, for $\alpha = 0.05$, is 5.991. The null hypothesis is rejected because 15.050 is greater than 5.991. We conclude that the birds do not distribute themselves among the three feeding locations in the same proportions in spring as in autumn.

6. Regression and correlation

Regression and correlation statistically relate two different sets of data considered to vary with one another. For example, both age and trunk diameter data might be collected from a group of trees, as follows:

tree	age (yr)	diameter (cm)
1	4	5.3
2	6	7.3
3	8	10.5
4	10	12.1
5	12	15.2

and the data may be plotted as a scatter of points on a two-dimensional graph as in figure 1B.3.

6.1. The Regression Equation   For the data in figure 1B.3, the diameter of a tree is assumed dependent on its age (and not vice versa), so diameter is termed the *dependent variable* (and conventionally plotted on the vertical axis), and age is the *independent variable* (on the horizontal axis). When a dependent and independent variable may be so designated we deal with a *regression* relationship and can derive a simple equation for the line that best passes through the data points.

The equation for a straight line is:

$$Y = a + bX, \qquad (21)$$

which often is called a *regression equation*. Y is the dependent variable, X is the independent variable, and a and b are the two regression statistics computed from the data. The statistic b is the *slope* of the line, which tells how much change in Y (tree diameter) exists for each unit change in X (each year of age). If the value of b is positive, then there is an increase in Y as X increases; if it is negative, Y decreases as X increases. If $b = 0$, then there is no

---

1. The smallest possible contingency table is that with two rows and two columns. For such a table, the *Yates correction for continuity* should be employed, as it was in section 4, above; simply substitute the numerator of equation 17 in place of the numerator of equation 19.

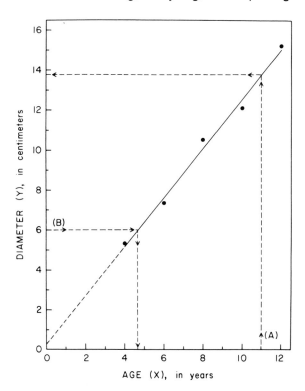

Figure 1B.3. A scatter plot of tree age and trunk diameter data, on which is drawn the least squares regression line ($Y=0.24 + 1.23X$). The broken line extension from the regression line shows how an extrapolation of the line to $X=0$ crosses the Y-intercept, 0.24cm (see section 1B.6.1). Also shown is a prediction of Y, given X (follow arrows from "A"); and a prediction of X, given Y (follow arrows from "B") (see sections 1B.6.2 and 1B.6.5, respectively).

change in Y for a change in X. The statistic a is called the *Y-intercept*, for it is the value of Y when X is mathematically set to zero (that is, if one extrapolates the regression line so that it crosses the Y axis at $X = 0$, it will cross it at $Y = a$). A regression line is uniquely defined by stating its a and b.

Computing the "best fit" regression line through the data points is performed by the method of "least squares," whereby one calculates:

$$b = SP/SS_X \qquad (22)$$

and

$$a = \bar{Y} - b\bar{X}. \qquad (23)$$

To compute b (called either the slope or the *regression coefficient*), one uses the sum of squares of the X values, SS_X, calculated from equation 3, and SP, a quantity called the *sum of the crossproducts of the deviations of* X *and* Y *from their means*. SP is defined as:

$$SP = \Sigma(X - \bar{X})(Y - \bar{Y}). \qquad (24)$$

The tediousness of subjecting large amounts of data to equation 24 is relieved by using a mathematically equivalent "machine formula" for which many calculators are well suited:

$$SP = \Sigma XY - \Sigma X \Sigma Y/n. \qquad (25)$$

For our data:

$$\Sigma XY = (4)(5.3 + (6)(7.3) + (8)(10.5)$$
$$+ (10)(12.1) + (12)(15.2)$$
$$= 452.4 \text{ yr-cm}$$
$$\Sigma X = 40 \text{ yr}$$
$$\Sigma Y = 50.4 \text{ cm}$$
$$n = 5$$
$$SP = 452.4 \text{ yr-cm} - (40 \text{ yr})(50.4 \text{ cm})/5$$
$$= 49.2 \text{ yr-cm}$$

Using equation 1:

$$\bar{X} = 40 \text{ yr}/5 = 8.0 \text{ yr}$$

and

$$\bar{Y} = 50.4 \text{ cm}/5 = 10.08 \text{ cm}.$$

Using equation 3, SS_X can be computed:

$$\Sigma X = 40 \text{ yr}$$
$$\Sigma X^2 = 4^2 + 6^2 + 8^2 + 10^2 + 12^2 = 360 \text{ yr}^2$$
$$n = 5$$
$$SS_X = 360 \text{ yr}^2 - (40 \text{ yr})^2/5$$
$$= 40 \text{ yr}^2$$

So the regression statistics for our data are:

$$b = 49.2 \text{ yr-cm}/40 \text{ yr}^2 = 1.23 \text{ cm/yr}$$

(that is, members of our tree population increase in diameter an average of 1.23 centimeters per year), and

$$a = 10.08 \text{ cm} - (1.23 \text{ cm/yr})(8.0 \text{ yr}) = 0.24 \text{ cm}.$$

6.2 Prediction   By inserting the computed values of a and b into equation 21, we have:

$$Y = 0.24 + 1.23X,$$

enabling us to predict diameters for a tree of a given age. To calculate the expected diameter of an eleven-year-old tree, for example, we simply insert $X = 11$ yr into our equation and predict:

$$Y = 0.24 + (1.23)(11) = 13.77 \text{ cm}.$$

Alternatively, predictions may be made graphically, as shown in figure 1B.3.

Care must be exercised in extrapolating—in predicting values of Y for values of X outside the range of ages originally used to calculate the regression line—for spurious results can thereby be obtained (Zar, 1984: 267).

The computed regression line may be drawn on a graph by connecting any two predicted points. For example, we have calculated a predicted Y of 13.77 cm for $X = 11$ yr, and we know that $Y = 0.24$ cm when $X = 0$ (i.e., the Y-intercept is 0.24 cm). So a line may be drawn along these two points, resulting in the regression line shown in figure 1B.3.

6.3. Significance Testing   The regression line slope, b, calculated from a statistical sample is an estimate of the slope within the entire statistical population (symbolized by the Greek beta, β). By testing the significance of b, we ask whether there is indeed a relationship between Y and X in the population. The null hypothesis states that there is not (H_o: $\beta = 0$); the alternate hypothesis states there is (H_A: $\beta \neq 0$). A t-test analogous to that in section 1B.3.1 is one appropriate procedure for testing the significance of a regression. In this case the test statistic, t, is computed as:

$$t = |b|/s_b, \qquad (26)$$

where s_b, the standard error of b, is:

$$s_b = \sqrt{s^2/SS_X}, \qquad (27)$$

SS_X is the sum of squares of X as used previously, and

$$s^2 = \frac{SS_Y - (SP)^2/SS_X}{n-2}. \qquad (28)$$

For the present data, SS_Y, the sum of squares of the sample Y values, is computed by subjecting the tree diameters to equation 3, resulting in 61.05 cm². Then:

$$s^2 = (61.05 \text{ cm}^2 - (49.2 \text{ yr-cm})^2/40 \text{ yr}^2)/(5-2)$$
$$= 0.178 \text{ cm}^2$$
$$s_b = \sqrt{0.178 \text{ cm}^2/40 \text{ yr}^2}$$
$$= \sqrt{0.00445 \text{ cm}^2/\text{yr}^2}$$
$$= 0.067 \text{ cm/yr}$$

and

$$t = \frac{|1.23 \text{ cm/yr}|}{0.067 \text{ cm/yr}} = 18.36$$

The critical value of t for this test is that for $n - 2$ degrees of freedom; using $\alpha = 0.05$, this value (from table 1B.1) is 3.18. As 18.36 is greater than 3.18, we reject H_0: $\beta = 0$ and conclude that there is a change in diameter with age in the population sampled.

6.4 Confidence Intervals   To calculate the precision of the computed rate of change of diameter with age, a confidence interval may be stated for β:

$$(1 - \alpha) \text{ confidence interval for } \beta = b \pm ts_b, \quad (29)$$

where t is the same critical value used in the t test (that for DF $= n - 2$). Therefore, for our data:

95% confidence interval for β = 1.23 cm/yr $\pm$
$$(3.18)(0.067 \text{ cm/yr})$$
$$= 1.23 \text{ cm/yr}$$
$$\pm 0.21 \text{ cm/yr}$$

and it may be asserted, with 95% confidence (i.e., with a 5% chance of error), that tree diameter in the population sampled increases with age at a rate between 1.02 and 1.44 cm/yr.

Confidence intervals for predicted Y values are computed as:

$$Y \pm ts_Y, \qquad (30)$$

where t has $n - 2$ degrees of freedom, and s_Y, the standard error of the predicted Y, is:

$$s_Y = \sqrt{s^2\left[\frac{1}{n} + \frac{(X - \bar{X})^2}{SS_X}\right]}. \qquad (31)$$

In equation 31, X is the value of the independent variable for which Y is being predicted. For our above (section 6.2) prediction of $Y = 13.77$ cm for $X = 11$ yr:

$$s_Y = \sqrt{0.178\left[\frac{1}{5} + \frac{(11 - 8.0)^2}{40}\right]}$$
$$= \sqrt{0.0757}$$
$$= 0.28 \text{ cm},$$

and the 95% confidence interval for the predicted Y is, by equation 30:

$$13.77 \text{ cm} \pm (3.18)(0.28 \text{ cm})$$
$$= 13.77 \text{ cm} \pm 0.89 \text{ cm}.$$

Thus we conclude, with 95% confidence, that in the population we sampled, the mean diameter of eleven-year-old trees is between 12.88 and 14.66 cm.

6.5. Inverse Prediction   "Inverse prediction" is used to estimate a population value of X corresponding to a given value Y. For example, we may measure a tree diameter of 6 cm and wish to predict how old the tree is. Rearranging the regression equation (equation 21) gives us:

$$X = \frac{Y - a}{b}, \qquad (32)$$

so we may compute:

$$X = (6 - 0.24)/1.23$$
$$= 4.68 \text{ yr}.$$

Alternatively, the prediction of X could be done graphically, as shown in figure 1B.3.

Regarding the precision of the computed age estimate, the 95% confidence interval for an inverse prediction may be computed as:

$$\bar{X} + \frac{b(Y - \bar{Y})}{c}$$
$$\pm \frac{t}{c} \sqrt{s^2\left[\frac{(Y - \bar{Y})^2}{SS_X} + c\left(1 + \frac{1}{n}\right)\right]}, \qquad (33)$$

where Y is the diameter for which age, X, is to be predicted, t has $n - 2$ degrees of freedom, and,

$$c = b^2 - t^2 s_b^2. \tag{34}$$

So, for 6 cm:

$$c = (1.23)^2 - (3.18)^2(0.00445)$$
$$= 1.47$$

and the 95% confidence interval for the prediction of 4.68 years is:

$$8.0 + \frac{1.23(6 - 10.08)}{1.47}$$
$$\pm \frac{3.18}{1.47} \sqrt{0.178\left[\frac{(6 - 10.08)^2}{40} + 1.47\left(1 + \frac{1}{5}\right)\right]}$$
$$= 8.0 - 3.41 \pm 2.30 \sqrt{0.178[2.18]}$$
$$= 4.59 \text{ cm} \pm 1.43 \text{ cm.}$$

Thus we can state with 95% confidence that the age of a tree 6 cm in diameter is between 3.16 and 6.02 years. (Note that contrary to other confidence intervals encountered the confidence interval for a predicted X is not symmetrical around the predicted value—in this case 4.68 yr.)

6.6. Correlation   If two variables are hypothesized to vary with each other, but one is not dependent on the other, then we are dealing with *correlation analysis*. For example, we may wish to correlate tree trunk diameter and tree height. The *correlation coefficient, r,* is a unitless number that may have a value in the range of -1.0 to $+1.0$. A positive correlation implies a direct relationship (e.g., large diameters are associated with tall trees), and the larger the r the stronger the relationship. A negative r would indicate an inverse relationship (e.g., tall trees had narrow trunks and short trees had thick trunks). A correlation coefficient of zero indicates no relationship between the two variables.

The calculation of r involves quantities we have already defined, where either of the two variables may be labeled X and the other labeled Y:

$$r = \frac{\text{SP}}{\sqrt{\text{SS}_X \text{SS}_Y}}. \tag{35}$$

To test the significance of r we are asking whether the population correlation coefficient (indicated by the Greek rho, ρ) is different from zero (i.e., we test H_0: $\rho = 0$ against H_A: $\rho \neq 0$). This may be done by:

$$t = |r|/s_r, \tag{36}$$

where

$$s_r = \sqrt{(1 - r^2)/(n - 2)}, \tag{37}$$

and the critical value of t is associated with $n - 2$ degrees of freedom.

7. Selected references

Dixon, W. J., and F. J. Massey, Jr. 1983. Introduction to statistical analysis. 3rd ed. McGraw-Hill Book Co., New York.

Green, R. H. 1979. Sampling Design and Statistical Methods for Environmental Biologists. John C. Wiley & Sons, New York.

Milton, R. C. 1964. An extended table of critical values for the Mann-Whitney (Wilcoxon) two-sample statistic. J. Amer. Statist. Assoc. 59:925–934.

Prince, S. D. 1986. Data anlalysis, 345–375. In P. D. Moore and S. B. Chapman (eds.), Methods in plant ecology. 2d ed. Blackwell Scientific Publications, Oxford, England.

Rohlf, F. J., and R. R. Sokal. 1981. Statistical tables. W. H. Freeman and Co., San Francisco.

Snedecor, G. W., and W. G. Cochran. 1980. Statistical methods. Iowa State University Press, Ames, Iowa.

Sokal, R. R., and F. J. Rohlf. 1981. Biometry: The principles and practice of statistics in biological research. 2d ed. W. H. Freeman and Co., San Francisco.

Steel, R. G. D., and J. H. Torrie. 1980. Principles and procedures of statistics: A biometrical approach. 2d ed. McGraw-Hill Book Co., New York.

Zar, J. H. 1984. Biostatistical analysis. 2nd ed. Prentice-Hall, Englewood Cliffs, N.J.

1. Introduction

Ecological research involves designing a study, collecting samples, measuring variables, analyzing data, and presenting the results in a formal report. The process of writing, evaluating, and rewriting research findings makes the author think more deeply about the study. The principal objectives of a research report are to present a record of one's work and to communicate the ecological ideas inherent in that work. Accurate, clear, and concise writing is essential to effective communication among researchers, teachers, and students. A scientific research report provides a writing experience different from that associated with a library term paper, for a research report is based on one's own data and personal involvement in an organized investigation.

2. Format and style

Generally, a biological paper has a title and byline (the later identifying the authors and their institutional affiliations), followed by such sections as Introduction, Materials and Methods, Results, Discussion, Summary, and Literature Cited (or References). Often an abstract at the beginning of the report will appear in place of or in addition to the summary. This format serves as a framework for preparing a more detailed working outline, which is a necessary first step in constructing a research paper.

Manuscripts are typed double spaced and with margins of one to one and one-half inches, and each page is numbered. Avoid the use of footnotes. Follow the conventions of section 8 below for referencing. A heading is customarily typed for each of the major sections of the report. Indented subheadings in a section may also be included for clarity. These subheadings generally are a product of the detailed working outline.

The style of a scientific paper varies, depending on the writer and his or her audience. The writing style of scientific papers often is poor, largely because the authors lack experience and training in writing. For the preparation of biological papers, the *CBE Style Manual* (Council of Biological Editors, 1983) is a standard reference for form and style; it is a book with which every serious biological scientist should become familiar. A good summary of report writing fundamentals, with an ecological emphasis, is provided by Scott and Ayars (1980). The following general guidelines gleaned from these sources should be helpful:

1. Wherever possible, use the first person (I or we) instead of awkward indirect statements (this author, these researchers).
2. Avoid long involved sentences and overuse of polysyllabic words. Long, run-on sentences often obscure your meaning, and frequent use of cumbersome words reduces the readability of the paper. Check for excessive use of commas and conjunctions (and, but, or). These often connect clauses that can be more clearly separated into two or more sentences.
3. Use the active voice instead of the passive voice. For example, "I measured the water temperature" is preferable to "The water temperature was measured by the author," as it uses fewer words and is unambiguous (i.e., it is clear who measured the temperature). And "I measured forty-four trees" is preferable to "forty-four trees were measured," because the latter statement does not tell us who performed the measurement.
4. Avoid excessive use of nouns as adjectives. Such use of nouns often is acceptable (*temperature* stratification or *tree* height), but it frequently is overused (e.g., *morning lake water temperature profile record sheet* format).
5. Be positive in your writing. Don't hide your findings in noncommittal statements. For example, "the data could possibly suggest" implies that the data actually may show nothing; simply state "the data show."
6. Avoid noninformative abbreviations such as "etc." and phrases such as "and so on" or "and the like."
7. Keep specialized jargon to a minimum. If (but only if) vernacular terminology is just as accurate, use it. Similarly, excessive use of Latin nomenclature should be avoided. If acceptable common names exist for organisms, introduce them together with the Latin names, and thereafter use the former. Otherwise, identify the Latin names. Whenever Latin genus or species names are written they are to be either italicized or underscored; higher taxonomic ranks—e.g., family, order, class, phylum—are not italicized or underlined.
8. Avoid repeating facts and thoughts. Decide in which portion of the report different statements are best placed, and do not repeat them elsewhere.

9. Be concise and succinct. Avoid verbosity in writing. For example, say "many species" rather than "a large number of species," and say "because" rather than "due to the fact that." Include all that is necessary, but don't pad the report with data irrelevant to the purpose or conclusions of the study.

3. Introduction section

In the introduction of the paper state the nature of the problem, objectives of the study and any hypotheses to be tested. Also, give a brief background for the study, which would typically include a brief review of the literature. Relate the problem and its significance to the general discipline of study. This part of the paper presents the background, justification, and relevance of your study.

4. Materials and methods section

Procedures in research reports are usually detailed enough for the reader to have an accurate idea of what was done in the study or to be guided to appropriate literature for this information. A good description of materials and methods used is one that would enable a reader to duplicate your investigative procedure. Keep to a minimum the details of standard and generally known procedures (such as how an item was weighed). Detailed published accounts, such as chemical formulations for reagents, may be omitted but should be referenced. In a field study, a general description of the study site is called for. If this description needs to be lengthy, then it may comprise a separate subsection (or a new section).

5. Results section

This portion of a report gives the facts found, even if they are contrary to hypothesis or expectation. Listings of raw data are rarely presented, except occasionally in a class activity or as an appendix to the report. Instead, data typically are summarized using means, frequency tables, percentages, or other descriptive statistics for presentation and analysis in some appropriate statistical manner (see section 1B). These data summaries may be incorporated into figures or tables if this results in additional clarity or helps illustrate a pattern or trend.

In general, the number of data collected should be indicated, and some measure of variability of the data should accompany statements of means (see section 1B). Statistics used, type of data analysis performed, and mode of presentation depend on the study and type of data collected. Statistical comparisons of different groups of data are often called for, as explained in section 1B.

The Results section is not just a data summarization or a collection of tables and figures; it should contain an explanation and description of the data. Tell the reader exactly what you found, what patterns, trends, or rela-

tionships were observed. For example, do not just say "The species-area curve is shown in figure 1." Tell the reader what is being presented, as "Figure 1 shows that the number of species in the habitat increases and then levels off as the area of the habitat increases."

Illustrations in the Results section may consist of graphs, photographs, or diagrams that visually depict your results. All such illustrations are individually numbered and cited in the text and referred to as a figure (e.g., "Dominance of sugar maple is shown in figure 4."). Labeling and citing tables of data in the text is done in the same manner as for graphs. If a graph will summarize the data as well or better than a table, then the graphical presentation typically is preferable. Each figure and table should contain an explanatory legend. In standard thesis and publication manuscripts the figure number, figure title, and legend are generally on a separate page from the illustration. Be sure the axes of all graphs are fully and correctly labeled with a scale marked off and the units of measurements given; units of measurement (preferably metric) must also be given for tabular data. (Appendix B provides conversion factors for common measurement scales.) Avoid the tendency to cram too much information into one graph or table, thus losing readability.

You may benefit from examining various portions of this book (e.g., section 2A) to observe how figures and tables are titled and are referred to in the text.

6. Discussion section

In the previous section of the paper the results are summarized and described. In this section they should be interpreted, critically evaluated, and compared to other research reports; and conclusions should then be drawn based on the study and its findings. Whereas the Results section presents the "news," the Discussion section contains the "editorial." Some research reports have a combined Results and Discussion section, and in some the conclusions are placed in a separate section or are included in a Summary and Conclusions section.

In the discussion, examine the amount and possible sources of variability in your data. Examine your results for bias and evaluate its consequences in data interpretation. Develop arguments for and against your hypotheses and interpretations. Do not make generalized statements that are not based on your data, known facts, or reason. Be sure to relate your findings to other studies and cite those studies. Draw positive conclusions from your study whenever possible.

7. Summary section

The end of your paper should contain a summary, which is a concise but exact statement of the problem, your general procedure, basic findings, and conclusions. It should not be just a vague hint of the topic covered, an amplified table of contents, or a shortened version of the report. In many scientific journals, an abstract of the paper at the beginning of the paper replaces a summary. Some research papers include a separate Conclusions section between the Discussion and Summary sections.

Example of a poor summary:

The food habits of various amphibians were studied in detail by the authors. The data were analyzed statistically and the findings were discussed at length. Certain similarities and differences were found between the species studied and the habitats in which they were found. Conclusions about feeding habits, habitat relationships, and niches were made for these species.

This summary or abstract is merely an expanded table of contents with verbs added to make complete sentences. Notice that no specific information is given to the reader.

Example of an acceptable summary:

Stomach contents of the red eft, red-backed salamander, and dusky salamander were identified. Analysis of overlap of food taxa shows that the feeding habits of only the latter two species were similar. As an example of niche segregation, the salamanders show less feeding overlap in habitats where they are living together.

8. Literature cited section

No comprehensive literature survey is required for a class research report; however, you are expected to use some sources other than a textbook (such as technical journals and reference works). These sources should be cited in the body of your report. Useful references are given at the end of each section in this manual, in textbooks, and in the Literature Cited or References sections of scientific papers. It is up to you to select the most useful references. All references given in your paper must appear in the Literature Cited section. Rarely (e.g., in an instructional report), it may be desirable to list references in addition to those cited in the paper. In this case the heading Literature Cited should be replaced by Bibliography, or Suggested References, or Selected References.

References may be cited in the text of your paper in one (but not both) of two forms: (1) by author and year or (2) by number. Citation by author and year is more common in biological writing; for example:

Smith (1980) stated that eastern grasslands are either tame or seral.

or

Eastern grasslands are either tame or seral (Smith, 1980).

If there are two authors of the reference, then they are referred to as "Smith and Jones"; if there are more than

two, then "Smith et al." is written (although all authors will be listed in the Literature Cited section). All references are then listed in the Literature Cited section in alphabetical order of the first author's surname. (If there are more than one reference for an author, they are listed chronologically for that author.)

If the reference numbering system is used, then the text citation would be of the following form:

Eastern grasslands are either tame or seral (21).

and the Literature Cited section would consist of a listing of references in numerical instead of alphabetical order.

For a book in a list of references, the general form is:

Smith, R. L. 1980. Ecology and field biology. 3rd ed. Harper & Row, New York.

where the author (all authors if more than one) is followed by the year of publication, the title, and the name and location of the publisher. Sometimes the number of pages is also indicated at the end of the citation (e.g., ". . .835p.").

For a journal article, the general form of citation is:

Greenwald, G. S. 1956. The reproductive cycle of the field mouse, *Microtus californicus*. J. Mammal. 37: 213–222.

where the author (all authors if more than one) is followed by the year of publication, the title, and the journal name, volume, and page numbers. In journal citations it has been customary to use standard abbreviations for the name of the journal (as above), but it is an increasing practice to spell out the entire name, especially if the audience is a general one that might not recognize the abbreviations.

You may benefit from observing the various chapters in this book to see how literature may be cited.

9. Some common problems

1. Use, evaluate, and interpret your data. Failure to do so is the most common problem students have in report writing. Many will calculate their results and make figures and tables, thereafter leaving these data to sit idly in the paper without any explanation or elaboration.
2. Do not ignore results because they differ from textbook generalizations. Your data are not incorrect just because they do not agree with some general principle or a conclusion in another report.
3. Use reference material only if pertinent to your data. Often, much irrelevant information is brought into reports.
4. Be careful about making small differences seem important. Different values are not necessarily significantly different. If you have not used statistical testing

(see Section 1B), you should at least consider in your subjective evaluation the amount of variability in your data.

5. Do not discard data because of variability and biases. There are some errors in nearly all scientific data. If recognized and accounted for in interpretation of results, errors of reasonable size need not discredit your data.

6. Round off final quantitative results to no more digits than can be reasonably justified. What sense does it make to compare two numbers such as 17.289761 and 19.82946? Do the last several digits have any special meaning? Reporting 17.3 and 19.8 may suffice in your case.

7. Label figures and tables properly and thoroughly and cite them in your text. Too often figures and tables are inserted in a report without identifying their contents or explaining their purpose to the reader.

8. Play around with your data before preparing the final graphs and tables. Get your mind working over the data, in order to seek patterns and trends. Try to organize the data in various ways, as different presentations may elucidate different patterns or trends. But be careful not to force a preconceived conclusion on the data.

9. Do not select or reject data in order to make desired results apparent. Any "fudging" of data is dishonest and unacceptable.

10. Do not perform calculations on data just for the sake of calculating. Have a reason for and draw conclusions from the calculations performed. Padding your report with excessive though honest numbers serves no useful function.

11. Document ideas, conclusions, and hypotheses with data, facts from the literature, and sound reasoning. Do not leave your ideas up in the air without support or they will fall with the first touch of the instructor's red pencil.

12. Relate your results and conclusions to accepted principles and concepts. Explain any discrepancies.

10. Selected references

Baker, S. 1984. The complete stylist and handbook. 3d ed. Harper and Row Publishers, New York.

Council of Biological Editors, Committee on Form and Style. 5th ed. 1983. CBE Style Manual. American Institute of Biological Sciences, Washington, D.C.

Rathbone, R. R. 1972. Communicating Technical Information in Scientific and Engineering Writing. Addison-Wesley Publishing Co., Reading, Mass.

Scott, T. G., and J. S. Ayars. 1980. Writing the scientific report, 55–60. In S. D. Schemnitz (ed.), Wildlife management techniques manual. 4th ed. Wildlife Society, Washington, D.C.

University of Chicago Press. 1982. The Chicago manual of style. 13th ed. University of Chicago, Chicago.

introduction

The *habitat* is the place where an organism or a group of organisms lives, and it is described by its geographic, physical, chemical, and biotic characteristics. *Environment* refers to the total set of conditions, biotic and abiotic, that surround and influence the biota and its habitat, including influences from outside the habitat. (For example, ozone in the upper atmosphere is an environmental factor that affects the amount of ultraviolet radiation in the habitat.) Some writers have used "habitat" and "environment" synonymously.

Another basic ecological concept is the *community,* the aggregation of interacting species in a habitat. Although the habitat has biotic and abiotic components, we should not confuse it with the concept of an *ecosystem,* which is a community plus its relationships with its abiotic environment. Habitat analysis measures and describes the settings in which organisms live, while ecosystem analysis studies a system of exchanges and interactions between a community and its abiotic environment. A related concept is that of the *niche,* the functional role of a species in an ecosystem.

1. Divisions of a habitat

The overall habitat of a community of organisms is the *macrohabitat.* It is divided into smaller units, or *microhabitats,* each of which is the portion of the habitat directly encountered by a population of a given species. Thus for example, we may consider the macrohabitat of a deciduous forest and the microhabitat of a population of oaks, warblers, or millipedes. We may also consider several ecologically related species as occupying a given microhabitat; for example, one may study the soil microhabitat or the microhabitat defined by a rotting log.

The habitat should be treated as a biophysical entity containing many dimensions. Collectively, they can provide a comprehensive and concise profile of where a population or community lives. We may consider four basic dimensions of a habitat: temporal, spatial, physical-chemical, and biotic. Each of these is then subdividable into other components.

The temporal dimension includes components related to daily and seasonal characteristics and components that are associated with time-related phenomena such as population growth and ecological succession. The spatial dimension includes geographical components such as location and topography as well as spatial patterns like stratification and zonation within habitats. The physical-chemical dimension includes three basic components: the *atmosphere* (air) the *lithosphere* (substrate), and the *hydrosphere* (aquatic component). Those portions of the atmosphere, lithosphere, and hydrosphere that contain life are collectively called the *biosphere.*

unit 2

analysis of habitats

2. Application of habitat studies

Habitat studies have become a focal point in ecology. They find many practical applications; e.g., assessment of environmental impacts; land use planning; management of fish, wildlife, and vegetation habitats for the benefit of desired species; and habitat reclamation.

Habitat studies should have clearly stated objectives so as to avoid extensive collection of irrelevant data. One then draws conclusions as to the type of habitat data and level of detail needed to meet the objectives. The objectives of a habitat study determine the data collection requirements; and the principal objective often is one of the following:

1. Basic ecological research: Proposing and testing hypotheses relating to ecological theories and principles
2. Ecological inventories: Collection of data and samples to be used for reference, as baseline—or predisturbance—data or historical documentation
3. Environmental planning: Use of ecological information for proposing potential sites for preservation, management, or other environmentally sound uses
4. Environmental impact assessment: Collection of information for assessing present or potential impacts from human activities
5. Ecological resource management: Collection of information needed for management of species populations and for reclamation of disturbed habitats

3. Study options

There is no single optimum systematic procedure for habitat analysis, and several options can be developed for a habitat study, depending on the objectives of the study, the geographical location, the type of habitat, the species of primary interest in that habitat, and the resources available to conduct the study (i.e., available time, personnel, supplies, equipment, and money). Ideally, a good habitat study would include the following:

1. An examination of pertinent documents and literature, including maps, aerial photographs, and local and regional studies. This is valuable for selection of study sites, preliminary description of the habitat, and collection of existing data that need not be duplicated.
2. Site survey or reconnaissance. Relevant data collected in the field.
3. Summary and analysis of documents and field data, including drawing conclusions related to the study's objectives.
4. Field verification of results. A follow-up site visit often is desirable to verify interpretation of aerial photographs, habitat maps, and other information.

A detailed analysis of a habitat cannot be performed adequately during only one or a few short field trips. Therefore, you may select one of several options from the sections in unit 2. A first level of analysis is to prepare a general habitat (macrohabitat) description by recording information about the geographic, climatic, geologic, and biotic factors most important to the ecological community (section 2A). A more detailed analysis of a specific habitat component, such as weather and soil and water quality (sections 2B, 2C, 2D, and 2E) can also be conducted. Thirdly, a microhabitat study can be designed to evaluate the environmental factors affecting one or more species. A microhabitat analysis would best be conducted in conjunction with studies in units 4, 5, and 6.

2a
macrohabitat analysis

1. Introduction

In a description of the macrohabitat one summarizes the dominant features of a habitat; this can serve several purposes. It is useful for regional studies, for categorizing habitats, for preliminary or baseline assessments of habitats, and for descriptive overviews associated with more detailed microhabitat analyses. This section may be completed as a terrestrial habitat analysis associated with exercises in units 4, 5, and 6, or in conjunction with more detailed habitat analyses in the rest of unit 2. See section 2D for analysis of aquatic habitats.

A macrohabitat analysis should include a brief description of the dominant temporal, spatial, physical-chemical, and biotic components of the environment, and each description may refer to the three distinct yet interrelated portions of the biosphere: atmosphere, lithosphere, and hydrosphere. In terrestrial habitats, a succinct description of the vegetation should also be included, or a more detailed biotic description may be obtained, as indicated in section 5A.

2. General procedures

To gather information efficiently, each of several teams of investigators may be responsible for obtaining information on one or more specific portions of the study. Four tasks may be completed: document research, field investigation, summary and analysis of information, and verification of results. By examining all available documents, including maps, aerial photographs, and relevant local studies on the site of interest, one can collect known information concerning the habitat, plan the field investigation, and become familiar with the study site. The field investigation may range from a walk-through site reconnaissance to a detailed collection of environmental data, depending on the objectives of the study and the amount of data already available. Prior to the field study, a rough map of the study site should be prepared from existing maps and photographs, so that annotations can be made on this map during the investigation. A habitat data form should be prepared in advance so that data collection is orderly and complete.

After the field investigation, all relevant information derived from the document and field study can be compiled and summarized. The information thus summarized can be in the form of a table, a detailed habitat map, or a written description. Where possible, the information obtained from maps and aerial photographs should be verified, preferably by a field check.

3. Temporal information

Recording of time is needed for all habitat analyses; this should include the date and time of day. The distribution and abundance of the physical-chemical components vary in both time and space, and this variation, in turn, may influence the distribution and abundance of biota. Time is also important because plants, animals, and microorganisms exhibit daily and seasonal cycles of activity, including reproduction and productivity. Physical-chemical variables may also change periodically. Temporal information can also be noted as phases or stages, such as season, successional stage, or growth phase.

4. Spatial information

Spatial information such as locality, topography, and distribution pattern is needed for habitat studies. Certain geographic information is easily obtained from topographic maps.[1] On the basis of these maps a locality can be specified by latitude, longitude, and section number. The habitat location should be described in detail, including the major political units from the largest to the smallest, such as country, state or province, county, and township. The specific locality is given as the distance (in kilometers) and compass direction from the nearest city or village. (Appendix B gives metric conversions.) Names of bodies of water and specific landmarks in or near the habitat should be recorded as aids in locating the site.

Topography is a description of the spatial arrangement of the habitat surface. Spatial characteristics such as elevation, slope, curvature, and direction of slope (i.e., aspect) affect critical physical factors such as drainage, soil properties, temperature, and light intensity. Topography depends on the geological land forms discussed below. Record the elevation in meters above sea level and give a general description of the land curvature. In ad-

1. Information on the availability of topographic maps for specific areas may be obtained from the Map Information Office, U.S. Geological Survey, Washington, DC 20242. A good start is to request the index of topographic maps for the state in question. Of additional interest might be the nautical charts prepared for U.S. sea coasts and large lakes, available from the Distribution Division (C44), National Ocean Survey, Riverside, MD 20840. Colleges, universities, and government agencies often maintain map libraries pertinent to local areas.

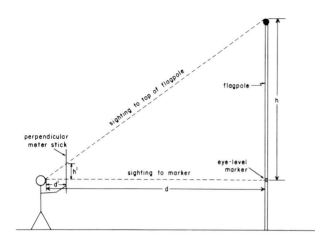

Figure 2A.1. Estimating the height of an object (as a tree or a flagpole) using a meter stick and elementary trigonometry.

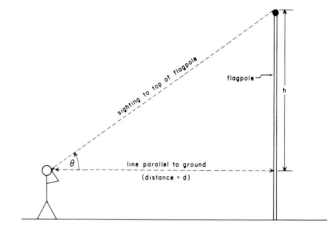

Figure 2A.2. Estimating the height of an object (as a tree or a flagpole) using an Abney level and elementary trigonometry.

dition, measure the area of the habitat site studied. An approximate area may be determined from outlining the study site on a topographic map or from calibrating measurements on an aerial photograph. The habitat area can be determined by partitioning the outlined site into a grid and counting the number of grid squares within the outline. More accurate measurements may be made by tracing the area, using a planimeter. Area on the photograph may be converted to actual land area (e.g., hectares) if the scale of the photograph is known. Aerial photographs can also provide valuable information concerning neighboring habitats and present land uses.[2] Remote sensing of habitats has become a sophisticated science, using recent technological improvements in cameras, infrared and color film, and computerized spectral analysis from space satellites. For details on the use of remote sensing in ecology, see Anderson, Wents, and Treadwell (1980).

The slope and aspect (i.e., direction) of the study area affect factors such as light, temperature, and soil moisture. The difference in elevation between two points may be expressed as relative to the horizontal distance between them (e.g., a slope of 15 m per 100 m). Measurement of elevation may use the same principles as shown in figures 2A.1 and 2A.2. In figure 2A.1, the observer holds a meter stick vertically and sights up the slope to a point as far off the ground as is the observer's eye. (This is conveniently done by sighting the head of a person standing upslope.) Then, the slope of the land is h'/d', where h' is the vertical distance on the meter stick between the eye height and the line of sight, and d' is the horizontal distance from the eye to the meter stick. Slope is often expressed as a percentage; for example, if the slope were 15 m/100 m, we could refer to it as a 15% slope; or if $h' = 10$ cm and

$d' = 50$ cm, then the slope would be 10/50 or 20%. Slope may also be expressed as an angle, by determining from trigonometric tables the angle that has a tangent of h'/d'. An alternative method, which gives a direct and more accurate angle of the slope, uses an Abney level (figure 2A.2). The angle is measured by sighting a point upslope that is as high above the ground as the observer's eye.

5. Lithosphere

Two subcomponents of the lithosphere should be evaluated: geological land forms and soil. The land form greatly affects physical factors such as drainage, soil properties, and microclimate. The land-form features depend on the physical and chemical properties of the substrate, source of the substrate (i.e., water, wind, glacial, or in situ deposits) and erosion.

Record general land forms (such as in table 2A.1). Approximate dimensions of major land forms, including elevation, diameter, or length, should be given. The principal type of geological substrate (such as granite, sandstone, shale, limestone, etc.), should be noted. Bare regions (such as rocky outcrops, cliffs, or eroded areas) should be recorded, along with their approximate sizes. Note also the nature and sizes of human-made features such as buildings, towers, power lines, bridges, fences, roads, railroads, or cemeteries.

Soil factors that are useful for characterizing macrohabitats include the major soil order, soil type, and parent material. In addition, a brief description of the upper soil horizons may be useful, such as a description of the litter layer and the depth of the adjacent horizons. More detail on soil horizons can be collected as described in section 2C.

Long-term interactions of climate, topography, and biota with parent substrate have resulted in a variety of soils in different regions. Two systems of classification of major soil types are in use today. The more recent system

Table 2A.1. *Some principal land forms.*

Alluvial (water body deposits)	Coastal
alluvial fan	delta
terrace	coastal plain
flood plain	outwash
meander	dune
levee	spit
delta	peneplain
	bench
Glacial (formed by glaciers)	reef
till sheet	fiord
kettle	cliff
outwash	beach
plain	bar
esker	tidal flat
terminal moraine	barrier island
lateral moraine	swale
medial moraine	
drumlin	**Arid**
kame	butte
	dune
Associated with mountains	mesa
ridge	canyon
talus	badlands (highly eroded)
volcano	playa (dried-up lake basin)
slope	fan slope
valley	plateau
dome	plain
peak	
plateau	
canyon	
cliff	
basin	
bluff	

is that recommended by the U.S. Department of Agriculture Soil Conservation Service (SCS) (Soil Survey Staff, 1960, 1967, 1975). However, the older system is still in wide use by ecologists and will only gradually be replaced by the new classification in future literature. Both systems classify soils in a hierarchical manner similar to that of biological classification. The older system has three orders subdivided into great soil groups and then into soil families, series, and types. The SCS system is a complete revision of the soil orders and great soil groups. In it there are ten soil orders divided into 44 suborders, and a large number of further subcategories. The suborders resemble the old great soil groups, although there are many differences. For a nontechnical discussion of them consult Wolfanger (1971).

The name of the soil type includes a description of the texture as well as an indication of the soil series (e.g., Miami silt loam). A series is comprised of soils of different textures but alike in color, depth, structure, and organic content of the horizons. It is generally named after the place or geophysical form in which it was first found and described. The soil type then represents the textural classification within a particular series. In the example above,

the soil series is Miami, and the soil type Miami silt loam. (The Miami series was first described near the Miami River in Ohio.) To determine the series name, consult an SCS soil map for the local county.

6. Atmosphere

A principal determinant of habitat characteristics is climate. Climates are often broadly categorized latitudinally as polar (or Arctic), cold temperate (boreal or subarctic), temperate, subtropical, and tropical. The Köppen system of classification considers temperature and precipitation as associated with vegetation types. The major divisions in this system are arid (subdivided into deserts and plains), microthermal and mesothermal (subdivided into those moist all year, those dry in winter, and those dry in summer), tropical (subdivided into tropical rainforest and tropical savanna), subarctic (boreal), and polar (subdivided into tundra and polar ice). Tables 2A.2 and 2A.3 summarize major climate categories in North America, considering temperature and moisture. Local climate can vary from the regional climate, and is affected by proximity to large bodies of water, large cities, and industrial areas, by elevation, and by topography. Therefore, regional climate descriptions should be further characterized to include these effects, and summary data from a local meteorological station should be given. Such data should include the mean monthly temperature (in degrees Celsius) for the coldest month and the warmest month of the year (usually January and July, respectively), date of last and first frost, mean annual precipitation (mm of rainfall), and monthly precipitation for the driest and wettest months as indications of seasonal precipitation.

It is essential to realize that such data may not represent accurately the conditions at your study site. Therefore, local data should be obtained from sources as close to that site as possible. When using average data, caution should be exercised in correlating these averages with ecological conditions, as extreme conditions may be more important than averages. Often extreme climatic conditions such as periodic droughts can have long-term ecological effects. Include in your data any periodic or recent extremes such as droughts, floods, and extremely early or late frosts.

Tables 2A.2 and 2A.3 can be used to characterize the climate for your region. Since climates may be characterized by a combination of seasonal changes in temperature and moisture, these tables provide the criteria for defining temperature and moisture categories for a given climate. Using this information, together with average monthly and annual data for temperature and precipitation in your region, compare your climate classification with that of climate maps based on Köppen's system or a modified Köppen system (Trewartha, 1968).

Table 2A.2. *Temperature categories for classification of climates.*

Thermal Group	Coldest Month (Mean Monthly Temperature, °C)	Warmest Month (Mean Monthly Temperature, °C)	Growing Season (Days Between Last and First Frost)
Polar (ice)	< −38	< 0	none
Arctic (tundra)	< −38	0 to 10	< 60
Subarctic (boreal)	< −38	10 to 22	60 to 90
Microthermal			
cold	−38 to 0	less than 4 mo between 10 and 22	90 to 120
cool summer	−38 to 0	more than 4 mo between 10 and 22	120 to 180
warm summer	−38 to 0	> 22	120 to 240
Mesothermal			
cool marine	0 to 18	less than 4 mo between 10 and 22	120 to 180
marine	0 to 18	more than 4 mo between 10 and 22	180 to 240
warm summer	0 to 18	> 22	> 240; also with periodic frosts
subtropical	0 to 18	> 22	usually 365; with infrequent frosts
Tropical	> 18	> 22	365
Highlands tropical mesothermal microthermal montane alpine glacier	Mountainous regions or plateaus generally higher than 1500 meters: Temperature and growing season highly variable. Day-night temperature differences often greater than 10°C. Therefore, thermal categories above only approximate analogous mountain zones. Highlands generally exhibit vegetation zonation.		

Criteria based on Trewartha (1968).

7. Community type

A *biome* is a large geographic area characterized by a common predominant climax community (see section 5D). Within a biome, however, several different community types may occur, most of them seral (i.e., intermediate successional stages). The major types of the communities should be recorded using accepted names such as those in table 2A.4. Record the dominant species of plants, those species that are important because of their controlling influence over the amount of light, heat, nutrients, soil, wind, and moisture in the habitat. (More information on biotic factors is presented in section 5A.) Note the successional stage of development by naming the seral stage or climax community. Any known historical events that have influenced the community type (e.g., recent burning, flooding, lumbering, grazing) should be recorded.

Habitat factors not only are associated with geographic location and the topography of the land but often display distinct spatial patterns. Two such patterns are zonation (typically horizontal) and stratification (vertical). Stratification in terrestrial habitats is most commonly seen in vegetation (figures 2A.3 and 2A.4) and soil horizons (section 2D).

Zonation results from a gradient in slope, moisture, soil type, soil chemistry, or light resulting in more or less distinct belts or zones of biotic communities. For example, a marsh habitat often shows the following zones related to water level: submerged vegetation, emergent vegetation, sedge meadow, and shrubs. Sometimes zones may be so broad as to be considered distinct habitats, such as mountain zonation and soil catenas (sequences of related soil types). In cases where the gradient of environmental change is very slight, only a gradual change in the biota may be visible, and demarked zones may not be clear.

A general description depicting existing zonal relationships should be given. Note any qualitative physical and biotic components such as gradients of light, temperature, water, soil moisture, slope, and vegetation. Describe the changes you observe in sequential order. For a detailed

Table 2A.3. *Moisture categories for classification of climates**

Moisture Group	Annual Precipitation (cm)	Monthly Precipitation (cm) Winter	Summer
Rainy			
wet	> 200	driest mo > 6	driest mo > 6
monsoon	> 150	driest mo > 6	driest mo < 6, but $a > 10 - r63.5$
wet and dry			
dry summer	50 to 200	driest mo > 6	driest mo < 6, and $a < 10 - r/63.5$
dry winter	50 to 200	$a < 10 - r/63.5$ and driest mo < 6	driest mo > 6
Humid			
moist	65 to 200 $r > 2T + 14$	< 3 times driest summer mo	driest mo > 3
dry summer	40 to 100 $r > 2T$	> 3 times driest summer mo	driest mo < 3
dry winter	40 to 100 $r > 2T + 28$	low and variable	> 10 times driest winter mo
Dry	precipitation is exceeded by evaporation		
semiarid	20 to 75		
uniform	$r < 2T + 14$ and $r > T + 7$	< 3 times driest summer mo	driest mo > 3
dry summer	$r < 2T$ and $r > T$	> 3 times driest summer mo	driest mo < 3
dry winter	$r < 2T + 28$ and $r > T + 14$	dry	> 10 times driest winter mo
arid	< 25		
uniform	$r < T + 7$	arid	arid
arid summer	$r < T$	> 3 times driest summer mo	arid
arid winter	$r < T + 14$	arid	> 10 times driest winter mo

Criteria based on Trewartha (1968).
*a = average monthly precipitation, r = average annual precipitation, and T = average annual temperature.

quantitative description, measure the above physical characteristics at regular intervals along the environmental gradient (see sections 2B, 2C, 2D, and 3B).

Stratification, the more or less distinct layering of vegetation, can be found in most habitats. In forests, for example, a description of stratification would include ground, herbaceous, shrub, understory, and canopy levels (figure 2A.3). In some forests, stratification may be complex enough to have more than one shrub or understory level, while in others, some strata may be missing. Plant life forms generally inhabit specific strata, as do many animal forms. The ground stratum may be divided into litter, surface, and subsurface layers. Surface plant taxa include mosses, lichens, and fungi. Herbs consist of many forms of annuals and perrenials. In the shrub stratum one finds bushes, shrubs, and young trees. Small canopy and noncanopy species are found in the understory while the canopy consists mainly of dominant tree species. In some habitats, the description of stratification may be rather subjective in the absence of clear distinction among shrubs, understory, and canopy. In grasslands, one generally describes the root stratum, ground stratum, forb stratum, and aerial grass stratum (figure 2A.4). For a general habitat analysis, a qualitative description of the stratification often is adequate. For more precise analysis, a quantitative index is described in section 2A.8 below.

Table 2A.4. *Major community types.*

Tundra   Cold and treeless; found in arctic regions or high mountain elevations; consists of low shrubs, forbs, lichens, and sedges.

Grassland   Grasses the dominant vegetation. Grasses are short in semiarid *plains* (called "steppes" in Eurasia), tall in semihumid *prairies*.

Field   Early successional stage of grasses and forbs common on abandoned farmland and other disturbed areas.

Meadow   Moist grassland.

Marsh   Herbaceous vegetation in standing water.

Swamp   Woody vegetation in standing water.

Bog   Standing water, usually with poor drainage, typically in northern latitudes, with sphagnum moss, sedges, heath shrubs, and peat formation.

Deciduous forest   Close stand of broad-leaved trees that shed their leaves during the cold or dry season.

Coniferous forest   Close stand of evergreen needle-leaved trees.

Broad-leaf evergreen forest   Trees of warm and humid regions that maintain foliage the entire year.

Scrub   Dense shrubs or small trees, often thorny or having small tough leaves.

Shrub   Dominant vegetation is tall shrubs. Semiarid shrublands are often called *chapparal*.

Woodland   Open growth of small trees, often evergreen, with well-developed growth of grasses.

Savanna   Grassland with scattered trees or groves of trees.

Desert   Hot and arid, with sparse thorny or scrubby vegetation (or, in extreme cases, no vegetation).

8. Habitat diversity

Diversity of species in a community (section 5B) is in part a function of diversity of the habitat. A description of horizontal habitat diversity would consider the variety and proportionality of land forms and plant life forms in the total habitat. For example, a homogeneous stand of coniferous trees would offer very low habitat diversity to animals, compared to a habitat containing coniferous trees, deciduous trees, bare ground, and standing water. It has been found in both terrestrial and aquatic habitats that the diversity of animal species may be strongly correlated with the structural diversity within the habitat.

Shannon's index of diversity (section 5B.2.3) is suitable as a quantitative measure of habitat diversity:

$$H' = -\Sigma p_i \log p_i, \qquad (1)$$

where H' is the diversity index and p_i is the proportion of the total habitat area covered by the ith category of coverage. For example, if 40% of a habitat area is covered by litter, 15% by rocks, 20% by sand, and 25% by standing water, the habitat diversity would be:

$$\begin{aligned}
H' &= -[0.40 \log 0.40 + 0.15 \log 0.15 \\
   &\quad + 0.20 \log 0.20 + 0.25 \log 0.25] \\
   &= -[0.40(-0.398) + 0.15(-0.824) \\
   &\quad + 0.20(-0.699) + 0.25(-0.602)] \\
   &= -[-0.159 - 0.124 - 0.140 - 0.151] \\
   &= 0.574.
\end{aligned}$$

The above calculation employs logarithms to the base ten (Appendix D, table D.2), but other bases could be used.

canopy

understory

shrub

herb
ground

Figure 2A.3. Stratification in a mixed deciduous forest.

Figure 2A.4. Stratification in a prairie.

H' makes a good measure for comparing different habitats. (See section 5B for further discussion of diversity indices and their interpretation.)

Vertical habitat diversity is also important as a determinant of species diversity of animals inhabiting several strata, such as birds and insects. A measure of stratum diversity would be H' (equation 1), computed where p_i is the proportion of the total foliage height occupied by each successive stratum. For example, consider a deciduous forest in which herbs are 20 cm (0.2 m) high, shrubs rise to 2.5 m, understory trees are 10 m tall, and canopy trees are 21 m tall. We would assign the following heights to the four strata: 0.2 m, 2.3 m (2.5 m − 0.2 m), 7.5 m (10 m − 2.5 m), and 11 m (21 m − 10 m). Therefore, p_i, the proportion of heights in each category, would be:

$$p_1 = 0.2 \text{ m}/21 \text{ m} = 0.010$$
$$p_2 = 2.3 \text{ m}/21 \text{ m} = 0.110$$
$$p_3 = 7.5 \text{ m}/21 \text{ m} = 0.357$$
$$p_4 = 11 \text{ m}/21 \text{ m} = 0.524$$

Using table D.2 in Appendix D, we can calculate:

$$\begin{aligned} H' = &-[0.010 \log 0.010 + 0.110 \log 0.110 \\ &+ 0.357 \log 0.357 + 0.524 \log 0.524] \\ = &-0.010(-2.000) + 0.110(-0.959) \\ &+ 0.357(-0.447) + 0.524(-0.281)] \\ = &-[-0.020 - 0.105 - 0.160 - 0.147] \\ = &\ 0.432. \end{aligned}$$

While this is a rather crude index of the vertical habitat diversity available to denizens of the habitat, it can be used for comparing different habitats. A better index of vertical habitat diversity would be one where p_i is a proportion of the foliage density (or screening efficiency—see section 5A—or some similar measure) in each stratum. MacArthur and MacArthur (1961) called such a measure *foliage height diversity* (FHD) and found it highly correlated with bird species diversity.

9. Suggested exercises

1. Prepare a description of a terrestrial habitat in terms of time, space, substrate, climate, and biota.
2. Describe the topographic differences between two habitats, examining areas having different slopes or different directions of slope. Relate your comparisons to observed differences in the biota.
3. Prepare a habitat map from aerial photographs and a topographic map. Include topographic and land form features and record distribution patterns of the biota. Verify the map by conducting a site reconnaissance.
4. Conduct a site survey of a habitat to collect information on slope, stratification, zonation, soil, and biota. Evaluate the interrelationships among these variables.
5. From quantitative estimates of substrate type, determine the horizontal habitat diversity of two different terrestrial or aquatic habitats.
6. From quantitative estimates of stratification, determine the vertical habitat diversity of two different habitats.

10. Selected references

Anderson, W. H., W. A. Wents, and B. D. Treadwell. 1980. A guide to remote sensing information for wildlife biologists, 291–303. In S. D. Schemnitz (ed.), Wildlife management techniques manual. 4th ed. The Wildlife Society, Washington, D.C..

Avery, T. E. 1977. Interpretation of aerial photographs. Burgess Publishing Co., Minneapolis.

Barbour, M. G., and W. D. Billings (eds.). 1988. North American terrestrial vegetation. Cambridge University Press, New York.

Barbour, M. G., J. H. Burk, and W. D. Pitts. 1980. Terrestrial plant ecology. Benjamin/Cummings Publishing Co., Menlo Park, Calif.

Barnes, J. R., and G. W. Marshall. 1983. Stream ecology: Application and testing of general ecological theory. Plenum Press, New York.

Clapham, W. B., Jr. 1983. Natural ecosystems. 2d ed. Macmillan Co., New York.

Fontaine, T. D., III, and S. M. Bartell (eds.). 1983. Dynamics of lotic ecosystems. Ann Arbor Science, Ann Arbor, Michigan.

Jenny, H. 1980. The soil resource. Springer-Verlag, New York.

Kimmins, J. P. 1987. Forest ecology. Macmillan Publishing Co., New York.

Lock, M. A. and D. D. Williams. 1981. Perspectives in running water ecology. Plenum Press, New York.

MacArthur, R. H., and J. W. MacArthur. 1961. On bird species diversity. Ecology 42:594–598.

Mitsch, W. J., and J. G. Gosselink. 1986. Wetlands. Van Nostrand Reinhold Co., New York.

Mosby, H. S. 1980. Reconnaissance mapping and map use, 277–290. In S. D. Schemnitz (ed.), Wildlife techniques manual. 4th ed. The Wildlife Society, Washington, D.C..

Moss, B. 1980. Ecoloy of fresh waters. Blackwell Scientific Publications, Boston.

Mueller-Dombois, D., and H. Ellenberg. 1974. Aims and methods of vegetation ecology. John C. Wiley & Sons, New York.

Shelford, V. E. 1963. The Ecology of North America. University of Illinois Press, Urbana, Ill.

Soil Survey Staff. 1960. Soil classification. A comprehensive system. 7th approximation. U.S. Soil Conservation Service, Washington, D.C.

Soil Survey Staff. 1967. Supplement to soil classification system. 7th approximation. U.S. Soil Conservation Service, Washington, D.C.

Soil Survey Staff. 1975. Soil taxonomy: A basic system for making and interpreting soil surveys. Agriculture Handbook 436. U.S. Soil Conservation service, Washington, D.C.

Strahler, A. N., and A. H. Strahler. 1984. Elements of physical geography. 3d ed. John C. Wiley & Sons, New York.

Thornthwaite, C. W. 1940. Atmospheric moisture in relation to ecological problems. Ecology 21:17–28.

Trewartha, G. T. 1968. An introduction to climate. McGraw-Hill Book Co., New York.

U.S. Geological Survey. 1969. Topographic maps. U.S. Geological Survey, Washington, D.C.

U.S. Geological Survey. 1971. Aerial photographic reproductions. U.S. Geological Survey, Washington, D.C.

Walter, H. 1985. Vegetation of the earth and ecological systems of the geobiosphere. 3d ed. Springer-Verlag, New York.

Weller, M. W. 1987. Freshwater marshes. University of Minnesota Press, Minneapolis.

Witkamp, M. 1974. Soils as components of ecosystems. Annu. Rev. Ecol. Systemat. 2:85–110.

Wolfanger, L. 1971. Soil orders and suborders of the United States and their utilization, 55–98. In G. H. Smith (ed.), Conservation of natural resources. John Wiley & Sons, New York.

1. Introduction

Climate, season, and weather affect the distribution and activity of both terrestrial and aquatic organisms. *Climate* refers to the general prevailing atmospheric conditions over the years in a given region. Climates are usually characterized by seasonal temperature, humidity, and precipitation. *Weather* refers to the momentary conditions of the atmosphere. Four major physical factors comprise the atmospheric component of a habitat: air moisture, temperature, wind, and solar radiation. Extremes, rather than averages, of these variables usually affect the distribution and abundance of organisms.

The chemical components of air are rather uniform over the earth and are only measured as a matter of concern in the analysis of air pollution and in soil microhabitats. Unlike in aquatic habitats, oxygen is abundant in aboveground terrestrial situations.

In this section we shall be concerned with the analysis of climatic factors. They largely determine the type of biotic community in an area and the distribution of individual species. We will pay similar attention to the analysis of weather conditions that largely affect the daily and seasonal behavior and abundance of species. A useful account of the ecological significance of atmospheric factors is given in Daubenmire (1974).

2. Climate

Climates are often broadly categorized by latitude, as polar (or arctic), cold temperate (or boreal), temperate, and tropical, with terms such as subtropical or subarctic denoting intermediate climates. See section 2A.6 for descriptions of climates based on temperature and precipitation.

For a more detailed picture of the climate for a given region, ecologists use two types of *climatographs*. In figure 2B.1, the mean monthly temperature and the mean monthly precipitation are plotted for each month of the

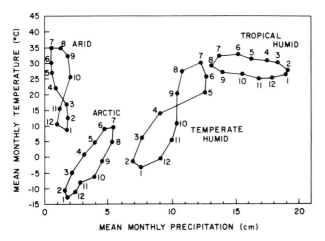

Figure 2B.1. Climatographs, describing climate in terms of mean monthly temperature and precipitation.

2b

atmospheric analysis

year, and the plotted points are connected sequentially to form an irregular polygon. Data for this type of graph are available from local weather stations or from government documents prepared by the National Oceanic and Atmospheric Administration (NOAA), Washington, DC, and the National Climate Center, Ashville, NC.

A second method of presenting the climate of a given region is to graph the mean monthly precipitation and mean monthly evapotranspiration as functions of time of year. Evapotranspiration includes loss of water to the atmosphere through both evaporation and plant transpiration. This measure gives a better picture of water availability to plants than does the temperature-precipitation graph of figure 2B.1. However, evapotranspiration data often are not available, and they must either be measured by the investigator or roughly approximated from temperature, humidity, and wind data (Rosenberg et al., 1983: Chapter 7).

Figure 2B.2 is an alternative presentation that helps to diagram water availability based on readily obtained temperature and precipitation data. As evapotranspiration is directly related to temperature, a plot of seasonal changes in temperature will be similar to a plot of evapotranspiration. A temperature of 10°C is considered roughly equivalent to 20 mm of monthly precipitation in terms of evapotranspiration (Walter, 1985). Consequently, points on figure 2B.2 where temperature and precipitation curves intersect represent a condition where the amount of water lost through evapotranspiration is about equal to the amount gained through precipitation. Thus in figure 2B.2, July and August would have a water deficit in the Pacific Northwest but would experience a water surplus for the mid-Atlantic coast of the United States.

3. Microclimate

Variation in the local climate due to such factors as elevation, slope, and shade can result in temperatures, humidities, and light intensities quite different from those of

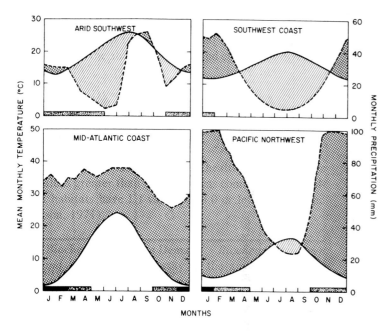

Figure 2B.2. Climatographs, emphasizing water availability. Since a mean monthly temperature of 10°C and a monthly precipitation of 20 mm are considered equivalent in terms of evapotranspiration, diagonal lines indicate periods of drought and cross-hatching indicates periods of water surplus. Stippled bars denote months with frost; solid bars indicate months with freezing temperature (after Walter, 1985).

surrounding areas. For example, the atmospheric conditions in a forest on a north-facing slope are quite different from those on a south-facing slope. Also, conditions near the ground are generally different from those a few meters above the ground. Therefore, when conducting a study of the microhabitat, one should determine the vertical profile of temperature, light intensity, relative humidity, and wind velocity (see section 2B.4 below). For such profiles, portable electronic instruments are conveniently used. Detectors having long leads are placed on extendable poles and elevated to the desired heights, and measurements are taken at 0.5- or 1-meter intervals. The detectors may also be attached to cords thrown over high tree branches and raised to the desired heights. One then graphs the measured variables as a function of height in the habitat.

A horizontal profile of these variables may be made where zonation or patchiness occurs within the habitat. A microhabitat study may also include analysis of these variables in specific locations, such as animal burrows, nests, and hollow trees or logs. These microhabitats are generally sheltered from large variations in the macrohabitat and represent rather moderate and stable microclimates for species that would not otherwise survive in the area.

4. Atmospheric measurements

Record atmospheric conditions at the time of sampling, since animal activities and plant functions may be dependent on them. Sampling of animals (as described in sections 3D through 3G) will often yield different results under different weather conditions. Therefore, always record air temperature, relative humidity, wind velocity and direction, relative amount of cloud cover, light intensity, and any occurrence of precipitation.

4.1. Light Intensity   The intensity and duration of solar radiation not only affects other atmospheric variables (such as temperature, relative humidity, and wind), but also the amount of energy available for production and the timing of seasonal cycles of plants and animals. (See Daubenmire, 1974, for a discussion of light as an ecological factor.)

Luminous flux is the amount of light energy per unit time. The *lumen*[1] is the unit of luminous flux, but, since the lumen is dependent on wavelength, light measurements are difficult to interpret as solar energy input (e.g., calories).

1. Common abbreviations for photometric units are: lumen, 1m; candela, cd; einstein, E; footcandle, fc; lux, lx.

Luminous intensity (measured in units called *candelas,* or *candlepower*) refers to the amount of light emitted by a source, measured relative to the so-called *standard,* or *international, candle.* The amount of light received one meter from a standard candle is a *lux;*[2] that received at a distance of one foot is a *footcandle.*[2]

A light meter may be calibrated to either lux or footcandles (1 footcandle = 10.76 lux; 1 lux = 0.0929 footcandles). When measuring luminance in a given habitat, determine its value at ground level under the vegetation and in open sunlight (outside the habitat, if necessary). Relative luminance may then be expressed as lux at ground level divided by the lux in open sunlight. This value may be expressed as percent transmittance in that habitat. We can estimate an *absorption coefficient,* analogous to the "extinction coefficient" of section 2D.3.3., equation 4, by designating the height of the tallest stratum (as determined in section 2A.7) as *d,* the luminance at ground level as I_d, and the luminance in the open (assumed to be the same as that above all foliage) as I_0.

In a markedly stratified habitat, a light sensor either attached to an extendable pole or hanging from a tall branch can be used to measure the screening efficiency in each stratum (sections 2A.7 and 6A.7). Then make a profile of light extinction for that habitat by graphing the percentage of light transmittance as a function of vegetation height.

Measurement of illuminance, in lux or footcandles, can allow us to describe some characteristics of a habitat (e.g., how much of the light striking the tree canopy reaches various lower strata). More pertinent to photobiological processes (e.g., photosynthesis) is the measurement of the flux of photons in the biologically important part of the light spectrum. Electronic instrumentation is available to obtain such data, measured in micromoles of protons per square meter per second (i.e., in units of microeinsteins/m²/sec).[3]

4.2. Temperature   Air temperature should be measured at ground level and compared to a measurement made in the open. A simple mercury thermometer may be used, but for a temperature profile of the habitat it is more convenient to use electronic telethermometers (employing thermistors) or thermocouples. A temperature probe may be placed at different heights as described in section 4.1 above, and the temperature at various intervals plotted as a function of height in the habitat. Care should be taken to shield temperature-measurement devices from direct sunlight.

4.3. Solar Radiation   For studying heat balance in an ecosystem, temperature and light intensity measurements are inadequate. As the ultimate source of heat in an ecosystem is solar radiation, an estimate of radiant heat is useful. Instruments that measure such environmental radiation are called radiometers. Those specifically measuring visible light are called pyranometers, or solarimeters; others may measure longer wavelengths (i.e., infrared) or all wavelengths. (A pyrheliometer is a radiometer specifically designed to measure the perpendicular radiation from the sun.)

A simple radiometer can be constructed that is usable for comparative field studies. The bulbs of two standard, calibrated laboratory thermometers, accurate to ± 0.1°C, are painted flat white and flat black, respectively, and sealed in an evacuated glass tube. The radiometer should be kept in a dark container until field measurements are to be made. The container with the radiometer should be allowed to equilibrate to air temperature. The radiometer is then removed from the container and exposed to the sky, positioned so that the incident solar radiation is perpendicular to the surface of the thermometers. After five minutes of exposure, record the temperatures of both thermometers. Temperature readings may be averaged if the mean radiation over the period of measurement is needed. Radiant heat may then be estimated as

$$H_r = ka(T_b^4 - T_w^4), \qquad (1)$$

where

H_r = radiant heat flux density in cal/cm²/hr
a = Stefan-Boltzman constant (4.96 × 10⁻⁹ cal/cm²/hr/°K⁴)
k = calibration constant determined for the radiation against a standard radiometer or source of radiation
T_b = temperature of the black bulb, in degrees Kelvin
T_w = temperature of the white bulb, in degrees Kelvin
H_r may be expressed in joules/m²/sec by using
a = 5.67 × 10⁻⁸J/m²/sec/°K⁴.

This simple procedure can be used to determine the daily solar radiation, to compare the available radiant heat in different ecosystems, or, with the appropriate light filters, to estimate the available daily solar energy at wavelengths needed for photosynthesis.

4.4. Wind Velocity   An electronic wind meter measures wind velocity most accurately. However, determinations may also be made using an anemometer or a simple wind meter in which the wind causes a small light ball to rise in a tube. Record the direction as well as the velocity of the wind. Wind velocity affects certain animal activities (e.g., insect flight) and the rate of evapotranspiration of water from the habitat.

2. Common abbreviations for photometric units are: lumen, 1m; candela, cd; einstein, E; footcandle, fc; lux, lx. In full sunlight, one might find measurements of 10,000 fc, or 107,600 lx, or 2000 μE/m²/sec (Barbour et al., 1980:303).
3. A mole (i.e., 6.02 × 10²³) of photons is called an einstein.

Table 2B.1. *Determination of percent relative humidity from dry bulb and wet bulb temperatures (°C) from a psychrometer.* Tabled values are for a barometric pressure of 743 mm Hg. For other barometric pressures, see table 2B.2 and the note at the bottom of that table.*

dry bulb (°C)	difference between dry and wet bulbs (°C)																								
	1	2	3	4	5	6	7	8	9	10	11	12	13	14	15	16	17	18	19	20	21	22	23	24	25
−10	60	31																							
−8	65	39	13																						
−6	70	46	23																						
−4	74	53	32	11																					
−2	78	58	39	21																					
0	81	64	46	29	13																				
2	84	68	52	37	22	7																			
4	85	71	57	43	29	16																			
6	86	73	60	48	35	24	11																		
8	87	75	63	51	40	29	19	8																	
10	88	77	66	55	44	34	24	15	6																
12	89	78	68	58	48	39	29	21	12																
14	90	79	70	60	51	42	34	26	18	10															
16	90	81	71	63	54	46	38	30	23	15	8														
18	91	82	73	65	57	49	41	34	27	20	14	7													
20	91	83	74	66	59	51	44	37	31	24	18	12	6												
22	92	83	76	68	61	54	47	40	34	28	22	17	11	6											
24	92	84	77	69	62	56	49	43	37	31	26	20	15	10	5										
26	92	85	78	71	64	58	51	46	40	34	29	24	19	14	10	5									
28	93	85	78	71	65	59	53	48	42	37	32	27	22	18	13	9	5								
30	93	86	79	73	67	61	55	50	44	39	35	30	25	21	17	13	9	5							
32	93	86	80	74	68	62	57	51	46	41	37	32	28	24	20	16	12	9	5						
34	93	87	81	75	69	63	58	53	48	43	39	35	30	26	23	19	15	12	8	5					
36	94	87	81	75	70	64	59	54	50	45	41	37	33	29	25	21	18	15	11	8	5				
38	94	88	82	76	71	66	61	56	51	47	43	39	35	31	27	24	20	17	14	11	8	5			
40	94	88	82	77	72	67	62	57	53	48	44	40	36	33	29	26	23	20	16	14	11	8	6		
42	94	88	83	78	72	67	63	58	54	50	45	42	38	35	31	28	25	22	19	16	13	11	8	6	
44	94	89	83	79	73	68	64	59	55	51	47	43	40	36	33	30	27	24	21	18	16	13	10	8	6

* From the more extensive table 1 of the U.S. Weather Bureau (1953). Interpolation may be made with an error of less than 1% relative humidity. (For example, the relative humidity associated with a temperature difference of 5°C for a dry bulb temperature of 9°C may be estimated as 42%, the midpoint of the values of 40% and 44% for air temperatures of 8° and 10°C, respectively.)

Determine the wind velocity on the ground and in the open. For a profile of wind velocity, record measurements at different heights, as explained in section 2B.4.1 above.

4.5. Precipitation Noting the type and intensity of any precipitation is sufficient during a single field trip. However, if one is sampling animals over a series of days, quantitative precipitation measurements are necessary. Simple rain gauges may be set out in the habitat and read daily. Precipitation data from local weather stations may suffice, but actual measurements in the habitat are often different, due to local variations. In dense vegetation much less rainfall may reach the ground surface than would fall on bare ground. Rain intercepted by plants may evaporate from them or be directed down their stems.

4.6. Humidity Atmospheric humidity, highly variable, is the amount of water vapor in the air. It has important biological effects on plant respiration, on rates of transpiration and evaporation, and on the amount of cooling of surfaces from which evaporation takes place. The amount of water vapor in air may be expressed as *vapor pressure,* the partial pressure of water vapor in the air. It can be stated as millibars (mb) or as an equivalent height of a mercury column, as in a barometer:

$$1 \text{ mm Hg} = 1.333 \text{ mb}; \quad 1 \text{ mb} = 0.7501 \text{ mm Hg} \quad (2)$$

$$1 \text{ in. Hg} = 33.86 \text{ mb}; \quad 1 \text{ mb} = 0.02953 \text{ in. Hg} \quad (3)$$

The maximum amount of water vapor that the air can hold when in equilibrium over liquid water is called *saturation vapor pressure* and is directly related to temper-

Table 2B.2. *Correction factors,* c, *for use with table 2B.1 to determine percent relative humidity from dry and wet bulb temperatures.**

Dry Bulb Temperature (T, in °C)	Correction Factor (c)	Dry Bulb Temperature (T, in °C)	Correction Factor (c)
−10	0.0304	18	0.00433
−8	0.0260	20	0.00383
−6	0.0224	22	0.00339
−4	0.0193	24	0.00302
−2	0.0166	26	0.00268
0	0.0144	28	0.00239
2	0.0125	30	0.00212
4	0.0109	32	0.00190
6	0.00945	34	0.00170
8	0.00827	36	0.00152
10	0.00724	38	0.00136
12	0.00634	40	0.00123
14	0.00559	42	0.01100
16	0.00492	44	0.00992

* If the dry bulb temperature is T (in degrees °C), the difference between dry and wet bulb temperature is ΔT (in °C), the relative humidity in table 2B.1 is RH, and the barometric pressure is P (in mm Hg), then the corrected relative humidity is:

$$RH_c = RH + c\Delta T(743 \text{ mm} - P). \qquad (4)$$

Values in this table were calculated from table 1a in U.S. Weather Bureau (1953). Note: The above correction is seldom needed (unless at high altitudes). It results in RH_c being different from RH by no more than 2% for P ranging from 714 to 770 mm Hg at 0°C, and with even less error at most other temperatures. The correction should routinely be used at altitudes over 600 m. If barometric pressure is not measured directly, it may be assumed to decrease about 9 mm Hg per 100 m altitude up to about 800 m, about 8 mm Hg per 100 m from 900 to 1700 m, and about 7 mm Hg per 100 m thereafter up to about 2500 m (from Golterman, 1969).

ature (see table 2E.1). The *vapor pressure deficit* is the difference between the saturation vapor pressure and the actual vapor pressure.

The commonest measure of atmospheric water vapor content is *relative humidity,* the actual vapor pressure expressed as a percentage of the saturation vapor pressure. It is important to note that rates of evaporation are not directly related to relative humidity, but are a function of the vapor pressure deficit and the air temperature.

Relative humidity is conveniently measured with a hygrometer or a sling psychrometer (or an electronic humidity meter standardized with one of the former). A psychrometer consists of two thermometers, one having a dry bulb and the second having a bulb wrapped with a wick continually moistened with distilled water. (If the air temperature is below freezing, then the wick will of course

be impregnated with ice.) Evaporation of water from the wet bulb cools the second thermometer; as the rate of evaporation (hence the cooling of the thermometer) is directly related to the vapor pressure deficit, the relative humidity can be estimated from the difference between the wet bulb and dry bulb temperatures.

Swing the psychrometer in a circle until the wet bulb temperature ceases to decline. (Be careful to keep a safe distance from people and other objects while swinging the instrument, as it is easy to injure a person or damage the thermometers.) Read the wet bulb temperature immediately after swinging; record the wet bulb and dry bulb temperatures and calculate the difference between the two. The relative humidity is then determined by consulting tables 2B.1 and 2B.2. For example, if the dry bulb temperature were 22°C and the wet bulb 18°C, the difference in temperature would be 4°C, and the relative humidity for 22°C would be 68%.

---

5. Suggested exercises

1. Construct climatographs for your region as illustrated in figures 2B.1 and 2B.2.
2. Collect data on light intensity and temperature in the different strata of a forest community, and plot these variables as a function of height.
3. Compare variables such as temperature, humidity, wind, and light in a forest to those in a nearby field or grassland.
4. On a clear day, compare the solar heat flux at ground level within a forest and outside the forest in direct sunlight. Compare the percent difference in air temperature between the two locations with the percent difference in solar radiation.

---

6. Selected references

Aikman, J. M. 1936. The radiometer: A simple instrument for the measurement of radiant energy in field studies. Proc. Iowa Acad. Sci. 43:95–99.

American Society for Testing Materials. 1981. Standard test methods for relative humidity by wet and dry bulb psychrometer, 868–877. *In* Annual Book of ASTM Standards, Part 26.

Barbour, M. G., J. H. Burk, and W. D. Pitts. 1980. Terrestrial plant ecology. Benjamin/Cummings Publishing Co., Menlo Park, Calif.

Daubenmire, R. F. 1974. Plants and environment: A textbook of plant autecology. 3d ed. John C. Wiley & Sons, New York.

Golterman, H. L., R. S. Clymo, and M. A. N. Ohnstad. 1978. Methods for physical and chemical analysis of fresh waters. Blackwell Scientific Publications, Oxford, England.

Rosenberg, N. J., B. L. Blad, and S. B. Verma. 1983. Microclimate: The biological environment. 2d ed. John C. Wiley & Sons, New York.

Unwin, D. M. 1980. Microclimate measurement for ecologists. Academic Press, New York.

U.S. Weather Bureau. 1953. Relative humidity—psychrometric tables. Celsius (centigrade) temperatures. U.S. Weather Bureau, Washington, D.C.

Walter, H. 1985. Vegetation of the earth and ecological systems of the geobiosphere. 3d ed. Springer-Verlag, New York.

2c

substrate analysis

1. Introduction

The portion of the lithosphere directly important to the ecologist is the top few meters of soil and aquatic sediments. Soil is a heterogeneous substance; it varies somewhat with season and interacts with climate and vegetation. Given below are some basic physical measurements desired in soil analysis. Section 2E describes chemical analyses.

2. Sampling methods

All substrate samples should be collected at random and taken in replicate (see section 1A). One commonly uses a soil corer (figure 2C.1) for soil samples. This consists of a hollow, half-open metal tube. The tube is pushed into the soil until its top is just at the ground surface, and then carefully pulled from the soil and examined. (The corer should be cleaned before taking the next sample.) For larger samples, small plots may be dug with a sharp, flat-tipped spade.

For sampling aquatic sediments, benthic grab samplers (described in section 3E.2.3) are often used. But to obtain cores of soft, lake sediments, you must use a specially designed corer. The simplest type is a metal tube with a heavy weight. The sampler is attached to a line and dropped into the water; the weight of the sampler drives it into the sediment. For deeper cores, specialized corers have an additional weight that is sent down the line repeatedly to drive it deeper into the sediment. After retrieving the corer, you extract the sample by pushing the core out with a rod or piston.

Samples should be stored in sturdy, tightly sealed plastic bags or tubes and analyzed as soon as possible, as some of the physical and chemical properties change with storage. For example, determine the pH of sediments in the field if possible.

Some types of analysis require that a sample be dried first to remove moisture, while others require the use of a fresh sample. In either case, final results are expressed relative to the dry weight of the soil sample, rather than to its fresh weight. Therefore, dry weights must be determined separately for samples requiring fresh material. Determination of the dry weight of a soil subsample and water percentage in the sample may proceed as for the dry weight of biological material (section 6A.2). The dry weight of the fresh sample is then estimated by multiplying the ratio of dry weight to fresh weight by the fresh weight of the sample.

Some procedures require that the soil or sediment be ground into fine particles to insure homogeneity of the material. To do this, take a 5- or 10-g substrate sample and pulverize it with a pestle in a mortar so that all particles pass through a 100-mesh screen, whereupon the material is weighed and analyzed.

3. Parent material

Parent material is the substrate from which the soil or sediment originated and may have been removed by wind, water, and gravity. Soil origin may be *residual* (formed in place), *alluvial* (deposited by water), *aeolian* (deposited by wind), *colluvial* (deposited by gravity), or *glacial* (deposited by glacier). For a determination of the parent material in your study area, consult local soil maps. These are available from the U.S. Soil Conservation Service and may be found in governmental and university map libraries.

4. Soil profile

Soils exhibit vertical zones, called *horizons* (figure 2C.2). The O horizon is the layer of deposited organic matter. The A horizon, or topsoil, is characterized by mineral soil having a granular, platelike, or crumb structure. The B horizon, or subsoil, collects the leached materials from the A layer and generally has a prismatic, blocky, or columnar structure. In arid regions, high evaporation rates inhibit percolation of water, and a "hardpan" may form in the B horizon. This is a hard deposition of salts impervious to water, roots, and burrowing animals. The C horizon contains weathered parent rock material unconsolidated into soil. Differences in color, structure, and chemistry within these major horizons are referred to by subdivision designations, such a A_1, A_2, B_1, B_2, etc. The basic soil horizons (and their subdivisions if possible) should be identified in a habitat analysis; measure the thickness of each. Describe the color, texture, and structure of the aggregates in each subdivision. Soil profile characteristics are essential in classifying basic soil types (see section 2C.9).

For a microhabitat analysis, a soil core will give enough information concerning the O, A, and often B horizons of the soil. Occasionally, complete soil profiles can be studied conveniently from a recent excavation in an area, but usually one has to dig a pit about 1.5 meters wide and 1.5

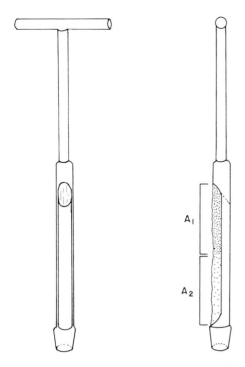

Figure 2C.1. A soil corer, empty on the left, and containing a core sample with A_1 and A_2 horizons on the right.

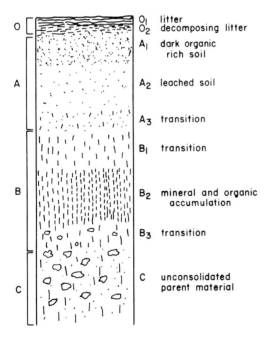

O	O_1	litter
	O_2	decomposing litter
	A_1	dark organic rich soil
A	A_2	leached soil
	A_3	transition
	B_1	transition
B	B_2	mineral and organic accumulation
	B_3	transition
C	C	unconsolidated parent material

Figure 2C.2. A generalized soil profile. (The layer of decomposing organic matter (O_2) is sometimes referred to as A_0, and the litter layer (O_1) as A_{00} or L.)

meters deep. However, dig such a pit in an area where it will not severely impact the habitat nor be a safety hazard. The pit may be safely covered with a lid, so that it can be used for demonstration for many years. For each viewing, remove a fresh slice of soil a few centimeters thick from the side of the pit with a sharp flat-tipped shovel to show an unweathered view of the profile. Each horizon may then be sampled with a soil corer by taking several samples parallel to the soil surface.

5. Soil moisture

For a general habitat survey, a relative measure of soil moisture content will suffice. A qualitative categorization would be: *dry soil* (crumbly, or hard and dry to the touch); *moist soil* (pliable, and moist to the touch); and *wet soil* (exuding water when squeezed, leaving the hand muddy).

Pocket-sized moisture meters are available for qualitative estimation of soil moisture. However, they basically measure conductivity and depend, therefore, not only on the moisture content of the soil but on salt content and pH. Therefore, such a meter should be standardized in the laboratory with a soil sample from the habitat of interest and calibrated against moisture content known from dry weight determinations, as explained below.

For precise determination of the percent moisture in the soil, obtain samples using a soil corer, seal the samples in separate plastic bags, and dry them in the laboratory for twenty-four hours at 105°C. See section 6A.2 for dry weight determination techniques. Fresh weight minus dry weight equals the amount of water in the soil and is expressed in grams of water per 100 grams of dry soil.

The amount of moisture in soil is related to the amount of rainfall, evapotranspiration, and drainage, and the water-holding capacity of the soil. The last factor is difficult to measure accurately, but it is related to soil texture and soil organic matter. For example, sand has a low water-holding capability, while silts, clays, and soils rich in organic matter have a high one.

6. Soil temperature

Soil temperature is a variable that affects the ecology of plants, animals, and microorganisms. Therefore, a profile of soil temperature is useful. A dial thermometer with a long metal stem can record shallow temperatures. If measurements of deep soil temperatures are needed, then thermistors may be buried at different depths. Allow the thermistor to equilibrate with the soil for at least half an hour before recording the temperature. For a profile of soil temperature, plot the temperature on the horizontal axis and the depth on the vertical axis of a graph, thus obtaining a plot resembling the temperature profile of a lake (figure 2D.2).

7. Soil organic matter

Soil organic matter is a major determinant of soil texture, moisture, *p*H, and nutrients. Chemical procedures for estimating organic carbon exist, but a simple approximation can be made by determining ash-free dry weight. This procedure is the same as described for biological samples in section 6A.3. Obtain a core sample of soil from the A and B horizons as described in section 2C.2, or sample each horizon as described in section 2C.4. The O horizon is best sampled from a plot using a clean hand trowel or flat-tipped shovel, collecting the organic matter down to the mineral soil. During sampling, determine the depth of each O subhorizon and separate the upper litter layer from the decomposed humus. The organic horizon samples can then be saved in plastic bags and analyzed in the laboratory for moisture, inorganic ash, and organic matter using methods in sections 6A.2 and 6A.3.

One of four types of organic matter generally is encountered: mull, moder, mor, and peat. Mor, which develops in coniferous forests and heathlands, is fairly thick and acidic, and it decomposes slowly. Mull forms on well-drained soils of deciduous forests and is relatively thin, is more neutral or basic, and decomposes rapidly. Moder is intermediate between mor and mull. Peat is undecomposed plant fibers and is formed in acidic marshes and bogs.

An analysis of the humus, the finely divided organic matter, is useful as this is the most biologically active portion of the soil. Humus by definition contains less than one-third plant fiber content by dry weight and has an ash content not exceeding 25% by dry weight. Plant fibers are separated from finer soil and humus in a number 100 (150-mm) sieve. If such fibers are more than 12.5 mm in length they may be classified as litter.

Prepare the soil litter as described for plants in section 6A. Air dry the soil samples and pulverize them as described in section 2C.2 above; then determine the oven-dry weight and ash-free dry weight. The percentage of organic matter is found from the difference between the oven-dry weight and the ash-free dry weight, divided by the oven-dry weight.

8. Soil density

The distribution and abundance of plants and soil organisms are affected by the density of the soil. Soil density (i.e., weight of soil per unit volume) is related to several factors: density of the mineral particles, amount of organic matter, compaction of the soil, burrowing activities of soil animals, and density of plant roots. A simple procedure to measure soil density is to dig a soil sample, determine its dry weight, and determine the volume of the sample by pouring a measured volume of dry sand in the hole from which the sample was removed.

Determine the volume of a clean, weighed jar (approximately 4 liters), by measuring the amount of water required to fill it. Dry the jar and fill it with dry, free-flowing sand. Weigh the jar and sand to the nearest gram and subtract the jar weight from the total weight to estimate the weight of the sand. The density of the sand is estimated as

$$D_1 = W_1/V_1, \tag{1}$$

where W_1 is the weight of the sand in the jar and V_1 is the volume of the jar.

Remove the organic layers from the top of the soil and dig a small hole that has a volume smaller than 4 liters. Be careful not to compact the soil bordering the hole, thereby making the volume of the hole larger than the volume of soil removed. Place the soil sample in a clean plastic bag. Carefully pour sand from the jar into the hole until it is just level with the surface. Weigh the jar and its remaining sand. If a field scale is not available, jars and sand may be weighed in the laboratory before and after the field collecting. Calculate the volume of the sand in the hole as follows:

$$W_3 = W_1 - W_2 \tag{2}$$

$$V_1 = W_3/D_1 \tag{3}$$

where W_3 is the weight of sand used to fill the sample hole, W_2 is the weight of the remaining sand in the jar, and V_2 is the estimated volume of sand in the sample hole.

To determine the density of the soil sample, first oven-dry the sample and determine its moisture content and dry weight as in sections 2B.2 and 6A.2. Then,

$$D_2 = W_4/V_2 \tag{4}$$

where D_2 = density of the soil, W_4 = dry weight of sample, and V_2 = estimated volume of sample. Soil density may be expressed as kilograms per liter; if other weight or volume units are used, the resultant density may be converted to kilograms per liter by consulting appendix B.

9. Soil fractions

For a general habitat survey, classify the soil textural types, by sight and touch, as *gravel, sand, silt, clay,* or *loam.* (Loam is a mixture of sand, silt, and clay.)

However, a more precise procedure is required for a detailed microhabitat study. There are several methods to estimate the fractions of sand, silt, and clay in substrate samples. That discussed here is one of the easiest; it is based on a physical principle ("Stoke's law"), that relates the velocity of a particle settling in a liquid to its size and density. Thus particle sizes can be estimated by knowing the density of the soil suspension at various settling times. This can easily be measured with a soil hydrometer[1] (Day, 1956, 1965; American Society for Testing and Materials, 1981).

1. The standard soil hydrometer referred to is the American Society for Testing and Materials (ASTM) hydrometer 152H, which is graduated in grams of soil per liter. This is available from scientific suppliers, such as Scientific Glass Apparatus Co., Bloomfield, NJ; Soiltest, Evanston, IL; and Nasco Agricultural Sciences, Ft. Atkinson, WI, or Modesto, CA.

The procedure is: Break up a sample of air-dried soil with a wooden roller, without grinding, to keep the natural particles unbroken. Shake the sample through a number 10 (2.0-mm) mesh sieve to separate and remove the gravel and larger components. Weigh and record these larger particles. Gravel-size particles can be separated further from rocks and stones by sieving through a number 4 (4.75-mm) sieve for fine gravel and a 75-mm sieve for coarse gravel. Sand fractions between 75μ and 2.0 mm are determined by sieving following the hydrometer testing of the sample that passed through the 2.0-mm sieve. Weigh a portion of the sieved sample of fine particles collected from the 2.0-mm sieve and determine the oven-dry weight as described in section 6A.2.

Use a 25- to 50-g sample for fine textured soils and 50 to 100 g for sandy soils. Place the oven-dried material in a 600-ml beaker. Prepare a 5% solution of sodium metaphosphate (also known as sodium hexametaphosphate, or Calgon), by dissolving 50 g in 1 liter of distilled water. This solution should be made up fresh each month and adjusted to a pH of 8 or 9 with sodium carbonate. Add 100 ml of the sodium metaphosphate solution and 400 ml of distilled water to the beaker with the sample. The sodium metaphosphate acts as a dispersing agent and neutralizes charges on the soil particles that might impede settling. Mix the suspension for 5, 10, or 15 minutes with an electric mixer. The longer two mixing times are used for silts and clays, respectively. Transfer the suspension to a calibrated 100-ml glass cylinder, rinse the beaker with distilled water, and add the rinsings to the cylinder from the beaker; bring the volume in the cylinder to one liter with distilled water. Mix the contents of the cylinder by capping it and inverting it 60 times; avoid shaking as that may cause foaming and air bubbles in the suspension. A special mixing plunger may be used instead.

Immediately after mixing, begin recording the time to the nearest second during the first hour and to the nearest minute thereafter. Carefully lower the hydrometer into the suspension to avoid disturbing it; allow at least 20 seconds before reading the hydrometer. Read the *top* of the meniscus, as grams of soil per liter in suspension, after 0.5, 1, 2, 5, 15, 30, and 60 minutes, and 4, 8, and 24 hours. Record the temperature and the precise time of each hydrometer reading. (To prevent temperature fluctuations, the cylinder may be kept immersed in a constant temperature bath.) The reading times may be varied as long as the time of the recording is accurately determined. Remove and rinse the hydrometer after each measurement (except after the 0.5-minute measurement). Let us call a hydrometer reading R. After taking the final hydrometer reading, a sieve analysis of the sand fractions can be made. Mix the suspension in the cylinder and transfer it to a number 200 (75-μ) sieve, rinsing the cylinder thoroughly. Wash

Table 2C.1. *Values of* c *(for equation 4) for calculating particle size for soil hydrometer (ASTM 152H) readings,* R, *taken at 25° C.*

R	c	R	c	R	c
0	52.0				
1	51.6	21	46.2	41	39.9
2	51.5	22	45.9	42	39.5
3	51.2	23	45.5	43	39.0
4	50.8	24	45.3	44	38.8
5	50.7	25	45.0	45	38.4
6	50.3	26	44.6	46	38.2
7	50.2	27	44.4	47	37.7
8	47.8	28	44.0	48	37.3
9	49.5	29	43.6	49	37.1
10	49.3	30	43.5	50	36.6
11	49.0	31	43.1	51	36.2
12	48.7	32	42.9	52	35.9
13	48.5	33	42.5	53	35.5
14	48.2	34	42.1	54	35.0
15	47.8	35	41.9	55	34.8
16	47.6	36	41.5	56	34.3
17	47.3	37	41.1	57	34.1
18	46.9	38	40.9	58	33.6
19	46.8	39	40.5	59	33.1
20	46.4	40	40.1	60	32.8

*Values of c were computed as described by Day (1956, 1965) using the effective hydrometer depths given by the American Society for Testing and Materials (1981).

the sample in the sieve with water until the washwater is clear. Then place the contents of the sieve into a drying pan or dish and dry at 110°C and determine the dry weight of the sample. Sieve the dry sand fractions through a series of sieves containing number 18 (1.0-mm), number 35 (500-μ), and number 60 (250-μ) size mesh. Weigh the sample fraction from each of these three sieves. The sum of these three weights should be approximately equal to the dried sieve contents.

The diameter *(d)* of the largest particle in suspension may be estimated at the time of each hydrometer reading as:

$$d = c/\sqrt{t}, \qquad (5)$$

where d is the particle diameter (in microns), t is the settling time (in minutes) until the time of reading, and c is a value given in table 2C.1 for a temperature of 25°C and a standardly assumed soil density of 2.65 g/cm³. (For some soils, e.g., those with much organic content, certain pretreatment is necessary before this procedure is employed, and the references in section 2C.11 below should be consulted.)

Table 2C.2. *Conversion factors for* c, *or* d *(in equation 1) for soil hydrometer readings taken at various temperatures.**

temperature (°C)	factor
20	1.060
21	1.048
22	1.035
23	1.023
24	1.011
25	1.000
26	0.989
27	0.978
28	0.967
29	0.957
30	0.947

* Calculated from water viscosities, as described by Day (1956, 1965).

For example, if after 65 minutes the hydrometer reads 21 g/l, then the maximum particle diameter remaining in suspension would be estimated to be:

$$d = 46.2/\sqrt{65} = 5.7 \text{ microns, or } 0.0057 \text{ mm.}$$

The values in table 2C.1 take into account the specific gravity of the sodium metaphosphate and the viscosity of water at 25°C. If this procedure is performed at another temperature, simply multiply the table 2C.1 value of c by the appropriate factor given in table 2C.2. Or, multiply the computed d by this factor; if the above example had been a hydrometer reading at 23°C, then calculate:

$$d = (5.7 \text{ microns})(1.023)$$
$$= 5.8 \text{ microns, or } 0.0058 \text{ mm.}$$

Now estimate the relative amounts of sand, silt, and clay in the substrate sample. Place 100 ml of the sodium metaphosphate solution in a clear cylinder and bring the volume to 1 liter with distilled water. Reading the *top* of the meniscus and take a hydrometer measurement (call it R_b) of this "blank" solution containing no soil. (If the readings on the soil suspensions are not all obtained at the same temperature, then a "blank" should be read for each temperature.) Then, the weight of soil left in the suspension at time t is:

$$W_t = R - R_b. \quad (6)$$

Therefore,

$$p = W_t/W_d \quad (7)$$

is the proportion (i.e., 100p is the percentage) of the dry weight (W_d) of soil originally placed in the cylinder still in suspension.

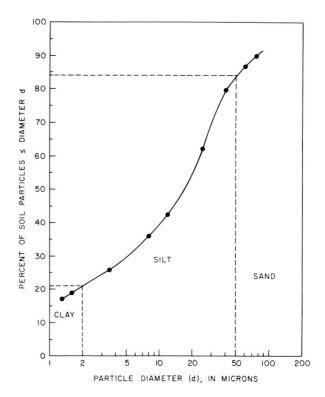

Figure 2C.3. Determination of the percent clay, silt, and sand composition of a substrate sample.

For each hydrometer reading on the soil suspension one should plot the percent of soil remaining in solution against the logarithm of the calculated maximum particle size left in suspension, as shown in figure 2C.3. The several data points are then connected by eye into a smooth curve. By consulting table 2C.3, you can see that the boundary particle sizes between clay, silt, and sand are 2 and 50 microns, and can locate these two points on the horizontal axis and determine where these sizes intercept the curve (the vertical dashed lines in figure 2C.3). These interception points on the curve are then read off of the vertical axis as percentages (horizontal dashed lines in figure 2C.3). For the results in figure 2C.3, for example, the sample contained 21% clay, 84% − 21% = 63% silt, and 100% − 84% = 16% sand. Then, according to figure 2C.4, this soil would be classified as silt loam.

10. Suggested exercises

1. Examine a soil profile from two different forests (e.g., a coniferous and a deciduous forest). What factors are responsible for the differences observed?
2. Examine the effects of topography on the soil moisture in different habitats.
3. Compare the soil fractions from two habitats using the hydrometer method.

Table 2C.3. *Size categories of soil particles.**

	U.S. Department of Agriculture (USDA) System			International Soil Science Society System		
		particle diameter				particle diameter	
category	mm	μ	category	mm	μ
clay	<0.002	<2	IV	<0.002	<2
silt	0.002–0.05	2–50	III	0.002–0.02	2–20
very fine sand	0.05–0.10	50–100			
fine sand	0.10–0.25	100–250	II	0.02–0.2	20–200
medium sand	0.25–0.5	250–500			
coarse sand	0.5–1.0	500–1000			
very coarse sand	1.0–2.0	1000–2000	I	0.2–2.0	200–2000

* From Soil Survey Staff (1951).

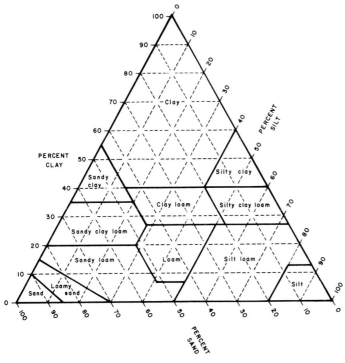

Figure 2C.4. Soil texture types based on percentages of sand, silt, and clay (Soil Survey Staff, 1951; as given in Millar et al., 1965).

4. Compare the soil temperature profiles of a forest and a nearby field or grassland, both at the same time of the same day.
5. Compare litter and humus in the O horizon in a deciduous forest with that in a coniferous forest.

11. Selected references

American Society for Testing and Materials. 1981. Particle-size analysis of soils, 191–202. In 1981 book of ASTM standards, part 19. American Society for Testing and Materials, Philadelphia.

American Society for Testing and Materials. 1981. Moisture, ash and organic matter of peat materials, 462–463. In 1981 book of ASTM Standards, part 19. American Society for Testing and Materials, Philadelphia.

Army Corps of Engineers. 1968. Student workbook: Soils. U.S. Army Corps of Engineers School, Fort Belvoir, Va.

Ball, D. F. 1986. Site and soils, 215–284. In P. D. Moore and S. B. Chapman (eds.), Methods in plant ecology. 2d ed. Blackwell Scientific Publications, Oxford, England.

Barbour, M. G., J. H. Burk, and W. D. Pitts. 1980. Terrestrial plant ecology. Benjamin/Cummings Publishing Co., Menlo Park, Calif.

Baver, L. D., W. H. Gardner, and W. R. Gardner. 1972. Soil physics. John C. Wiley & Sons, New York.

Berger, K. C. 1965. Introductory soils. Macmillan Co., New York.

Black, C. A. (ed.) 1965. Methods of soil analysis, Part 1. American Society of Agronomy, Madison, Wis.

Black, C. A. 1968. Soil-plant relationships. John C. Wiley & Sons, New York.

Black, C. A. (ed.) 1982. Methods of soil analysis, Part 2. American Society of Agronomy, Madison, Wis.

Brody, N. C. 1974. The nature and properties of soils. 8th ed. Macmillan Publishing Co., New York.

Buol, S. W. 1980. Soil genesis and classification. 2d ed. Iowa State University Press, Ames, Iowa.

Daubenmire, R. F. 1974. Plants and environment: A textbook of plant autecology. 3d ed. John C. Wiley & Sons, New York.

Day, P. R. 1956. Report of the committee of physical analysis, 1954–1955, Soil Science Society of America. Soil Sci. Soc. Amer. Proc. 20:167–169.

Day, P. R. 1965. Particle fractionation and particle-size analysis, 545–567. In C. A. Black (ed.), Methods of soil analysis, Part 1. American Society of Agronomy, Madison, Wis.

Etherington, J. R. 1975. Environment and plant ecology. John Wiley & Sons, New York.

Jenny, H. 1980. The soil resource: Origin and behavior. Springer-Verlag, New York.

Millar, C. E., L. M. Turk, and H. D. Foth. 1965. Fundamentals of Soil Science. John C. Wiley & Sons, New York.

Soil Survey Staff. 1951. Soil survey manual. U.S. Department of Agriculture Handbook 18. Washington, D.C. (Cited in Day, 1965).

Soil Survey Staff. 1960. Soil classification. A comprehensive system. 7th approximation. U.S. Soil Conservation Service, Washington, D.C.

Soil Survey Staff. 1967. Supplement to soil classification system. 7th approximation. U.S. Soil Conservation Service, Washington, D.C.

Soil Survey Staff. 1975. Soil taxonomy. A basic system for making and interpreting soil surveys. U.S. Department of Agriculture Handbook 436. Soil Conservation Service, Washington, D.C.

Wolfanger, L. 1971. Soil orders and suborders of the United States and their utilization, 59–98. In G. H. Smith (ed.), Conservation of natural resources. John C. Wiley & Sons, New York.

1. Introduction

Limnology is the study of fresh waters, including both their physical and biological aspects; *oceanography* considers the physical and biotic properties of marine and estuarine environments. Many of the basic principles and concepts concerning terrestrial habitats have parallels in aquatic habitats, although numerous details and patterns are unique to the latter. The aquatic habitat can be divided into certain basic dimensions, such as time, space, and physical and chemical components. Unlike the terrestrial ecologist, the aquatic ecologist generally emphasizes physical and chemical factors instead of biological factors when describing the habitat. In aquatic systems these factors are often more complex than in terrestrial environments, and vegetation has a relatively minor role in modifying the physical characteristics of the habitat. This section will deal with methods for analyzing physical factors such as light, temperature, current, and conductivity. Section 2E presents techniques for analyzing chemical factors.

There are two basic types of freshwater habitats: *lentic* (calm) waters and *lotic* (running) waters. *Lakes* and *ponds* are lentic habitats. Lakes are deep and generally stratified with respect to temperature, oxygen, and nutrients; ponds are shallow bodies of water without seasonal stratification and whose waters mix regularly from top to bottom. A common system of classifying lakes refers to relatively young, deep, cold, and nonproductive lakes as *oligotrophic;* relatively shallow, warm, and productive lakes as *eutrophic;* and lakes having intermediate characteristics as *mesotrophic.* Ponds may be *temporary,* especially in dry climates; *vernal* ponds are those that fill in the spring and dry up in the summer.

Lentic habitats shallow enough to be inhabited by much vegetation are often called *wetlands,* although there is no agreement on their nomenclature (Mitsch and Gosselink, 1986; Chapters 3 and 17). Common North American terminology emphasizes the following wetland types: A *bog* is characterized by having no significant inward or outward water flow (i.e., no source of water other than precipitation), an accumulation of partly decayed organic matter as peat, a low (acidic) *pH,* and mosses such as sphagnum. Other wetlands have relatively neutral *pH* and typically do not accumulate peat. A *marsh* has emergent (and, sometimes, floating) herbaceous vegetation (and may be saline in coastal regions); a *fen* is a peat-accumulating marsh; and a *swamp* has trees or shrubs as the predominant vegetation. Some additional regional terms are found: Shallow marsh-like ponds may be called *potholes* (or *playas*) in the southwest United States; a *muskeg* is a peat-forming wetland in Canada and Alaska, a periodically flooded forest area is a *bottomland,* a *wet meadow* is a grassland with waterlogged soils but without permanent standing water, and a *slough* is a term variously applied to swamps, marshes, or ponds in different parts of the

analysis of aquatic habitats

United States. In Europe, a *swamp* is dominated by reed grass, and peatlands may be called *moors* or *mires.* Wetland areas have been greatly reduced in extent by human activities.

A lentic body of water often may be deep enough to exhibit distinct zonation. The *littoral zone* is the shallow portion along the shore, where light penetrates with sufficient intensity to sustain a significant photosynthetic rate down to the bottom. Rooted vegetation is commonly found in this region. In the open water beyond the littoral zone a depth exists—the *compensation depth*—at which light penetration is so poor that the photosynthetic rate is just equal to the respiratory rate. Above the compensation depth is the *limnetic* region of the lake; below, the *profundal* zone. The littoral and limnetic waters often are collectively termed the *euphotic* zone, that portion of the lake where photosynthetic rate exceeds respiratory rate.

Streams are lotic, being flowing bodies of water. *Creeks* are small streams that are narrow, shallow, and may consist of relatively still areas (*pools*), areas of rapid shallow flow over gravel or rock (*riffles*), and areas of deeper flows (*channels*). *Rivers* are wide and deep streams, and may have more violent *rapids,* instead of riffles. Some small streams flow only seasonally or intermittently during periods of rainfall. The terrestrial borders of a stream are said to be the *riparian* habitat, and the *floodplain* is that land that is periodically subject to flooding.

2. Temporal and spatial information

When studying an aquatic habitat, record the date, time of day, and the name of the observers. Recorded spatial information (noted in subsection 2A.4) should include specific locality, topography, and drainage characteristics.

As most freshwater drains from or into some other body of water, the major drainage system (the *watershed*) should be identified, along with the name of the water body. The watershed is the total land area that drains into the waterway; it incorporates the energy and material exchanges of the terrestrial and aquatic ecosystems within it and is named by the major river system that eventually collects the water from it. The major watersheds are of the rivers that eventually enter the ocean, such as the Mississippi, St. Lawrence, Columbia, Colorado, and Hudson rivers. The *drainage density* for a watershed is

$$\frac{\text{total stream length}}{\text{total area drained}} . \qquad (1)$$

A topographic description of the study area (see section 2A.5) should include the type of water body, such as creek, river, pond, lake, or reservoir. A map or aerial photograph of the water is desirable. If none is available, the mapping methods described by Lind (1985) or Wetzel and Likens (1979) may be employed. Record surface features such as the slope and form of the surrounding terrain and shoreline, form of stream channel, and formations, such as riffles, rapids, falls, and islands. Record the size of the water body and its approximate center depth. If water, substrate, or biological samples are taken, the distance from shore and the depth of the sampling should be noted.

For lakes, the surface area may be estimated from a topographic map or aerial photograph (see section 2A.5). An important variable in limnological studies, particularly those dealing with lakes, is the ratio of the surface area to the volume of the lake. The larger the surface area relative to its volume, the greater will be the amount of gas exchange and mixing due to winds. If the volume and surface area are known, then we can define

$$\text{mean depth} = \text{volume/surface area.} \qquad (2)$$

But the surface area and especially the volume of lakes are usually difficult to measure, so a simple ratio of the width of the water body divided by the center depth can be used as a rough index of the surface area-volume ratio. In elongated lakes, the length of the lake may be considered instead of the surface area, particularly if the long axis is parallel to the direction of the prevailing winds.

It may be of interest, especially in deep lakes, to express the pressure at particular depths. This may be done as:

$$P = 1 + 0.0967d \qquad (3)$$

(Wetzel, 1983:159), where P is the combined atmospheric and hydrostatic pressure, in atmospheres (1 atmosphere = 760 mm Hg), at a water depth of d meters.

3. Physical environment

A description of the physical factors affecting the aquatic environment includes information on atmospheric conditions and substrate, as well as water. Atmospheric conditions control the climate, season, and daily weather conditions, which of course affect the amount of incipient light at the surface, volume of water, temperature, and water currents, and, subsequently, the distribution of organisms in the body of water. As biotic sampling results

Figure 2D.1. The Kemmerer sampler, with which water and plankton samples may be obtained. The sampler pictured has a capacity of 1.2 liters. (Photograph courtesy of the Wildlife Supply Company, Saginaw, Michigan.)

may vary with short-term changes in weather conditions, record the following atmospheric information: climatic zone, air temperature, wind velocity and direction, cloud conditions, and type and intensity of precipitation (see section 2B).

The substrate of the water body provides habitat for a distinctive animal aggregation called the *benthos* (see section 3E.1). Therefore, one should record the type of bottom materials: clay, silt, sand, gravel, or rock. Methods for physical analysis of the sediment are given in section 2D. Streams with swift currents may lack sediment, having a bottom of bedrock or large rocks and boulders. Such rock may be recorded as sandstone, shale, limestone, granite, or other specific type. The slope of the bottom, the depth of any silt, and the occurrence of riffles, rapids, channels, and pools should also be recorded. Samples of substrate, other than rock, and the benthos therein, may be obtained by the methods of section 3E.2.

For a general analysis of an aquatic habitat, record in the field the following basic water measurements: surface water temperature, current velocity, turbidity, and conductivity. For a general chemical survey (see section 2E), hardness, dissolved oxygen, alkalinity, and *p*H are properties measurable in the field and often included in a general habitat description. The Kemmerer water sampler (figure 2D.1) can collect a known volume of water (as well as the organisms suspended therein). The sampler is lowered to the desired water depth and closed by dropping a "messenger" (a metal weight) down the supporting line. In lentic habitats also record water temperature, dissolved oxygen, and *p*H measurements taken from just above the bottom.

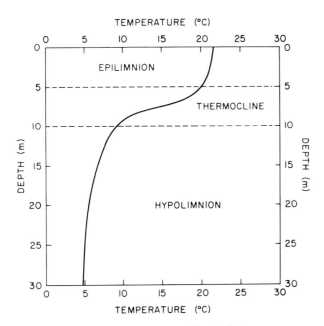

Figure 2D.2. The temperature profile of a lake.

3.1. Temperature   In lakes and ponds water temperature varies with depth and location. Importantly, it affects not only the distribution of organisms but the density of the water and the solubility of minerals and gases. For a general analysis of the habitat, measure the water temperature a few centimeters below the surface and just above the bottom, record this from a number of locations, and calculate the mean surface and bottom temperatures. For a more detailed study of a lake, take temperatures at one-meter intervals at a number of different depths to make a *temperature profile* of the pond or lake. For this purpose, a maximum-minimum thermometer or a thermistor with a long extension is useful. Temperatures of water samples from different depths can be measured immediately after the sample is taken, but this will be accurate only if a large volume of water is collected and measured very rapidly. Some commercial water samplers contain a thermometer readable through a plastic window in the sampler. To graph a temperature profile, it is customary to place water temperature on a horizontal axis and depth on the vertical axis, with the water surface (zero depth) at the top (figure 2D.2). A lake may be *stratified* thermally, having layers of water at distinctly different temperatures. If it is, there is often a short range of depths—the *thermocline*—in which the water temperature changes very abruptly. The water above the thermocline is the *epilimnion;* that below it is the *hypolimnion.* Some authors use the term *metalimnion* for the transitional layer between the epilimnion and hypolimnion and restrict the term *thermocline* to that portion of the metalimnion where the temperaure change is most abrupt.

3.2. Current   Use a current flow meter to measure the current velocity in streams at a number of locations. If such a meter is unavailable, the velocity can be approximated through use of a Pitot tube, an L-shaped glass tube marked off in linear units. The base of the L is placed in the stream with the tube's opening facing the current and its upper arm perpendicular to the surface. Pressure created by the current will cause water to rise in the tube. The height of the water column above the water surface is related to the stream velocity, which can be estimated using the following equation:

$$v = 0.977\sqrt{2gh} \qquad (4)$$

(Welch, 1948), where v is the velocity of the water (cm/sec), g is the gravitational constant (981 cm/sec²), and h is the height of the water (in cm) in the tube. One may also determine surface velocity by measuring the time it takes for a floating object to travel a known distance downstream (but care must be taken to use an object that does not project much above the water's surface and is thus influenced by wind). Take a number of readings across the stream at the same depth, and calculate the mean surface velocity. This procedure is not recommended when turbulence is great or current is slow. Keep in mind that you are measuring the surface velocity and not the mean velocity of a cross-sectional area of the stream, which is more difficult to obtain. Velocity varies with distance from the shore and depth of the stream.

Discharge, another measurement, is the volume of water flowing past a given section of a stream per unit time. It may be calculated as the mean velocity of the stream times its mean cross-sectional area. The mean cross-sectional area is approximated from the mean width of the stream times the mean depth.

3.3. Turbidity   An optical property of water, *turbidity* causes light to be scattered or absorbed in the water, resulting in a decrease in water transparency. It is a function of at least three variables: (1) dissolved chemicals, such as tannins, acids, and salts; (2) suspended particles, such as silt, clay, and organic matter; and (3) density of microbiological organisms.

Turbidity should be measured because the depth of light penetration affects the distribution and intensity of photosynthesis in the body of water. (See section 2D.1 for a description of the compensation depth and the euphotic zone.) One scale of turbidity measurement employs the *Jackson turbidity unit* (JTU). This unit, on a logarithmic scale, considers the height of a column of water that extinguishes the light emitted by a standard "candle."[1] (A height of 2.30 cm that extinguishes the candle image represents 1000 JTU.) Difficult to interpret ecologically, this scale of turbidity measurement is useful mainly in comparing different sites or times.

---

1. This refers to luminous intensity as described in section 2B.4.1.

A common and sounder measurement of turbidity is the *extinction coefficient:*

$$E = 2.30 \log(I_o/I_d)/d, \qquad (5)$$

where E is the extinction coefficient, d is the depth at which the measurement is taken, I_d is the light intensity at that depth, and I_o is the light intensity at zero depth, or just below the surface. (Use table D.2 in Appendix D to obtain logarithms.) This coefficient may be measured with a waterproof light meter lowered to the desired depths. Ideally, these measurements should be taken at about the same time of day and under fairly clear skies. The extinction coefficient is a measure of the amount of light absorbed per unit depth of the water, and can therefore be related to the photosynthetic potential of that body of water.

Another method for measuring turbidity is that using a colorimeter or a spectrophotometer. A water sample is shaken well to avoid settling and is placed in a colorimeter tube to its marked level. The percent transmittance (T) is compared to that of distilled water. The wavelength of the spectrophotometer is set at 450 nm, for this blue-green wavelength is an optimal one for photosynthesis. (Most of the light at the other photosynthetically optimal wavelength of 650 nm in the red-orange region is rapidly absorbed by water and thus has little role in photosynthesis below the first meter of depth.) Since the percent transmittance is $100(I_d/I_o)$, an extinction coefficient may be estimated by substituting $100/T$ for (I_o/I_d) in equation 5; the value of d represents the inside diameter or light path distance of the colorimeter tube. Because artificial white light or a specific wavelength is being used, the extinction coefficient will not be identical to that of a direct field measurement given above, which employs sunlight. A nephelometer is an instrument that measures light scattering in water, higher turbidity causing greater scattering. Nephelometer turbidity units are not the same as Jackson turbidity units; they are useful as comparative measures among various water samples.

A third but more subjective method of turbidity measurement uses the *Secchi disc.* For limnological studies, this disc typically is 20 cm in diameter, having four quadrants, two opposing ones painted black and the other two either white or unpainted. The disc is suspended from the center by a cord or chain and is usually lowered from a boat into the water. It is lowered slowly until no longer visible; the depth at this point is recorded. The disc is lowered further and then slowly raised until it just becomes visible; then the depth at this point is recorded. Calculate the mean of these two determinations and repeat the procedure at a few other locations. This method is a quick and easy method for relative comparisons of degrees of light penetration, but exercise care when interpreting the results as the method is difficult to standardize between individual observers and between different overhead light conditions.

3.4. Conductivity   The inverse of electrical resistance, *conductivity,* is another useful physical measurement in aquatic habitats. The greater the conductivity, the greater the amount of ions in the water. Thus conductivity is an indirect measure of salinity, which reflects the osmotic concentration of solutes. And osmotic concentration is an important physical property of water related to the water and salt balance of organisms. Since polluted waters have a higher conductivity than natural waters, this measure is often used as an index of pollution. The unit of conductivity is mho/cm and represents the amount of current that can be conducted between two electrodes one centimeter apart. Commercial conductivity meters are convenient, but you may also use a standard resistance test meter with platinum electrodes spaced one centimeter apart. Because conductivity is dependent on temperature, a correction for this variable must be made. See section 2E.5 for details on measuring conductivity.

4. Biological components

Biological components in aquatic environments are not as important as physical and chemical factors for rapid habitat descriptions in the field. Unlike terrestrial habitats, where plants dominate the community and strongly influence the physical environment (see section 2B), aquatic habitats are less conspicuously affected by organisms. Their effect is largely on the concentrations of dissolved nutrients and gases. Here, the task of the ecologist is to sample and tabulate quantitatively the more common plant and animal forms (see section 3E). Except in ponds, marshes, and swamps, most aquatic plants are suspended algae and make up the part of the community termed *phytoplankton.* Enumeration of certain "indicator" species is common practice in water pollution studies (see section 5.1 below).

In the littoral zone of most ponds and marshes and often along river edges, a well-developed pattern of vegetation occurs, described as free-floating plants (e.g., duckweeds), rooted floating plants (e.g., pond lilies), submerged plants (e.g., stoneworts, hornworts), and emergent plants (e.g., arrowhead, sedges, rushes, and cattails).

5. Water pollution

Few bodies of water remain free of human contamination. Contaminants, or pollutants, have drastically altered the ecology of many lakes and streams. Therefore some measure of the degree of pollution should be included in an aquatic habitat description.

Some pollution involves introduction of excess amounts of naturally present substrates (e.g., organic matter, nitrates, phosphates). Other pollutants (e.g., most pesticides) are substances foreign to natural habitats. The major sources of pollution are: *industrial* (chemical, organic, and thermal wastes), *municipal* (largely sewage consisting of human wastes, other organic wastes, and detergents), and *agricultural* (animal wastes, pesticides, and fertilizers). Different sorts of pollution may have vastly different effects on an ecosystem. For example, some characteristics of organically polluted waters include low dissolved oxygen, high biochemical oxygen demand (BOD), high turbidity, and high concentrations of such nutrients as phosphates, nitrates, and ammonia. However, acid mine drainage may be associated with water that is rich in oxygen, clear, low in nutrients, and low in organic carbon, but if introduced into the above waters could have devastating ecological effects.

5.1. Biological Indicators   Some organisms serve as indicators of organically or nutrient enriched waters, such as fecal coliform bacteria, "blooms" of blue-green algae, sludge worms (Tubificidae), and the so-called *rat-tailed maggots* of some syrphid flies. Organisms not present in such an environment are either intolerant of it or depend for food on organisms intolerant of it. Indicator organisms are described by Gaufin (1973), Goodnight (1973), Hart and Fuller (1974), Palmer (1962, 1969), and Patrick (1973). These authors, as well as Warren (1971) and Wilber (1969), discuss the use and misuse of indicator organisms in water pollution studies. Often the greater the density of these organisms the greater the degree of organic pollution. Also, biological indicators can signal the occurrence of pollution even if the pollutant is temporarily absent at the time of measurement.

However, be cautious about conclusions drawn from the presence or absence of indicator organisms. The presence of a pollution-tolerant species is not always an indication of pollution as these species occur naturally under less disturbed conditions. Likewise, the absence of such clean water forms such as stonefly naiads, mayfly naiads, caddisfly larvae, or damselfly naiads may be due to habitat conditions other than pollution. Also, organisms that indicate specific types of pollution may differ in different geographic regions or different types of habitats. Differences between the species composition of two areas can be quantitatively described (see section 5C).

5.2. Species Diversity   A popular comparative measure of water pollution and other habitat disturbances is the species diversity index (Wilhm, 1967; Wilhm and Dorris, 1968). In general, the more polluted a body of water the lower is the diversity index, but the use of such an index is difficult to standardize because a variety of factors other than pollution will affect it. The use of artificial substrate samplers (section 3E.2.5) helps alleviate many standardization problems.

Section 5B summarizes several measures of species diversity, and such measures may be used to assess taxonomic diversity even when species identifications are not made. Here we describe a rapid and very simple method (Cairns et al., 1968) to obtain a relative measure of diversity without requiring any taxonomic knowledge. Mix thoroughly the collection of organisms by shaking them in a container of water or preservative, and then examine them, one at a time, *at random.* (A subsample from a suspension may be placed on a microscope slide and the slide examined systematically, from left to right, from top to bottom. Or, a well-mixed collection of macroinvertebrates may be placed in a shallow pan marked with lines or a grid for systematic examination.)

In examining each organism, decide only whether it looks like the previous organism examined (on the basis of shape, size, color, and other obvious characteristics). If so, it is a member of the same "run"; if not, it is said to belong to a new "run." For example, a series of organisms observed at random might look like this (where different letters depict taxa subjectively judged to be different):

$$\underline{A}\ \underline{B\ B}\ \underline{A}\ \underline{C\ C\ C}\ \underline{B}\ \underline{A\ A}\ \underline{B}\ \underline{C\ C}\ \underline{D}$$

Here, a total of fourteen individuals appears in a sequence forming nine runs (a run indicated by an underline).

A *sequential comparison index* may then be expressed as:

$$\text{SCI} = \text{number of runs}/n, \qquad (6)$$

where n is the number of specimens examined. For the above example:

$$\text{SCI} = 9/14 = 0.64.$$

Obviously, the greater the variety (diversity) of organisms in the collection, the higher will be the computed SCI. The greatest possible diversity would be when each individual was unlike each preceding individual (SCI = 1.0); and the lowest possible diversity would be indicated by all of the n specimens being judged identical (SCI = $1/n$). A disadvantage of this index is the variability with which different observers declare organisms to be the same or different.

If the collection contains a large number of organisms, a performance curve (section 1A.3) may be used to determine how many individuals should be counted. Count 50 specimens and calculate the SCI; count another 50 and calculate the SCI for all 100; then proceed to the next 50, and so on. Each time a value of SCI is computed for the cumulative number of organisms, plot it. Counting may be terminated when the performance curve levels off.

The counted organisms should then be returned to the collection and the latter thoroughly mixed once again. A second SCI should be determined for that same collection in the same manner as the first. Calculate the mean of the two replicate determinations of SCI. Cairns and Dickson (1971) have found that a mean of two replicates is sufficient to analyze most ecological assemblages. For more precision, however, six to eight replicates should be obtained for each collection.

The sequential comparison index (SCI) may be transformed into a somewhat more refined index of diversity with little additional effort. While determining the SCI one can also keep track of the total number of different taxa (four in the above example), and calculate a diversity index as:

$$DI = SCI \times \text{number of taxa}. \qquad (7)$$

For the present hypothetical data,

$$DI = (0.64)(4) = 2.56.$$

Field experience has shown that "healthy" streams have DI values greater than 12.0, whereas communities in polluted habitats have DI values of 8.0 or less (Cairns and Dickson, 1971).

5.3. Biochemical Oxygen Demand   *Biochemical oxygen demand* (BOD) is a bioassay of the amount of biodegradable organic carbon in water. Two samples of water are taken in glass-stoppered bottles. Then, the amount of dissolved oxygen is determined in one of them, as described in section 2E.6. The second sample is stored for five days at 20°C, after which its dissolved oxygen is determined. The difference in the concentration of oxygen in the original sample and the stored sample represents the amount of oxygen (in mg/1, or parts per million [ppm]) consumed by microorganisms while decomposing organic material:

$$BOD = (C_1 - C_2)/c, \qquad (8)$$

where C_1 and C_2 are the original and final dissolved oxygen concentrations, respectively, and c is the dilution factor. For polluted water, the sample should be diluted 1:10 or 5:10 (resulting in c values of 0.1 and 0.5, respectively), depending on the expected concentration of organic matter.

BOD represents a laboratory measurement, so extrapolation of this value to the actual oxygen demand of a body of water is questionable. However, it has become standard procedure for comparing the relative amounts of organic enrichment of streams, lakes, or waste waters. In nature BOD may range from a trace to 5 ppm oxygen consumed over a five-day period. Ten to 20 ppm oxygen would indicate a high level of organic pollution, and some waste waters may have BOD values over 100 ppm.

Measurement of BOD can be biased by free chlorine in the water, supersaturation of oxygen, large concentrations of acids or bases, reduced inorganic compounds in the water (sulfite, ammonia, nitrite), and reduced iron. The types of available microorganisms can also affect the results.

5.4. Physical and Chemical Factors   Biological indicators, diversity indices, and bioassays do not reveal the exact identity of pollutants. For this, a chemical and physical analysis of the water should be made. The chemical analyses of section 2E can determine quantities of some pollutants.

In habitats where nutrient enrichment is suspected of causing algal blooms, phosphate and nitrate concentrations should be determined. However, if the algal bloom is far advanced, most of the soluble nutrients would be in the algal biomass and an analysis of soluble phosphate and nitrate may reveal low concentrations. Salt contamination, if suspected, is determined from measurement of conductivity and an analysis of chlorides. And acid mine drainage results in a low *p*H and high amounts of sulfates.

Low values of dissolved oxygen when accompanied by a high BOD will often result in greater concentrations of ammonia. High values of BOD are also accompanied by high turbidity and conductivity. However, high turbidity and conductivity are also associated with siltation, the major contaminant in many streams, lakes, and reservoirs. But siltation will not necessarily be associated with a high BOD or low dissolved oxygen.

Thermal pollution is easily detected by measuring the temperature of various parts of a lake or stream. This form of habitat alteration, unlike that caused by siltation or organic wastes, is less visible and may have more subtle effects on the diversity, productivity, and species composition of a body of water. Slight increases in the temperature of a water body may increase the rate of nutrient cycling, alter the reproductive efficiency of certain fishes, and even encourage algal blooms.

6. Aquatic habitat profile

How does one use all of these seemingly disjunct pieces of data from a habitat? One can simply but confusingly graph each of the physical and chemical variables as a function of space or time, as in a temperature profile of a lake in relation to depth. Such illustrations are useful for evaluation of individual physical and chemical variables, but it is desirable to have a holistic impression of the habitat. Kaill and Frey (1973) have attempted to summarize habitat data into an environmental profile, based on measurements taken at dawn and dusk. The following procedure, while using a lake as the example, may also be applied to streams and terrestrial habitats. In the latter case, use physical and chemical soil data and atmospheric measurements.

To prepare an environmental profile, construct a histogram of the measured variables for each ecological situation (location or time) being studied (figure 2D.3). In these histograms the logarithm of the value is plotted on the vertical axis, and the environmental factors are sequenced along the horizontal axis. (Commercially available 3- or 4-cycle semilogarithmic graph paper is very convenient for this purpose.) The logarithmic scale is used so that very small and very large numbers may be placed on the same graph. The order of the environmental variables is arbitrary but should be consistent from one profile to another. For convenience physical measurements may be placed together and chemical determinations grouped

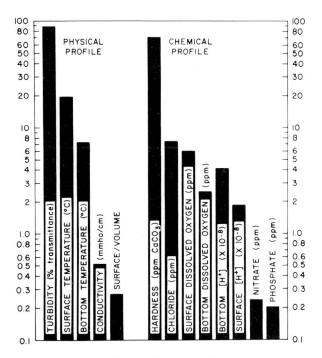

Figure 2D.3. A habitat profile of a pond.

together. (Biotic measures, such as abundance and species diversity, are not included in the habitat profile.)

Since the extinction coefficient is a logarithm, it is convenient to graph the percent transmittance instead. The plotting of pH presents a problem for this variable normally falls in a very narrow range of 6.5 to 8.5. Thus a small but important difference would appear insignificant on the graph, so it is recommended that pH be converted to the hydrogen ion concentration [H⁺]:

$$pH = \log (1/[H^+]);$$ (9)

therefore,

$$[H^+] = 1/(\text{antilog } pH).$$ (10)

Logarithms and antilogarithms may be obtained from table D.1 in Appendix D. For example, if the pH of a water sample is 7.62, we would use equation 10 to compute:

$$[H^+] = 1/(\text{antilog } 7.62)$$
$$= 1/(4.17 \times 10^7)$$
$$= 0.240 \times 10^{-7} \text{ or } 2.40 \times 10^{-8}.$$

The selection of units of measurement is done to allow the bars in the graph to fall within easily plotted limits. For example, conductivity values less than 0.1 mmho/cm may be plotted as µmho/cm; hardness values exceeding 100 ppm CaCO₃ may be plotted as ppm × 10⁻². Acidity is represented as [H⁺] × 10⁻⁸ (so a pH of 8 would appear on the profile graph as 1, a pH of 7 would appear as 10, a pH of 6 as 100, and so on). If habitat profiles are to be compared, they should each employ the same units of measurement.

Measurements such as temperature, turbidity, conductivity, dissolved oxygen, and pH should be made near the surface and near the bottom of a lake, and both surface and bottom values should be graphed. The measurement of these variables only near the surface can lead to a poor representation of the lake.

7. Suggested exercises

1. Compare the habitat profiles of two different ponds, lakes, or streams. Explain the differences between the physical and chemical variables observed.
2. Compare the habitat profiles of a polluted and an unpolluted body of water.
3. Compare the habitat profile of a riffle and a pool in a stream.
4. Select environmental variables such as temperature, turbidity, and oxygen, and determine the profile of these as a function of depth in a lake or pond (see figure 2E.2).
5. Sample a polluted stream, both upstream and downstream from the source of contamination, attempting to sample in habitats with similar currents, depths, and substrates. Identify the major taxa of algae or benthic invertebrates and note the relative abundances of typical clean-water or polluted-water taxa.
6. In a stream pollution study such as above, calculate a measure of taxonomic diversity at each sampling location. Determine how this changes with distance from the source of contamination.

8. Selected references

Barnes, J. E., and G. W. Minshall. 1983. Stream ecology: Applications and testing of general ecological theory. Plenum Press, New York.

Brown, A. L. 1971. Ecology of fresh water. Harvard University Press, Cambridge, Mass.

Cairns, J., Jr., D. W. Albaugh, F. Busey, and M. D. Chanay. 1968. The sequential comparison index—a simplified method for non-biologists to estimate relative differences in biological diversity in stream pollution studies. J. Water Poll. Contr. Fed. 40:1607–1613.

Cairns, J., Jr., and K. L. Dickson. 1971. A simple method for the biological assessment of the effects of water discharges on aquatic bottom-dwelling organisms. J. Water Poll. Contr. Fed. 43:755–772.

Cairns, J. C., Jr., and K. L. Dickson (eds.). 1973. Biological methods for the assessment of water quality. American Society for Testing and Materials, Philadelphia.

Cairns, J., Jr., G. R. Lanza, and B. C. Parker. 1972. Pollution related structural and functional changes in aquatic communities with emphasis on freshwater algae and protozoa. Proc. Acad. Natur. Sci. Philadelphia 124:79–127.

Cole, G. A. 1983. Textbook of limnology. 3d ed. C. V. Mosby Co., St. Louis.

Fontaine, T. D., III, and S. M. Bartell. 1983. Dynamics of lotic ecosystems. Ann Arbor Science, Ann Arbor, Mich.

Gaufin, A. R. 1973. Use of aquatic invertebrates in the assessment of water quality, 96–116. In J. C. Cairns, Jr., and K. L. Dickson (eds.), Biological methods for the assessment of water quality. American Society for Testing and Materials, Philadelphia.

Goldman, C. R., and A. J. Horne. 1983. Limnology. McGraw-Hill Book Co., New York.

Goodnight, C. J. 1973. The use of aquatic macroinvertebrates as indicators of stream pollution. Trans. Amer. Microscop. Soc. 92:1–13.

Greenberg, A. E., J. J. Conners, and D. Jenkins (eds.). 1985. Standard methods for the examination of water and wastewater. 16th ed. American Public Health Association, Washington, D.C.

Hart, C. W., Jr., and S. L. H. Fuller (eds.). 1974. Pollution ecology of freshwater invertebrates. Academic Press, New York.

Hutchinson, G. E. 1957. A treatise on limnology. Volume 1. Geography, physics, and chemistry. John Wiley & Sons, New York.

Hutchinson, G. E. 1967. A treatise on limnology. Volume 2. Introduction to lake biology and the limnoplankton. John Wiley & Sons, New York.

Hutchinson, G. E. 1973. Eutrophication. Sci. Amer. 61:269–279.

Hutchinson, G. E. 1975. A treatise on limnology. Volume 3: Limnological botany. John Wiley & Sons, New York.

Hynes, H. B. N. 1970. The ecology of running waters. University of Toronto Press, Toronto.

Kaill, M. W., and J. K. Frey. 1973. Environments in profile, an aquatic perspective. Canfield Press, San Francisco.

Lind, O. T. 1985. Handbook of common methods of limnology. 2d ed. Kendall/Hunt Publishing Co., Dubuque, Iowa.

Lock, M. A., and D. D. Williams. 1981. Perspectives in running water ecology. Plenum Press, New York.

Longhurst, A. R., and D. Pauly. 1987. Ecology of tropical oceans. Academic Press, New York.

Macan, T. T. 1974. Freshwater ecology. Longmans, Green and Co., London.

Meadows, P. S., and J. I. Campbell. 1987. An introduction to marine science. 2d ed. John Wiley & Sons, New York.

Merritt, R. W., and K. W. Cummins (eds.). 1983. An introduction to the aquatic insects of North America. 2d ed. Kendall/Hunt Publishing Co., Dubuque, Iowa.

Mitsch, W. J., and J. G. Gosselak. 1986. Wetlands. Van Nostrand Reinhold Co., New York.

Moss, B. 1980. Ecology of fresh waters. Blackwell Scientific Publications, Boston.

Palmer, C. M. 1962. Algae in water supplies. Public Health Service Publ. No. 657. U.S. Department of Health, Education, and Welfare, Washington, D.C.

Palmer, C. M. 1969. A composite rating of algae tolerating organic pollution. J. Phycol. 5:78–82.

Parsons, T. R., M. Takahashi, and B. Hargrave. 1984. Biological oceanographic processes. 3d ed. Pergamon Press, New York.

Patrick, R. 1973. Use of algae, especially diatoms, in the assessment of water quality, 76–95. In J. C. Cairns, Jr., and K. L. Dickson (eds.), Biological methods for the assessment of water quality. American Society for Testing and Materials, Philadelphia.

Reid, G. K., and R. D. Wood. 1976. Ecology of inland waters and estuaries. D. Van Nostrand Co., New York.

Reise, K. 1985. Tidal flat ecology: An experimental approach to species interactions. Springer-Verlag, New York.

Ruttner, F. 1963. Fundamentals of Limnology. 3d ed. University of Toronto Press, Toronto.

Stumm, W., and J. J. Morgan. 1981. Aquatic chemistry: An introduction emphasizing chemical equilibria in natural waters. 2d ed. John C. Wiley & Wiley and Sons, New York.

Tait, R. V. 1981. Elements of marine ecology. 3d ed. Butterworth Publishers, Woburn, Mass.

van der Valk. 1989. Northern prairie wetlands. Iowa State University Press, Ames, Iowa.

Warren, C. E. 1971. Biology and water pollution control. W. B. Saunders Co., Philadelphia.

Welch, P. S. 1948. Limnological methods. McGraw-Hill Book Co., New York.

Weller, M. W. 1987. Freshwater marshes. University of Minnesota Press, Minneapolis.

Wetzel, R. G. 1983. Limnology. 2d ed. Saunders College Publishing. Philadelphia.

Wetzel, R. G., and G. E. Likens. 1979. Limnological analysis. W. B. Saunders Co., Philadelphia.

Wilber, C. G. 1969. Biological aspects of water pollution. Charles C. Thomas, Springfield, Ill.

Wilhm, J. L. 1967. Comparison of some diversity indices applied to a population of benthic macroinvertebrates in a stream receiving organic wastes. J. Water Poll. Contr. Fed.

Wilhm, J. L., and T. C. Dorris. 1968. Biological parameters for water quality criteria. BioScience 18:477–481.

1. Introduction

Chemical components of the environment, though generally less visible than biotic and physical factors, significantly affect the abundance and distribution of species. The analytical methods given here are based on those given in *Standard Methods for the Examination of Water and Wastewater* (Greenberg et al., 1985), the American Society for Testing Materials (ASTM) *Annual Book of ASTM Standards* (ASTM, 1981), *A Manual of Chemical and Biological Methods for Seawater Analysis* (Parsons et al., 1984), and *Methods of Soil Analysis: Chemical and Microbiological Properties* (Black et al., 1965). One may also consult Wetzel and Likens (1979). We have avoided selecting chemical analyses requiring highly sophisticated equipment or long periods of time, but we have included procedures based on simple titrations and spectrophotometry.[1] If the reader is fortunate enough to have electronic instruments to collect chemical data, then perusal of the following methods should offer testimony to the extent of that good fortune.

Because the chemical components in soil are extracted into an aqueous solution before analysis, the basic analytical determinations described below can be used for either soil or fresh water. For salt water or highly polluted waters, some of the techniques given here may require slight modifications, and Greenberg et al. (1985) or Parsons et al. (1984) should then be consulted. *Care should be exercised in handling the chemicals indicated, particularly the acids and alkalis, and disposal of all solutions should be in accordance with environmental regulations and good laboratory practice.*

The following procedures may be performed as part of a general chemical analysis of a habitat. Or, one may select specific chemical factors, such as phosphate or nitrate concentrations, and compare different environmental situations. Techniques for measuring dissolved oxygen or total dissolved carbon dioxide may form the basis for determining other factors, such as biochemical oxygen demand (section 2D.5.3) or aquatic productivity (section 6B).

Of the many chemical factors that can affect an environment, the dominant ones are oxygen, hydrogen ion concentration (measured by *pH*), salinity, calcium, magnesium, nitrate, ammonia, phosphate, chloride, sulfate, carbon dioxide, carbonate, and bicarbonate. When measuring any of these, remember that you are making a determination at a given time. Therefore, your determination may not represent the magnitude of a given factor at that site yesterday, last week, last summer, or some time in the future. Unless a monitoring program is performed over an extended period of time, your measurement will be interpretable only within the framework of your particular study.

1. Prepackaged water and soil analysis kits are available for the chemical determinations given (for example, those of the Hach Chemical Co., Ames, IA, and the LaMotte Chemical Products Co., Chestertown, MD).

2e

chemical analysis of habitats

Since most ecological studies are comparative in nature, relative measures of the chemical constituents often are as informative as absolute measures. One often compares the results of chemical analyses to standards set by the U.S. Environmental Protection Agency (EPA) or by state and local environmental regulatory agencies. However, legal standards often have significance only locally, and they may have a public health, recreational, or aesthetic foundation rather than an ecological one. Nonetheless, some ecologists use these standards as reference points in water pollution studies to draw conclusions about environmental quality.

Basic principles of sampling (section 1A) must be followed and sufficient replicate samples taken when physically sampling water or soil in the habitat. Techniques for collecting soil and water are given in sections 2C and 2D, respectively. Preparation and analysis of the sample should be done as soon after sampling as possible, ensuring that care of the material is taken during storage and transport. Certain measurements, such as *pH*, alkalinity, carbon dioxide, and dissolved oxygen, are best performed in the field, for these quantities change rapidly due to temperature changes and biological activity in the sample.

1.1. Solutions Thoroughly wash all glassware and then rinse in distilled water. Be especially cautious when using detergents, since they can contaminate the glass with phosphorus, sulfur, or nitrogen compounds. Use distilled water for all reagents; boiling it will drive off dissolved CO_2 and O_2. Special care must be taken when preparing and handling toxic chemical solutions.

Concentrations in aqueous solutions are commonly expressed as parts per million (ppm), parts per thousand ($^o/_{oo}$), or percent (% = parts per hundred). These represent, respectively, the weight of solute per weight of solvent, as mg/kg (or μg/g), g/kg, and g/100 g, respectively. As one liter of water weighs one kilogram, it is common to measure the volume of water rather than its weight (i.e., mg/1, g/1, and g/100 ml, respectively). Units such as milliequivalents (meq), normality (N), and molarity (M) are also used by environmental chemists but mostly to express concentrations of chemical reagents used in titration analysis, rather than concentrations in nature.

1.2. Standards   Standards are solutions of known concentrations to which we compare the results of chemical analyses. The accuracy of final results depends heavily on the accuracy of standards, so great care must be taken when preparing them. Use only fresh reagent-grade chemicals and only freshly prepared solutions, unless stock solutions are known to be very stable. Prepare a range of concentrations of the standards so that your sample values fall within this range.

To calculate the concentration of a standard stock solution, first find the proportion of the desired substance relative to the formula weight of the compound used in making the solution. For example, when preparing a 1000-part-per-million (ppm) stock solution of phosphate, KH_2PO_4 is commonly used. The formula weight of KH_2PO_4 is the sum of the atomic weights of all atoms in the compound (see Appendix C for atomic weights): $39.102 + 2(1.0080) + 30.9738 + 4(15.9994) = 136.0894$. Since a 1000-ppm stock of phosphate solution contains 1 gram of PO_4 per liter, then,

$$W = FW_1/FW_2, \quad (1)$$

where W is the number of grams of KH_2PO_4 to be added to 1 liter of water to yield a 1000-ppm PO_4 standard, FW_1 is the formula weight of KH_2PO_4, and FW_2 is the formula weight of PO_4. The formula weight of PO_4 is 94.975, so

$$W = 136.0894/94.9714 = 1.433.$$

Therefore, 1.433 g of KH_2PO_4 is added to 1 liter of water to give 1000 ppm PO_4. Additional standards may then be made by diluting measured aliquots (fractional portions) of the stock solution. For example, a 100-ppm solution would be made by adding 1 ml of the 1000-ppm stock to 9 ml of water, a 10-ppm standard would be 1 ml of the stock added to 99 ml of water, and a 1-ppm standard would be 1 ml of 100-ppm standard added to 99 ml of water.

A standard may be used to determine the concentration of a substance in a prepared sample. For example, if the substance is detectable spectrophotometrically, plot the absorbance reading from the spectrophotometer against the concentration for a series of standards. Then fit a line through the plotted points. (If the data have very little departure from a straight line, the line may be fit visually; if not, then use the regression methods of sections 1B.6.1 and 1B.6.5.) It is common to assume that the standard curve passes through the origin of the graph; if so, then the standard curve coefficient is:

$$b = A_s/C_s, \quad (2)$$

where A_s is an absorbance value from the standard curve, and C_s is the concentration associated with A_s. Using equation 2, the unknown concentration in the prepared sample (C_p) is:

$$C_p = A_p/b, \quad (3)$$

where A_p is the absorbance reading obtained on the prepared sample.

1.3. Titration   Titrations determine what volume of a known concentration of a known substance (the *titrant*) is needed to react with the substance of interest in the prepared sample. Hence titrations are also referred to as volumetric methods. The concentration of a substance in a sample (C_p, in mg/l) is calculated from the normality of the titrant expressed as equivalents per liter and the amount of titrant used:

$$C_p = (N_t)(FW)(1000)V_t/nV_p, \quad (4)$$

where N_t is the normality of the titrant, FW is the formula weight in grams of the substance analyzed, n is the valence of this substance, and V_t is the volume of titrant needed to react completely with the substance in the sample of volume V_p.

The concentration of the titrant (N_t) generally should be calibrated by titrating it against a standard solution. The value of N_t is, then,

$$N_t = (N_s)(V_s)/V_t, \quad (5)$$

where N_s is the normality of the standard and V_s is the volume of the standard. This is a general equation usable for any of the titrametric methods given below. However, for ease of calculation it is more convenient to estimate first the concentration of the titrant as an equivalent weight (C_t):

$$C_t = N_t FW(1000)/n. \quad (6)$$

For example, if one were to express results as mg $CaCO_3$ per liter using a titrant having a concentration of 0.05N, the equivalent weight of the titrant would be:

$$C_t = (0.05)(100.0892)(1000)/2$$
$$= 2502.2 \text{ mg } CaCO_3/\text{liter.}$$

Then,

$$C_p = C_t V_t/V_p. \quad (7)$$

This last equation is simple to use, for C_t is known and C_p is then readily calculated from the volumes of the titrant and the prepared sample. For further simplification, prepare the concentration of the titrant (C_t) so that 1 ml of titrant is equivalent to 1, 2, 5, or 10 mg of the substance analyzed.

1.4. Estimating Sample Concentration

If a sample was diluted or concentrated, then the concentration in the original solution is:

$$C_o = C_p V_p / V_o, \qquad (8)$$

where V_o is the volume of the original aqueous sample prior to dilution or concentration. For example, if a 20-ml sample were diluted to 100 ml and the prepared sample contained 1.5 ppm PO_4, then,

$$C_o = (1.5)(100)/20 = 7.5 \text{ ppm}$$

is the concentration of the original sample. If no dilution or concentration was made on the original sample, then

$$C_o = C_p.$$

For soil extracts, the concentration in mg/kg (or $\mu g/g$) dry weight of soil is:

$$C_w = (C_o)(V_o)/W_d, \qquad (9)$$

where C_w is the concentration per unit dry weight, and W_d is the dry weight of the sample. For most soil extractions given below, fresh samples rather than dried samples are used. Therefore, the dry weight must be estimated by multiplying the fresh weight by the ratio of dry weight to fresh weight determined from a separate sample (see section 6A.2 for determination of dry weights).

1.5. Analytical Error

As discussed in sections 1A and 1B, all measurements derived from sampling have some amount of random error (i.e., lack of precision), expressible by the standard deviation, standard error, or coefficient of variation. (Chemists sometimes call the latter quantity the "relative standard deviation.") However, one must also consider bias, or lack of accuracy, due to analytical technique. The above considerations for calculation of sample concentration from titrations and absorbances apply only if:

1. contamination is negligible;
2. losses of the substance of interest are negligible;
3. interferences, as from other chemicals, turbidity, and color, are negligible; and
4. the standard curve represents the true concentration of the substance and is linear.

Loss or contamination of the chemical constituent can occur during sampling and preparation of the sample. Also, bias in standardization may result from careless preparation of standard solutions, use of impure chemicals, physical and chemical differences between the standards and the prepared sample, and nonlinearity of the standard curve. Fortunately, the latter is seldom a problem in the methods given below.

Always perform a given analysis on *blanks,* a useful procedure for correcting for contamination and certain types of background interferences during an analytical procedure. In a *blank determination* you duplicate the procedure exactly from beginning to end, either on distilled water, including the color step, or on a sample with only the color step omitted. The latter type of blank may be used to set 100% transmittance on the spectrophotometer. If it is not, then obtain readings from the blank, the standard, and the sample, and the concentration of the substance in the prepared sample is estimated as:

$$C_p = C_s(A_p - A_b)/(A_s - A_b), \qquad (10)$$

or, for titrations:

$$C_p = (C_t)(V_t - V_b)/V_p, \qquad (11)$$

where A_b and V_b, respectively, are the absorbances of a distilled water blank and the volume of the titrant used for the blank. Because the standard is run through the same procedure as the sample and the blank, any loss of sample substance may be corrected for if we can assume the same relative loss of the material in the standard. Use of the blanks will then correct for any background contamination or background interferences that accumulate during the analytical procedure.

2. Alkalinity and carbon dioxide

Alkalinity is the capacity to neutralize a strong acid and depends on the concentrations of buffers as well as bases in the medium. The dominant buffering system in most habitats is the carbonate-bicarbonate system. Additionally, many organic acids and weak inorganic acids can affect alkalinity, but these are generally of minor importance in fresh waters. Because dissolved CO_2 is of major importance in aquatic systems, the determination of alkalinity is often accompanied by the determination of dissolved CO_2. *Total* CO_2 represents the amounts of carbon dioxide (CO_2), bicarbonate (HCO_3^-), and carbonate ($CO_3^=$) dissolved in water. It is common practice to express alkalinity as an equivalent in ppm $CaCO_3$ because the calcium ion is the major associate of carbonate and bicarbonate in fresh waters. However, the carbonate and bicarbonate and hydroxide ions are the predominant determinants of alkalinity, not calcium. Do not confuse this measurement with hardness (see section 2E.3), also commonly expressed as an equivalent of ppm $CaCO_3$.

The relative concentrations of CO_2, $HCO_3^=$, and $CO_3^=$ are related to pH, as in figure 2E.1 and equation 12:

$$CO_2 + H_2O \rightleftharpoons H^+ + HCO_3^- \rightleftharpoons 2H^+ + CO_3^= \qquad (12)$$

Addition of hydrogen ions to the system results in a shift in the above reactions to the left, thus producing more CO_2. Addition of hydroxide ions removes the hydrogen ions and shifts the reaction to the right, thus producing more HCO_3^- and $CO_3^=$. Between a pH of 6.0 and 7.5, relatively large amounts of acid or base are required to produce a small change in pH. Thus water high in bicarbonate helps to maintain the pH range commonly found in nature. At

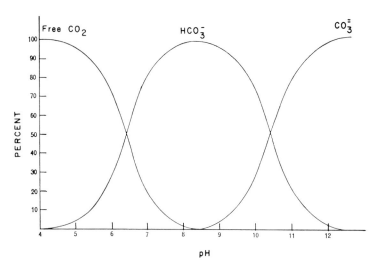

Figure 2E.1. The relative amounts of CO_2, HCO_3^-, and $CO_3^=$ are dependent on pH.

a pH above 12.5, all CO_2 is in the form of carbonate, and above 8.3 all alkalinity is due to OH^-, HCO_3^-, and $CO_3^=$. The concentrations above pH 8.3 are said to be in the range of *phenolphthalein alkalinity* (PA), since this indicator dye has a titration end point of about pH 8.3. At this point bicarbonate is the prevailing form of CO_2 (see figure 2E.1). *Total alkalinity* (TA) refers to the total bicarbonate, carbonate, and hydroxide that can neutralize acid, and thus includes PA. As can be seen in figure 2E.1, the end point where all HCO_3^- is converted to CO_2 is around a pH of 4.5 to 5.0. Methyl orange or a mixture of methyl red and bromocresol green is used, for their end points are within this range (ASTM, 1981; Greenberg et al., 1985).

2.1. Materials   Titration burette, 250-ml beaker, 100-ml water sample, phenolphthalein indicator (0.5 g/50 ml of 95% ethanol plus 50 ml water with a few drops of dilute NaOH added until faintly pink), methyl orange indicator (0.05 g/100 ml water) or mixed methyl red (20 mg) and bromocresol green (100 mg/100 ml ethanol), 1060 ppm Na_2CO_3 standard (1.06 g anhydrous sodium carbonate is equivalent to 1.0 g calcium carbonate and has a higher solubility), 0.1N or 0.05N H_2SO_4 (depending on the range of alkalinity expected), 0.0227N NaOH, 0.1% sodium hexametaphosphate. All reagents should be made with CO_2-free distilled water by boiling the water for 15 minutes and then storing it in airtight containers. The pH of the distilled water should not be less than 6.0. The H_2SO_4 titrant should be standardized with Na_2CO_3 standard to ensure accurate normality using equations 5 and 6.

2.2. Sample Preparation   Water samples: Alkalinity and CO_2 should be determined immediately, because respiration and photosynthesis affect their measurement. If samples cannot be analyzed in the field, store them in air-free, airtight bottles on ice and analyze them at room tem-

perature as soon as you return to the laboratory. Filter the water only if turbid, for filtration exposes the sample to air.

For a soil sample, extract soluble $CO_3^=$ and HCO_3^- in distilled water, as follows: Weigh 100 g of sifted fresh (undried) soil and place in a 250-ml flask. Add 100 ml of distilled water, shake vigorously, allow to settle, and shake again. Filter and add 4 drops of 0.1% sodium hexametaphosphate. Determine the dry weight of the soil on a separate sample.

2.3. Analysis   Add 2 drops of phenolphthalein to 25, 50, or 100 ml of prepared sample (use the smaller volumes if high alkalinities are expected). If red then titrate over a white surface with 0.05 or 0.1N H_2SO_4 until faintly pink and record the milliliters of titrant used. (The 0.1N H_2SO_4 is used for expected high alkalinities.) At this point the pH is about 8.3, and all $CO_3^=$ has been converted to HCO_3^-. After the determination of PA, or if the sample is not red after the addition of the indicator, add 2 or 3 drops of methyl orange or mixed indicator. Then titrate slowly with the acid until the solution just reaches its end point (orange for methyl orange at a pH of 4.6 or pink for the mixed indicator). At this point all the bicarbonate has been converted to CO_2 (see figure 2F.1).

To determine dissolved CO_2, reverse the titration. If the sample turned red in the phenolphthalein alkalinity test, then there is no dissolved CO_2. If the sample did not turn red, then take a fresh 100-ml sample and add 2 drops of phenolphthalein. Titrate with 0.0227N NaOH slowly until the sample just turns red. One ml of 0.0227N NaOH represents 1.0 mg of dissolved CO_2.

Total CO_2 is often of more importance than free CO_2, for CO_2 is in equilibrium with bicarbonate and carbonate, two important sources of carbon in photosynthesis. For determination of total CO_2 (CO_2 + HCO_3^- + $CO_3^=$), first adjust the pH of a fresh sample to 8.3 by titrating the sample with an acid if PA is greater than zero or with a base if PA is zero. This procedure converts all $CO_3^=$ or CO_2 to HCO_3^-. The amount of HCO_3^- is then determined as described for total alkalinity. For application of this method to measurement of CO_2 production or consumption in an ecosystem, see section 6B.2.

2.4. Calculations   Phenolphthalein alkalinity and total alkalinity are estimated using equations 6 and 7 where C_p is the total alkalinity, in ppm $CaCO_3$, $n = 2$, FW = 100.0892, and V_p is the volume of the prepared sample (25, 50, or 100 ml). For 0.1 and 0.05N H_2SO_4, 1 ml of titrant is equivalent to 5 and 2.5 mg $CaCO_3$, respectively. If the same sample was used for PA and TA, then V_t for TA must also include the volume of the titrant used for the determination of PA. For example, if 5 ml of acid were used to determine PA, and 9 ml of titrant were used to reach the TA end point, then $V_t = 13$ ml.

The milligrams of free CO_2 is the number of milliliters of 0.0227N NaOH used to titrate the sample to a pH of 8.3. The concentration of dissolved CO_2 and total CO_2, in ppm, is estimated from equations 6 and 7, where FW = 44.0098 and $n = 1$. The amount of OH^- at a pH of 8.3 is assumed to be small compared to the amount of HCO_3^-. Equation 8 corrects for dilution or concentration if either were done, and equation 9 expresses the alkalinity of soil samples where V_p is 100 ml. Use equation 11 for blank correction.

2.5. Sources of Error   Care must be taken to have reagents free of CO_2. Temperature also affects the results, for CO_2 is more soluble at colder temperatures. Samples can be biased by respiration, impure reagents, and long exposure to air. Loss of CO_2 to the air or through photosynthesis can also occur. Alkalinity and CO_2 measurements can be further biased by a high degree of conductivity, salinity, weak acids, turbidity, or strong coloring agents in the water. Between 10 and 500 ppm $CaCO_2$, this technique has an accuracy of ± 3 ppm $CaCO_3$ and a precision of ± 10%. Values less than 5.0 ppm as $CaCO_3$ are less reliable (Golterman et al., 1978).

3. Calcium, magnesium, and hardness

Calcium and magnesium in soil and water are important mineral nutrients, and they affect pH and the solubility of carbonate and phosphate. These two cations also have an antagonistic effect on many materials that would otherwise be toxic. Calcium and magnesium in water cause "hardness," the ability of water to precipitate soap. These metal ions are easily determined using compleximetric titration. An indicator dye such as Eriochrome Black T forms a red-colored complex with these cations. In the absence of calcium and magnesium ions, the dye is blue. For other dyes that can be used see ASTM (1981), Greenberg et al. (1985), and Black et al. (1965). Ethylenediaminetetraacetic acid (EDTA) is a chelating agent (a chemical that forms a soluble complex with metal ions). As EDTA is added to the sample containing the metal and the indicator dye, it effectively binds the metal as an EDTA complex. When the metal ion is thus removed from solution, the dye turns blue, and the amount of the metal originally present is directly proportional to the amount of EDTA required.

One may determine the amount of calcium by first precipitating the magnesium out of solution at a pH of 12.0, and then performing the analysis. Subtract the amount of calcium from the measurement of total calcium and magnesium, and you will have an estimate of the magnesium in the original sample. Calcium is generally the dominant bivalent metal ion in soil and fresh water, and is largely associated with carbonate. Thus the hardness, or total calcium and magnesium concentration, is commonly expressed as the equivalent of ppm $CaCO_3$, even though one is not actually measuring $CaCO_3$. This measurement should not be confused with alkalinity (see section 2E.2), which is also commonly expressed as ppm equivalent of $CaCO_3$. Some environmental chemists prefer to express hardness as ppm equivalent of calcium.

3.1. Materials   250-ml Erlenmeyer flasks, burettes, filter paper, suction funnel, 100-ml graduated cylinder, 50-ml beaker, pH meter, distilled water, EDTA (3.723 g $Na_2EDTA:2H_2O$/1000 ml, 1.0 ml titrant = 1 mg $CaCO_3$), Eriochrome Black T (0.4 g mixed with 100 g NaCl; keep dry) or other suitable calcium-magnesium indicator, $CaCO_3$ or $CaCl_2$ standard ($CaCl_2$ is more soluble than $CaCO_3$ and 1.11 g $CaCl_2$/liter is equivalent to 1000 ppm $CaCO_3$), buffer solution (16.9 g NH_4Cl/143 ml concentrated NH_4OH plus 1.25 g Mg salt of EDTA, dilute to 250 ml, pH = 10), 1N sodium acetate. The EDTA titrant must be checked against standards to assure its correct normality. The normality may be calculated using equation 5.

3.2. Sample Preparation   For water samples, filter 100 ml of water.

For soil samples, place 3 to 5 g of fresh soil (do not dry it) in a 100-ml flask. Add 30 ml of 1N sodium acetate, stopper, and shake vigorously for one minute; allow soil to settle then decant and save the supernatant. Repeat twice more on the same sample so that the soil has been rinsed with a total of 90 ml of the acetate extracting solution. Filter the supernatant and rinse the filter with 10 ml of distilled water. The volume of the sample (V_p) is then 100 ml. The calcium and magnesium in the extract represents in part what soil chemists call *exchangeable cations*.

3.3. Analysis   Place 25 ml of prepared sample in a 50-ml beaker, add 1 ml of buffer solution to adjust pH to 10.0. Add a few milligrams of the indicator-salt mixture (or a few drops of other indicator solution), which gives a reddish color if Ca and Mg are present. Titrate slowly with the EDTA until a distinct blue color appears (the solution may show an intermittent purple stage). Record the volume of EDTA needed to titrate the sample to its blue end point. If more than 15 ml of EDTA titrant is required, then use a smaller sample (e.g., 10 ml) or dilute the sample (e.g., 10 ml of sample + 90 ml of distilled water). If less than 1 ml of titrant is required, use 50 or 100 ml of sample. Analysis must be completed within five minutes to reduce the chance of calcium precipitation. To analyze calcium alone, place 25 ml of prepared sample in a 50-ml beaker, and add enough 10% NaOH to bring the pH to 12.0. At this pH, $Mg(OH)_2$ is precipitated. Add a few drops of indicator and slowly titrate with EDTA until a blue color is obtained and record the volume of the EDTA used.

3.4. Calculations
Estimate total calcium and magnesium or calcium alone, as ppm $CaCO_3$, using equation 7. The concentration of Mg is the difference between the total Ca and Mg concentration and that for Ca alone. Use equation 8 to correct for any dilutions made. The concentration of exchangeable Ca and Mg in the soil sample is found from equation 9.

3.5. Sources of Error
Results may be biased if there are relatively high concentrations of other bivalent metal ions such as Ba^{++} or a high concentration of phosphate. (If these interferences occur, then use a more complex preparation of the sample.) Precision of this technique is about 2.9%, with precision decreasing below 5 ppm $CaCO_3$ (Greenberg et al., 1985).

4. Nitrogen

Nitrogen is one of the major elements required for life. Although abundant as an atmospheric gas in the form of molecular nitrogen (N_2), it must first be converted to ammonia, nitrate, or some organic form before it can be used by most organisms. Nitrate is typically present in habitats, with nitrite in much lower quantities. The cycling of nitrogen in the ecosystem is a complex process described in ecology and microbiology textbooks.[2] Techniques given here for nitrogen determination involve the analysis of ammonia as the nitrogen-containing compound, thus nitrate must be converted to ammonia before analysis.

Other techniques for nitrate determination are available (ASTM, 1981). With the exception of the cadmium reduction procedure (Greenberg et al., 1985), they are more time-consuming, and require complex preparation and the use of dangerous reagents. If a spectrophotometer is available for use in the ultraviolet wavelength of 220 nm, then nitrate can be analyzed directly (Greenberg et al., 1985).

2. However, some ecology texts incorrectly state or imply that nitrates (NO_3) are the direct product of nitrogen fixation. In fact, nitrogen fixation by microorganisms results in the incorporation of atmospheric nitrogen into ammonia and organic nitrogen compounds.

4.1. Materials
Spectrophotometer (set to read at 410 nm), spectrophotometer tubes, 250-ml flasks, suction filter, filter paper, 100-ml beakers, graduated or volumetric pipettes, 10% $ZnSO_4$ solution, 6N NaOH (240 g/l), EDTA solution (50 g Na_2EDTA:$2H_2O$/100 ml of 2.4N NaOH), finely granulated Devarda's alloy (Cu, Zn, and Cd), NH_4Cl stock solution (3.819 g/l = 1000 ppm as N, or 3.146 g/l = 1000 ppm as NH_3), KNO_3 stock solution (7.218 g/l = 1000 ppm as N, or 1.630 g/l = 1000 ppm as NO_3), Nessler's reagent (100 g HgI_2 and 70 g KI per 200 ml, 160 g NaOH/500 ml, mix both solutions and dilute to 1000 ml; *the solution is toxic*).

4.2. Sample Preparation
Analyze the sample as soon as possible as ammonia and nitrate are biologically active and can change rapidly. For water samples, add 1 ml of 4000 ppm $HgCl_2$ or 1 ml of concentrated H_2SO_4 to the sample bottles as a preservative. Filter the sample.

For soil samples, exchangeable ammonium and nitrate ions are extracted by shaking 20 g of soil in a 250-ml flask with 100 ml of 2N KCl. Add 0.5 g of $ZnSO_4$ to the extract and shake; allow the suspension to settle and filter the extract.

4.3. Nitrate
This method gives reliable results if the concentration of nitrites is much lower than the concentration of nitrates. Care must be taken to duplicate this procedure exactly on blanks and standards, for you must time the reduction of NO_3^-. This method is based on the principle that NO_3^- is reduced to NH_3 in a basic solution in the presence of a metal catalyst (Golterman et al., 1978). In a 250-ml flask place 50 or 100 ml of the water sample or soil extract and adjust the pH to 10.5 using 6N NaOH. Gently heat the sample to remove any ammonia present, but do not boil. After 15 minutes, cool and add 0.2 g of Devarda's alloy. Bubbles of hydrogen gas generate during this reaction. Gently mix so all the nitrate in the sample has a chance of reducing to NH_3. Accurately time the reaction for 15 minutes. Ammonia vaporizes rapidly at temperatures above 25°C, so keep the flasks from heating by placing them periodically on ice, but do not chill them. Carefully decant the sample into a clean flask, leaving the alloy behind. Determine the concentration of ammonia using Nessler's procedure as discussed in section 2E.4.4 below.

Some ammonia is lost in this procedure, but careful timing of the reduction using the nitrate standards will correct for this loss. A blank using distilled water must also be run along with a sample and standards. For more precise results the reduction should be carried out in a distillation flask (Golterman et al., 1978; Greenberg et al., 1985).

4.4. Ammonia
This procedure (Greenberg et al., 1985) is used for the determination of ammonia and for the final portion of the analysis of nitrate (section 2E.4.3). Add 1

ml of $ZnSO_4$ solution to 100 ml of the prepared sample. Adjust the *p*H to 10.5 using 6N NaOH while gently mixing. Let stand for 15 minutes, filter, and then add 1 drop of EDTA solution. The $ZnSO_4$ flocculates colloidal and suspended materials, and the EDTA holds bivalent cations such as those of Ca and Mg in solution to retard turbidity during analysis. Add slowly 1 ml of Nessler's reagent to 25 ml of this sample and allow color to develop for 10 minutes. Place a portion of the developed sample in a colorimeter tube and record the absorbance at 410 nm. Repeat the procedure exactly for the standards and blank. If the samples become turbid when Nessler's reagent is added, the concentration of NH_3 may be too high. In that case, take a second 25-ml aliquot of the remaining 75 ml of prepared sample, dilute it to 100 ml, and add 1 ml of Nessler's reagent. Then proceed from the color development step, as above.

4.5. Calculations   Use equation 3 to estimate the ppm NH_3 or NO_3^- in the prepared sample after a standard curve coefficient (*b*, in equation 2) has been determined. If no dilution was made, then this value is the concentration in the water sample. Otherwise, use equation 8 to correct for any dilution.

For soil samples, use equation 9 to estimate the μg N/g dry weight. If the concentration in terms of μg NH_3 or μg NO_3^- is preferred, then the results are determined directly from the ppm NH_3 or ppm NO_3^- of the standards. For conversion of ppm NH_3 or ppm NO_3^- to ppm N, the following relationships may be used:

$$\text{ppm N} = (14/17) \text{ ppm } NH_3; \qquad (13)$$

$$\text{ppm N} = (14/48) \text{ ppm } NO_3^-. \qquad (14)$$

In the above calculations, the volume of the sample (V_o) from equation 9 represents the 100 ml of sample extract, not the 25-ml aliquot.

4.6. Sources of Error   Accurate timing of the reactions is essential. Any substances that cause turbidity, such as Ca, Mg, $Fe^=$, or SH^-, when treated with Nessler's reagent will bias results. Pretreatment with $ZnSO_4$ and EDTA will remove most of the common interferences. A precision of 5% to 40% is obtainable, depending on the presence of other materials in solution. Under optimal conditions, the minimum detectable concentration of NH_3 is 20 μg/l (Greenberg et al., 1985).

5. Salinity and chlorides

Salinity refers to the total amount of soluble salts in water or soil. Collectively, it generally includes such common ions as Ca^{++}, Mg^{++}, K^+, Na^+, Cl^-, $SO_4^=$, HCO_3^-, and $CO_3^=$, either naturally or added to the environment as pollutants. One basic measure of salinity is the ability of water to conduct electric current, for ions are excellent conductors.

The unit of *specific conductance* is mho/cm, the inverse of resistance (measured in ohms) divided by the distance between two electrodes. In natural waters and soils, measure in 10^{-3} mho (millimhos, or mmho) per centimeter, or 10^{-6} mho (micromhos, or μmho) per centimeter. Freshly distilled water, for example, has a specific conductance of 0.5 to 2.0 μmho/cm, and typical fresh waters often fall in the range of 50 to 500 μmho/cm. As conductivity increases as temperature increases, values of specific conductance are corrected to a standard temperature of 25°C (see section 2E.5.4).

A conductivity measuring device consists of a simple resistance meter and a pair of clean platinum electrodes spaced 1 centimeter apart. If a standard resistance meter is used instead of a conductivity meter, simply invert the reading in ohms.

It is also common to report salinity values as millimolar equivalents or ppm equivalents of KCl or NaCl, instead of in terms of mho/cm. The millimolar equivalent of conductivity is approximately 0.0068 times the specific conductance measured in μmho/cm, and this value times the molecular weight of KCl (namely 74.555) or NaCl (58.4428) represents the concentration in ppm. However, for more precise results, conductivity determinations can be calibrated against a standard curve for KCl or NaCl. Bear in mind that the calculated value does not represent the actual KCl or NaCl concentration in the water, but rather the total conductivity of the ionic solution expressed as an equivalent of that salt.

Because chloride contributes greatly to water salinity, this anion is generally determined separately. In soft waters, concentrations of chlorides greater than 250 ppm give a salty taste to the water. Salinity is also important in its osmotic effects on organisms and its effect on solubility of oxygen (see section 2E.6). Salinity commonly increases due to contamination from highway deicing salts (principally NaCl and $CaCl_2$) and NaCl from residential and industrial water softening.

Chloride is easily determined by titrating a sample with mercuric nitrate, which then forms mercuric chloride. Excess mercuric ions not reacting with the chloride form a purple complex with the indicator diphenylcarbazone at a *p*H of 2.3 to 2.8. An alternative titration may be performed with silver nitrate, using potassium chromate as the indicator, the resultant silver chromate being red in color. Either the mercuric or silver titration is acceptable (ASTM, 1981; Greenberg et al., 1985).

5.1. Materials   Burettes, 250-ml flasks, NaCl standard (0.0141N = 0.824 g/l; 1 ml = 500 μg Cl^-), 0.01M KCl (= 1414 μmho/cm at 25°C), 0.1N HNO_3, 0.1N NaOH, indicator acidifier reagent (250 mg s-diphenylcarbazone, 4.0 ml concentrated HNO_3, 30 mg xylene cyanol FF/100 ml ethanol), 0.014N $Hg(NO_3)_2$ titrant (2.5 g $Hg[NO_3]_2$:H_2O/100 ml water plus 0.25 ml concentrated HNO_3; dilute to 1 liter).

5.2. Sample Preparation   For water samples, filter 100 ml of water if turbid. For determination of conductivity, the measurement may be made directly in the water.

For soil, place 20 g fresh soil in a 250-ml flask with 100 ml distilled water. Stopper and shake for one minute, allow to settle, repeat shaking four more times, and filter the suspension. Add 2 drops of 1% sodium hexametaphosphate to inhibit precipitation of calcium carbonate.

5.3. Analysis   To determine specific conductance, insert the conductivity electrode into a sample, adjust the temperature correction dial and standardize the meter, and then read the specific conductivity in μmho/cm or mmho/cm.

For determination of actual chloride concentration, proceed as follows. If the sample is highly alkaline, adjust the pH to 8.0. Add 1.0 ml of indicator acidifier reagent to 100 ml of prepared sample. The sample should give a blue-green color indicating a pH of approximately 2.5. If a high concentration of chloride is expected, or more than 15 ml of titrant is required, then use only 25 or 50 ml of sample or dilute the sample. Titrate with mercuric nitrate to a definite purple end point.

5.4. Calculations   Equation 15 may be used to estimate the specific conductance in μmho/cm at 25°C:

$$C_{25} = (1413)(C_p)/C_s, \qquad (15)$$

where C_{25} is the specific conductance at 25°C, C_p is the specific conductance of the prepared sample measured at a given temperature, and C_s is the specific conductance measured for the 0.01M KCl standard at the latter temperature.

Salinity expressed as milliequivalents of KCl or ppm KCl is:

$$\text{meq KCl} = (0.0068)C_{25} \qquad (16)$$

$$\text{ppm KCl} = (74.55)(0.0068)C_{25}. \qquad (17)$$

Since C_p and C_s are seldom measured at the same temperature in the field, corrections for this difference must be made by adding the corrected difference to the colder solution before using equation 15. For 0° to 10°C the correction factor is 0.03 μmho/cm per degree difference, for 10° to 20°C, it is 0.025 μmho/cm per degree, and for 20° to 30°C it is 0.02 μmho/cm per degree. For example, if C_s was measured at 20°C, and C_p at 6°C, then C_p corrected to 20°C would be (4)(0.03) + (10)(0.025) + C_p μmho/cm.

Titrametric determination of chloride requires equation 7. One milliliter of 0.014N Hg(NO$_3$)$_2$ is equivalent to 500 mg Cl⁻. Therefore the number of milliliters of titrant times 500 represents the number of milligrams of chloride in the prepared sample. Use equation 8 to correct for dilutions, and equation 9 to estimate soil concentration

in mg/g dry weight. Chlorinity (in parts per thousand—e.g., g/l) may be converted to salinity (in parts per thousand) by the following formula, which assumes a salt chemical composition similar to that of sea water:

$$\text{salinity} = (1.80655)(\text{chlorinity}) \qquad (18)$$

(Johnston, 1964).

Another way to report seawater-associated salinity is to express:

$$\frac{\text{percent}}{\text{seawater}} = \frac{\text{salinity of sample}}{\text{salinity of seawater } (= 34\ ^o/_{oo})} \times 100\%, \qquad (19)$$

or,

$$\frac{\text{percent}}{\text{seawater}} = \frac{\text{chlorinity of sample}}{\text{chlorinity of seawater } (= 19\ ^o/_{oo})} \times 100\%. \qquad (20)$$

Waters with a salinity up to about 0.5 $^o/_{oo}$ (about 1.5% seawater) are considered "fresh," those with salinities between about 0.5 and 30 $^o/_{oo}$ are considered "brackish," and those in the salinity range of about 30–40 $^o/_{oo}$ are referred to as "seawater."

5.5. Sources of Error   The above titrametric procedure for chlorides has a coefficient of variation of 3.3% for repeated determinations by experienced analysts (Greenberg et al., 1985). A pH close to 2.5 is essential for an accurate determination. For each 0.1 pH unit deviation an error of 1% may occur. Interfering ions include Zn^{++}, B^{+++}, I^-, $Cr_2O_7^=$, Fe^{++}, and $SO_3^=$, but the concentrations of these ions are generally low, except in highly contaminated habitats.

6. Dissolved oxygen

Besides being necessary for aerobic respiration, dissolved oxygen (DO) in water affects the oxidation-reduction state of many other chemical variables, such as nitrate and ammonia, sulfate and sulfite, and ferrous and ferric ions. Also, a low dissolved oxygen content is often an indicator of organic pollution, and the degree of such pollution may be examined by measuring biochemical oxygen demand (section 2D.5.3). Unlike in the atmosphere, oxygen in aquatic environments is highly variable and generally low, and is affected by many factors, such as temperature, salinity, respiration, and photosynthesis.

Portable oxygen meters simplify the measurement of dissolved oxygen. However, the standard Winkler method of chemical analysis is still used for accurate DO determination and to calibrate the electronic instruments. Each oxygen meter model has specific instructions for operation and calibration, so consult its operating manual before

Table 2E.1. *Saturation concentrations of dissolved oxygen in water of various temperatures and salinities, at an atmospheric pressure of 760 mm Hg.‡*

Tempera-ture (° C)	Saturation concentration* (mg/l)	b† (mg/l per °/oo)	p‡ (mm Hg)
0	14.16	0.08405	4.6
1	13.77	0.08153	4.9
2	13.40	0.07908	5.3
3	13.05	0.07671	5.7
4	12.70	0.07440	6.1
5	12.37	0.07218	6.5
6	12.06	0.07002	7.0
7	11.76	0.06795	7.5
8	11.47	0.06595	8.0
9	11.19	0.06402	8.6
10	10.92	0.06217	9.2
11	10.67	0.06039	9.8
12	10.43	0.05869	10.5
13	10.20	0.05706	11.2
14	9.98	0.05551	12.0
15	9.76	0.05404	12.8
16	9.56	0.05263	13.6
17	9.37	0.05130	14.5
18	9.18	0.05005	15.5
19	9.01	0.04887	16.5
20	8.84	0.04777	17.5
21	8.68	0.04674	18.7
22	8.53	0.04579	19.8
23	8.38	0.04491	21.1
24	8.25	0.04410	22.4
25	8.11	0.04338	23.8
26	7.99	0.04272	25.2
27	7.86	0.04214	26.7
28	7.75	0.04164	28.3
29	7.64	0.04121	30.0
30	7.53	0.04085	31.8
31	7.42	0.04057	33.7
32	7.32	0.04036	35.7
33	7.22	0.04023	37.7
34	7.13	0.04018	39.9
35	7.04	0.04020	42.2
36	6.94	0.04029	44.6
37	6.86	0.04046	47.1
38	6.76	0.04070	49.7
39	6.68	0.04102	52.4
40	6.59	0.04141	55.3

* For pure water; from Truesdale et al. (1955).

† To determine oxygen saturation (in mg/l) at salinity S (in °/oo), subtract $(b)(S)$ from the saturation concentration given for pure water (Truesdale et al., 1955). (This calculation is accurate at least within the range of 0–30 °/oo.) For example, the saturation concentration of oxygen at 10°C and 15 °/oo salinity is:

$$10.92 - (0.06217)(15) = 10.92 - 0.93 = 9.99 \text{ mg/l.}$$

‡Saturation water vapor pressure (from Weast, 1980: D-180). The oxygen concentration at saturation at an atmospheric pressure of P is the saturation concentration from the table times the factor $(P- p)/(760 - p)$. For example, the oxygen

measuring DO. Most O_2 meters are calibrated using a water sample saturated with oxygen. Table 2E.1 gives the values for water saturated with oxygen at different temperatures and salinities, and can be used to standardize an oxygen meter. The determination of DO, either with a meter or by the Winkler method given below, can be used in a habitat analysis (section 2E), for BOD analysis (section 2D.5.3), or the analysis of oxygen consumption of aquatic organisms, photosynthesis in algae, or primary productivity (section 6B).

The Winkler method (ASTM, 1981; Greenberg et al., 1985) is a titrametric procedure involving a series of reactions prior to the titration. Mn^{++} ions are oxidized rapidly to Mn^{+++}, and in a basic solution the latter form a precipitate in the form of $Mn(OH)_2$ and $MnO(OH)_2$. The amount of oxidation is, of course, related directly to the amount of dissolved oxygen. In the presence of iodide ions in dilute sulfuric acid, the manganese hydroxide is converted to manganous sulfate and in the process oxidizes the iodide ions to molecular iodine (I_2). Therefore, the concentration of iodine is directly related to the concentration of oxygen in the original water sample. The I_2 concentration can easily be determined by titration with a thiosulfate solution using starch as the indicator of the end point.

6.1. Materials   Burette, 300-ml glass-stoppered bottles, 250-ml flasks, $MnSO_4$ solution (400 g $MnSO_4$:H_2O/500 ml water, filtered and diluted to 1000 ml), alkali-iodide-azide reagent (500 g NaOH, 135 g NaI/1000 ml water plus 10 g NaN_3/40 ml), concentrated H_2SO_4, starch solution (5 g starch/800 ml, boiled and diluted to 1000 ml; use only clear solution; add a few drops of toluene as a preservative), 0.1N sodium thiosulfate solution (24.82 g $Na_2S_2O_2$:$5H_2O$/1000 ml plus 1 g NaOH as preservative), 0.025N sodium thiosulfate titrant (250 ml of stock solution/1000 ml water; 1 ml of titrant = 200 μg oxygen). Keep all reagents sealed from the air, and boil all water just prior to the preparation of reagents to remove oxygen.

concentration at saturation (in pure water) at 10°C and 750 mm Hg is:

$$10.92 \left(\frac{750 - 9.2}{760 - 9.2} \right) = 10.92(0.987) = 10.78 \text{ mg/l.}$$

The correction for atmospheric pressure will not be great for low temperatures and altitudes, but should be used if the temperature and/or altitude is high. If barometric pressure is not measured directly, it may be considered to decrease at about 9 mm Hg per 100 m altitude up to about 800 m, about 8 mm Hg per 100 m from 900 to 1700, and about 7 mm Hg per 100 m thereafter up to about 2500 m (from Golterman et al., 1978).

Interestingly, the solubility of oxygen, unlike that of nitrogen, carbon dioxide, and other gases, is practically constant with water depth (Eckert, 1973; Fenn, 1972). The effect of hydrostatic pressure is very slight, predicted to be a less than 0.1% decrease in O_2 solubility at 1000 meters depth (at 23°C) (Mancy, 1966).

6.2. Sample Preparation Special care in sampling and preparation of the water samples is essential for accurate determination of DO, and any exposure of the sample to air will bias your results. Always collect the water under the surface, and allow it to flow into the sample bottle very slowly to prevent mixing it with air. Record the water temperature at the time of collection.

Glass-stoppered bottles are best because the stoppers eliminate air spaces. If a water sampler is used, such as the Kemmerer sampler (figure 2D.1), care must be taken when filling the glass bottle from the sampler. Attach a tube to the sampler so the water can fill the bottle from the bottom and allow about 300 ml of water to spill out over the top to remove water contaminated during filling.

Immediately after taking the sample, add 2 ml of $MnSO_4$ solution, followed by 2 ml of alkali-iodide-azide reagent. Add these solutions below the surface of the water sample to avoid contamination with air. Stopper the bottles, and gently mix by inverting them several times. A precipitate of $Mn(OH)_2$ and $MnO(OH)_2$ will form. Allow the precipitate to settle and gently shake again. Keep the prepared samples cool and dark, and complete the analysis within 4 to 8 hours.

6.3. Analysis In the laboratory (or in the field if possible), carefully add 2 ml of concentrated sulfuric acid under the surface of the prepared sample, restopper, and mix. The brown precipitate should dissolve, but if it doesn't, then add a few more drops of acid. Complete the analysis within 45 minutes.

Transfer the contents of the sample bottle to a 500-ml flask or take a 100- or 200-ml aliquot in a 250-ml flask. Add 1 ml of starch indicator solution. Titrate with 0.025N sodium thiosulfate until the blue color disappears, and record the volume of titrant used.

6.4. Calculations If a 200-ml sample is used, then 1 ml of titrant equals 1 ppm O_2, for 1 ml of titrant equals 200 μg O_2. For other sample volumes use equations 6 and 7. However, the initial 4 ml of reagents diluted the original 300 ml of sample, so the actual volume of sample (V_p) in equation 7 is 300, but the volume of the original sample (V_o in equation 8) is 300 ml − 4 ml or 296 ml. If different sample volumes were used for the titration, then substitute those values in equation 7.

The amount of dissolved oxygen is dependent on water temperature and salinity, so ecologists often indicate the percent saturation of oxygen in the water:

$$\%O_2 = (DO' - DO)/DO', \qquad (21)$$

where DO' is the dissolved oxygen concentration for oxygen-saturated water at a given temperature and salinity, and DO is the measured dissolved O_2 in the sample. Table 2E.1 gives values of DO' for different temperatures and salinities.

6.5. Sources of Error The greatest error is that of careless sampling and sample preparation. Also, significant biases may result from high turbidity, free chlorine, and reduced ions such as nitrite and ferric iron. These ions may be common in polluted waters and in water low in dissolved oxygen. Also, avoid sunlight during sample preparation. For distilled water, this method has a precision of about ± 0.02 ppm O_2 (Greenberg et al., 1985).

7. Hydrogen ions and pH

The hydrogen ion concentration of soil or water is one of the most important chemical components of the habitat. It not only affects directly the diversity and distribution of organisms, but also determines the nature of many chemical reactions that occur in the environment.

Hydrogen ion concentration, $[H^+]$, is conventionally expressed in terms of pH, where,

$$pH = \log (1/[H^+]). \qquad (22)$$

On this scale a pH of 7 represents a neutral solution where the concentrations of H^+ and OH^- are equal; values less than 7 represent acid conditions; and values greater than 7 represent basic conditions.

Measurements of pH should be made either in the field or on freshly-collected samples; biological activity and other chemical changes can rapidly alter a sample by as much as 0.5 to 1.0 pH unit.

If a portable pH meter is not readily available, then use the following colorimetric method. Although convenient for use in the field, colorimetric determination of pH can be biased by temperature, turbidity, and chemical interferences.

For water samples, the measured pH is representative of the actual pH of the water. For soil samples, however, the pH is determined for an extract and is only an approximation of the soil conditions. As standard practice, results of the pH determination on a soil extract are given as the soil pH, but such results should be stated as the pH of a 1:1 soil-water extract or whatever ratio of soil to water is used.

7.1. Materials Battery-operated pH meters are most convenient for field and laboratory use. For colorimetric determination: 100-ml graduated cylinder, test tubes, balance, suction filter, filter paper, $BaSO_4$ powder, indicator solutions, pH color comparison chart. Prepare the following common pH indicator solutions by dissolving 40 mg of indicator in 100 ml of water or alcohol. Using dilute acid or base, adjust the pH to the midpoint of the range shown: thymol blue (pH range of 1.2–2.8), bromocresol green (3.0–5.4), methyl red (4.4–6.0), bromocresol purple (5.2–6.8), bromothymol blue (6.0–7.6), cresol red (7.2–8.8) (Golterman et al., 1978).

7.2. Sample Preparation   For water samples, filter if turbid.

For soil, place 25 g of soil in a flask, add 25 ml of distilled water, shake well, and allow to settle. If the sample cannot be readily weighed in the field, fill a graduated cylinder with distilled water to 25 ml. Add soil until the volume reaches 40 ml. By assuming an average soil density of 1.65 g/ml, 15 ml of soil is approximately 25 g. Filter the sample in the laboratory. Or, in the field, add approximately 0.5 g of $BaSO_4$ to the suspension, shake, and allow to clear; this aids in reducing the turbidity of the extract.

7.3. Analysis   If a pH meter is used, first calibrate the meter at the temperature of the sample using a standard pH buffer. Thoroughly rinse the electrode with distilled water after each measurement of buffer or sample. If measured in the laboratory, remember that the pH of the sample will differ from that of a field-determined sample, for the pH of natural water changes after collection.

For colorimetric analysis, use 25 ml of water or soil extract, to which is added 1 ml of indicator solution for the expected pH range. Then compare the color of the sample to standard color comparison charts appropriate to the indicator used. If the expected pH range is not known, use a 10 ml subsample subjected to a wide-range indicator (as pH indicator paper) to determine the approximate pH, and then use a second 10-ml sample with a narrow-range indicator.

7.4. Sources of Error   Temperature can affect the measurement of pH. It is common to correct pH to a standard temperature of 20°C. For most field studies, an increase of 0.01 pH unit occurs for each degree decrease in temperature; conversely, a decrease of 0.01 pH unit occurs for each degree of temperature increase. Thus if the temperature in the field is 15°C, and the pH reading is 7.61, the pH corrected to 20°C would be:

$$pH = 7.61 - 5(0.01) = 7.56.$$

For environmental temperatures not far from 20°C, this correction will be insignificant. The pH measurement of soil depends on the amount of time elapsed since sample collection, the soil moisture, soluble salts, temperature, and amount of water used for the soil extraction. These factors can affect the results by as much as 0.5 pH unit. One is generally justified in expressing pH to no more accuracy than 0.1 pH unit.

8. Phosphate

Phosphorus, one of the nutrients critical for growth, is limiting in many environments. Unlike nitrogen, phosphorus does not have a large immediate storage reservoir like the atmosphere. Phosphate ($PO_4^{\equiv}$) is the common form of usable phosphorus and is relatively insoluble. Soluble phosphate has a short residence time in the environ-

ment and can be tied up for long periods of time in plant biomass or as insoluble salts in sediments or soil. An excess of phosphate can cause algal blooms and hence secondary pollution problems. This may result in an acceleration of natural eutrophication of ponds and lakes.

We speak of two fractions of phosphate in soil and water: soluble (filterable through a 0.45-μ filter) and insoluble (nonfilterable through a 0.45-μ filter). Within these fractions there are three forms of phosphate commonly measured: *organic phosphate*, largely of biological origin; *condensed phosphate*, mainly from polyphosphate detergents and fertilizers; and *orthophosphate*, from natural minerals, fertilizers, and natural breakdown of condensed and organic phosphates. *Total phosphate* refers to all the forms of phosphate found in a sample. Condensed phosphates are sometimes called polyphosphates, but they also include pyrophosphates (P_2O_7) as well as metaphosphates (P_nO_n).

Three treatments of the sample can be used to roughly estimate ortho-, condensed, and organic phosphates. *Soluble phosphate* is largely orthophosphate in a filtered untreated sample. *Acid hydrolyzable phosphate* approximately represents both ortho- and condensed phosphates. *Total phosphate* includes all three forms of phosphate and is determined through acid oxidation of either a filtered or unfiltered sample depending on whether the soluble or insoluble fraction, respectively, is desired. Acid hydrolysis and acid oxidation prior to the analysis converts condensed and organic phosphates, respectively, to orthophosphate. Therefore, the final chemical analysis for all three forms is the same. Subtraction of the amount of soluble or filterable phosphate from the hydrolyzable phosphate gives an estimate of condensed phosphate.

The basic analytical method for determining orthophosphate is the molybdate procedure (ASTM, 1981; Black et al., 1965; Greenberg et al., 1985). Ammonium molybdate forms a blue phosphate-molybdate complex in the presence of stannous chloride, and the intensity of the color is proportional to the concentration of the phosphate. The procedures given below are for determining soluble and acid-hydrolyzable phosphate. Because acid oxidation is time-consuming and requires a fume hood and distillation apparatus, we have not included total phosphate analysis in our methods. (Consult Black et al., 1965; Golterman, 1969; or Greenberg et al., 1985).

8.1. Materials   Spectrophotometer, suction filter, filter paper (0.45-μ recommended), 250-ml beakers, graduated cylinder, ammonium molybdate solution (50 g $(NH_4)_6Mo_7O_{24}$/100 ml water plus 400 ml 10N H_2SO_4, diluted to 1 liter), stannous chloride stock (10 g $SnCl_2$: $2H_2O$/25 ml concentrated HCl, diluted to 100 ml; prepare diluted $SnCl_2$ solution from stock on day of analysis (1 ml stock/100 ml water), K_2PO_4 standard stock solution (100 ppm PO_4, with few drops of toluene for preservative), charcoal powder.

8.2. Sample Preparation *Phosphate determination in water:* If a water sample cannot be analyzed soon after sampling, add 1 ml of 40-ppm $HgCl_2$ to it in the field as a preservative. Try not to use phosphate detergents when washing sample bottles and glassware to avoid contamination. Wash all glassware in 5% HCl. For soluble phosphate, filter 100 ml of sample. For hydrolyzable phosphate (soluble and insoluble), add 0.5 ml of concentrated HCl and 0.5 ml of concentrated H_2SO_4 to 100 ml of sample, boil gently for 30 minutes, and filter.

Phosphate determination in substrate: For water-soluble phosphate, shake 100 ml of distilled water and 10 g of soil. Allow to settle, shake again, and then filter. For hydrolyzable phosphate, add 40 ml of 0.1N HCl and 40 ml of 0.05N H_2SO_4 to 10 g of soil along with 200 mg of charcoal powder. Shake sample and boil gently for 30 minutes. Allow sample to settle, and filter. Rinse filter with 20 ml of water and bring final volume to 100 ml.

8.3. Analysis Mix 1 ml of ammonium molybdate solution with 10 ml of prepared sample and with 10 ml of a standard. Then add one or two drops of diluted $SnCl_2$ solution to the sample and standard and allow the blue color to develop for 5 minutes. Precise results are obtained only if color development time is kept constant. Measure the absorbance on a spectrophotometer at 650 nm. Repeat the sample preparation and analysis steps for the standards (0.1, 0.2, 0.4, 0.6, and 1.0 ppm PO_4). If the absorbance on the sample is too high for this range of standards, then dilute the sample 1:10.

8.4. Calculations Use equation 3 for estimating the concentration of PO_4 in the prepared sample. Equation 8 corrects for dilution of the sample, and equation 9 estimates the concentration in the soil.

8.5. Sources of Error Relatively high concentrations of arsenates, silicates, and reduced substances such as Fe^{++} and SH^- interfere with the analysis. Detection limits are 3 μg P/l (Greenberg et al., 1985).

9. Sulfate

Sulfate ($SO_4^=$) is a common anion in water and soil, ranging from a few to over a thousand parts per million. Concentrations in excess of 250 ppm in water have a cathartic effect, especially when associated with magnesium ions. High sulfate concentrations are often associated with acid conditions, especially if calcium, magnesium, and alkalinity are low. Sulfate is a pollutant found in acid mine drainage, fossil-fuel combustion products, and pulp and paper wastes. Sulfate is also a natural product of the microbial regulated sulfur cycle. It is commonly measured

by precipitating $BaSO_4$ with $BaCl_2$ in an acid solution. The degree of the resultant turbidity is measured spectrophotometrically and compared to standard sulfate solutions (ASTM, 1981; Black et al., 1965; Greenberg et al., 1985).

9.1. Materials Spectrophotometer set at 420 nm, 250-ml flasks, conditioning reagent (30 ml glycerol plus 30 ml concentrated HCl in 300 ml distilled water, with 75 g NaCl and 100 ml 95% ethanol added), barium chloride crystals, 100 ppm Na_2SO_4 stock solution (147.9 mg/1000 ml = 100 ppm SO_4), acetate extracting solution (39 g ammonium acetate/1000 ml of 0.25N acetic acid), sulfate-free activated charcoal.

9.2. Sample Preparation For water, filter 100 ml of sample to remove turbidity.

For soil samples, add 20 g of soil to a 250-ml flask, add 50 ml of acetate extracting solution, add 0.25 g charcoal, shake 1 minute every 10 minutes for a half hour, and then filter.

9.3. Analysis Add 100 ml of prepared sample or standard to a 250-ml flask, add 5 ml of conditioning reagent, and mix by swirling the flask. Add 0.3 g of $BaCl_2$ crystals and swirl for exactly 60 seconds. Immediately add a portion of the cloudy sample to a colorimeter tube and record the absorbance at 420 nm at 4 minutes from the time the $BaCl_2$ was added.

Because color and turbidity of the sample will bias the results, prepare a blank by using a prepared sample. Follow the same analysis procedure, but do not add the $BaCl_2$ crystals. Prepare a standard curve for 0, 1, 5, 10, 20, and 40 ppm SO_4. If the sample exceeds 40 ppm, then dilute it 1:10 prior to the analysis.

9.4. Calculations Use equation 3 to estimate the concentration in the prepared sample. Equation 10 corrects for the distilled water blank run through the above procedure. However, after determining the standard curve, set the spectrophotometer to 100% transmittance using the sample blank without $BaCl_2$. Samples measured thereafter will automatically be corrected. Equations 8 and 9 correct for sample dilution and for the concentration in the original soil sample, respectively.

9.5. Sources of Error The minimum detectable limit with this procedure is 1 ppm SO_4. Measured by the coefficient of variation, a precision of 9% to 10% can be expected by this method (Greenberg et al., 1985). Turbidity and color of the aqueous sample are the major sources of bias. As only $BaSO_4$ forms a precipitate in an acid solution, chemical interferences are limited.

10. Selected references

Allen, S. E., H. M. Grimshaw, and A. P. Rowland. 1986. Chemical analysis, 285–344. In P. D. Moore and S. B. Chapman (eds.), Methods in plant ecology. Blackwell Scientific Publications, Oxford, England.

American Society for Testing Materials. (ASTM). 1981. Water. In 1981 annual book of ASTM standards, part 31. American Society for Testing Materials, Philadelphia.

Black, C. A., D. D. Evans, J. L. White, L. E. Ensminger, and F. E. Clark (eds.). 1965. Methods of soil analysis. Part 2. Chemical and microbiological properties. American Society of Agronomy, Madison, Wis.

Eckert, C. A. 1973. The thermodynamics of gases dissolved at great depths. Science 180:426–427.

Fenn, W. O. 1972. Partial pressure of gases dissolved at great depths. Science 176:1011–1012.

Golterman, H. L., R. S. Clymo, and M.A.N. Ohnstad. 1978. Methods for physical and chemical analysis of fresh waters. Blackwell Scientific Publications, Oxford, England.

Greenberg, A. E., J. J. Conners, and D. Jenkins (eds.). 1985. Standard methods for the examination of water and wastewater. 16th ed. American Public Health Association, Washington, D.C.

Johnston, R. 1964. Recent advances in the estimation of salinity. Oceanogr. Mar. Biol. Annu. Rev. 2:97–120.

Lind, O. T. 1985. Handbook of common methods in limnology. 2d ed. Kendall/Hunt Publishing Co., Dubuque, Iowa.

Mancy, K. H. 1966. Analysis of dissolved oxygen in natural and waste waters. Public Health Service Publication no. 999-WP-37. U.S. Department of Health Education and Welfare, Cincinnati.

Parsons, T. R., Y. Maita, and C. M. Lalli. 1984. A manual of chemical and biological methods for seawater analysis. Pergamon Press, New York.

Truesdale, G. A., A. L. Downing, and G. F. Lowden. 1955. The solubility of oxygen in pure water and seawater. J. Appl. Chem. 5:53–62.

Weast, R. C. (ed.). 1980. Handbook of chemistry and physics. 61st ed. CRC Press, Cleveland.

Wetzel, R. G. 1983. Limnology. 2d ed. Saunders College Publishing, Philadelphia.

Wetzel, R. G., and G. E. Likens. 1979. Limnological analysis. W. B. Saunders, Philadelphia.

2f

habitat assessment

1. Introduction

There is a need for assessing the quality or value of ecological resources because land and water, which are limited resources, have diverse uses and encounter increasing demands by contemporary society. The field of environmental impact assessment is relatively new, being stimulated by the National Environmental Policy Act of 1969 (NEPA). In recent years, biologists have developed three approaches to assessing ecological resources: the ecosystem approach, which uses methods described in unit 6; the community inventory approach, which uses methods of units 3 and 5; and the habitat approach, which uses methods in unit 2. The approach used depends on the purposes of the ecological assessment. Ideally, elements of all three would be used for determining impacts from large-scale construction projects or for assessing habitats for purposes of ecological resource management.

The habitat assessment strategy is commonly used for several reasons. It can be employed to obtain an overall evaluation of a habitat at relatively low cost, it is useful for identifying potential environmental impacts at early stages of project planning, it facilitates identifying alternatives that will minimize adverse impacts, and it often is a prerequisite for wildlife and fisheries management studies. In addition, habit assessment provides information needed to evaluate ecosystem and inventory studies. Whether they are assessing ecosystems, communities, species, or habitats, ecologists generally conduct three types of assessments: assessment of value or importance of the resource, assessment of potential adverse or beneficial impacts, and assessment of actual adverse or beneficial impacts. The latter requires environmental and biological monitoring of a habitat over a long period of time. For example, baseline monitoring studies would be conducted prior to construction with potential impact, preoperational monitoring would be done during construction and prior to operation of the facility, and operational monitoring would take place during the first several years of the facility's operation. The second type of as-

sessment, commonly called an environmental impact assessment, involves predicting possible adverse impacts from the evaluation of baseline information coupled with knowledge of the activity causing the impact. In this section we will discuss methods useful for determining the value or importance of a habitat and estimating adverse impacts. It is common practice in habitat assessments to consider the concept of the habitat broadly; and in terrestrial habitats, biotic components (see section 5A) are often emphasized over the physical and chemical components of the habitat.

One pitfall of many ecological assessments is the collection of excessive and irrelevant data. To avoid this problem, follow these guidelines:

1. Develop a clear objective for the study. Keep in mind that the principal goal of an ecological study is an evaluation of biotic changes in relation to environmental changes and not the study of the environment in and of itself.
2. Define the ecological unit to be studied (population, community, or ecosystem). (For example, is the primary ecological resource of interest an endangered species of bird, a unique forest community, or the fishing productivity of a river?)
3. Delimit the scope of the habitat analysis. How large an area and how many sites must be studied? What physical or chemical factors need to be measured? (For example, are soil analyses needed? If so, delimit the type and quantity of the analyses.)
4. Specify a plan for efficient study, and prepare an experimental design and a sampling program (see section 1A). Identify study tasks clearly and prepare a schedule for their implementation.
5. Determine the procedures to be used in your analysis. What sampling and measurement procedures will be used?

2. Setting assessment criteria

A given habitat may be assessed in a distinct manner, depending on the criteria used to set values on habitats. A real estate developer has a different set of values than a conservationist. A forester will evaluate a woodland in a fashion different from a wildlife manager. Although decisions on the future of an ecological resource are often described from a subjective valuation of that resource, it is imperative that an objective basis be established for determining the importance of a particular site and its habitat components. Some criteria are relatively easy to define, because they have been established as federal, state, or local laws with which one must comply. In some instances criteria are derived from a consensus of professional opinion or from judicial decisions. The standards or established rules by which we can measure or test a habitat by comparison are the criteria that, once established, can be used systematically to assess a habitat in an objective manner.

Table 2F.1. *Some commonly used habitat assessment factors.*

Ecological Factors		
Physical and Chemical Environment	**Biotic Environment**	**Biological Indicators**
Habitat area	Habitat diversity	Species diversity
Soil factors	Vegetation cover	Productivity
Atmospheric factors	Breeding sites	Biomass
	Protective cover	Density
Hydrological factors	Stratification	Dominant species
Geological factors	Plant distribution	Indicator species
Chemical factors	Habitat type	Sensitive species
	Zonation	Rare species
Limnological factors	Plant form	Eutrophication species
Migrating species' ranges		Food chains
		Disease vectors
		Pest species
		Stage in succession

Human Factors		
Economic	**Legal**	**Esthetic**
Timber value	Endangered species	Unique species
Commercial fisheries	Protected species	Beautiful species
Crop production	Wetland regulations	Scenic environments
Fur production	Regulated game species	Pristine or virgin habitat
Domestic animal production	Wildlife refuge	Unique habitat features
Mineral resources	Air quality standards	Degraded habitat
Soil conservation	Water quality standards	Wilderness and open space
Water conservation	Solid waste regulations	Historical or archeological sites
Land value	Zoning laws	
Hazards to human health	Toxic materials regulations	

Table 2F.1 lists several factors that can be used to develop criteria for evaluating the worth of a habitat. Selection of or emphasis on a particular category will depend largely on the purpose of the evaluation. These factors can be ranked systematically, using one of the quantitative

Table 2F.2. *Criteria useful for ecological assessments.*

Each of these questions is answered *yes* or *no* for each factor under consideration.
A. Static factors.
   Is this factor:
   1. regulated by law?
   2. an important limiting factor?
   3. a sensitive habitat component, species, or ecosystem?
   4. a dominant or controlling factor?
B. Dynamic factors.
   Would a significant reduction or increase in this factor:
   1. have an adverse effect on the regional abundance or survival of a valued species?
   2. have an adverse effect on economic, aesthetic, or recreational values?
   3. result in an adverse change in the composition of the dominant plant or animal species?
   4. result in an adverse irreversible or long-term effect?
   5. result in a significant adverse effect on other factors?
   6. result in significant degradation of the physical or chemical factors of the habitat?

methods in section 2F.3; then those considered most critical for the study objectives can be selected. Some factors may simply be omitted because they do not apply to the habitat in question or because there are insufficient available data. Note that the factors listed are not necessarily environmental or biological, for the value of a habitat is often determined by the importance of the species and the community it supports or by its economic and human uses.

Other factors for setting criteria can be added to table 2F.1, while certain categories of factors can be emphasized or expanded. For example, in a basic ecological study, economic, legal, and esthetic factors would not be used to a significant degree. On the other hand, in an applied ecological study, economic values may be relatively important in setting criteria. In a full environmental impact statement, legal and aesthetic values are considered as well as ecological, human health, and economic factors.

Factors selected for ecological assessments should be:

1. consistent with the study objectives (i.e., basic vs. applied ecology, or impact assessment vs. habitat management);
2. determined prior to the collection of data;
3. broad and inclusive, including as many factors as reasonable;
4. balanced not biased toward one discipline or set of factors;
5. practical, that is, measurable or observable with reasonable expenditure of effort;
6. critically evaluated for validity, bias, precision, and specificity.

Table 2F.2 presents criteria useful for evaluating assessment factors. Once a set of critical factors is selected, criteria may be established for the factors. The simplest criterion is presence or absence of a factor, but one must

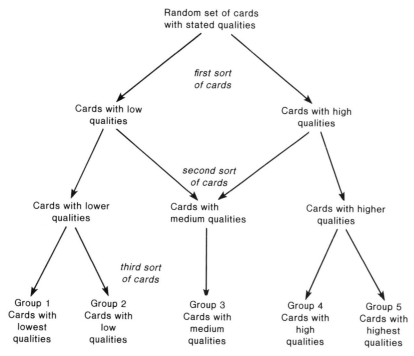

Figure 2F.1. A diagram of the Q sort method for categorizing habitat qualities.

be cautious, because presence is self-confirming whereas the absence of a factor seldom can be confirmed; also, the quantity of a factor may be more significant than its occurrence. Examples of presence or absence might include: presence of an endangered species, absence of a critical food source, presence of nest building sites, presence of an indicator species.

An alternative to the presence/absence criterion is the determination of those situations in which a factor of interest is present in a magnitude greater or less than some prespecified quantity. One relatively common procedure is to use legal or voluntary environmental quality standards. For example, standards exist for determining minimum drinking water quality and water quality for swimming and for fishing streams. Values also exist in the literature for primary production in the prairies or forests of a given region, and these can be used to determine criteria for minimal productivity.

Criteria can also be derived especially for a given study. For example, one can use the species area curve (figure 1A.1) to establish a minimal area of habitat needed to maintain a specific number of bird or plant species. It might be determined that a minimum of 5 hectares are required to support 9 out of 10 species found in that type of habitat. Therefore, a habitat of 5 or more hectares would be important.

Standards are simple ground rules agreed on in order to minimize bias and maximize objectivity. For example, if one states that a certain forest is of prime importance and should be protected, or that a stream is an excellent game fish habitat, on what basis can these evaluations be made? A researcher should avoid conclusions such as "prime importance" or "excellent" when evaluating a habitat. Based on specific criteria, however, one can conclude that the forest is ranked second among other forests. Or a particular creek has a rating for sport fish of 8 on a scale of 1 to 10. Setting objective criteria is necessary to ranking, rating, and scoring discussed below.

3. Quantifying assessments

Once criteria, or ground rules, have been established for evaluating certain habitats, an objective method is desirable for quantifying the information needed for assessment. Physical measurements or quantitative ecological measures of values in table 2F.1 often are difficult or impractical. If so, three basic procedures can be used to quantify information of a qualitative nature: sorting (i.e., categorizing), ranking, and rating.

3.1. Sorting   Sorting is a procedure for categorizing non-numerical information and placing this information on a nominal scale. Usually, three to ten categories are established, with five being usual. One common method, called the Q sort, is illustrated in figure 2F.1. It is useful both for establishing and ranking categories.

Specific habitat factors and their respective qualities are written on index cards without any indications of the habitat with which they are associated. The cards are randomized (i.e., shuffled) to mask their relationships with one another and then are sorted into two piles, as in figure 2F.1 and table 2F.3, according to whether the sorter judges

Table 2F.3. *An example of sorting habitat qualities.*

A. *Randomize a list (i.e., a deck of cards) of descriptive variables:*

Loam, deer, sparse cover, climax, gravel, wetland, old field, clay, pheasant, endangered species, sand, forest, pasture grasses, dense cover, subclimax, rare species, pioneer plants, degraded, pristine, field mice, rabbits, moderate cover, disturbed.

B. *First sort of qualities:*

(1) Low Quality	(2) High Quality
sparse cover	loam
gravel	deer
old field	climax
sand	clay
pasture grasses	pheasant
pioneer plants	endangered species
degraded	forest
disturbed	dense cover
moderate cover	subclimax
field mice	rare species
rabbits	pristine
	wetland

C. *Second sort of qualities:*

(1) Low Quality	(2) Medium Quality	(3) High Quality
sparse cover	clay	loam
pioneer plants	pheasant	deer
sand	subclimax	endangered species
degraded	rare species	climax
field mice	old field	forest
pasture grasses	gravel	wetland
	disturbed	dense cover
	rabbits	pristine
	moderate cover	

them to be in either of two prespecified categories (e.g., high or low importance). Each of the two piles is then shuffled and sorted into two groups. This procedure may then be repeated as shown if more categories are desired. After the sort, numerical values (e.g., 1 through 5) are assigned to the groups of factors in each habitat to denote relative importance or quality. The totals for each habitat are then used to provide a relative quantitative assessment of the

habitat. This is shown in table 2F.4, where, for example, the cattail marsh was assessed to have greatest ecological value.

This procedure is applied most easily in situations where more refined quantification of habitat factors is impractical or not critical to the assessment. It is most useful where several choices are clear and relatively easy to make, but is most difficult to apply when the compared habitats are very similar. As the sorting procedure depends on subjective judgments, different persons can produce different, even conflicting, results. However, if criteria are established for the categories beforehand, then such personal differences in judgment are minimized. For example, if higher soil productivity is agreed to be of greater value than low soil fertility, then loam would sort in a high group and sand in a low group. Bias also can be reduced by having several persons sorting the same cards and averaging the results.

3.2. Ranking Although it is useful to sort and compare groups of items, ranking each item according to some criterion of importance provides more detailed information. Such ranking permits a more systematic and objective assessment of each item of interest according to an ordinal measurement scale. (On an ordinal scale of measurement, one only knows, for example, that item 3 has greater value than item 2, not how much more.) Ranking can also be used to select items of significance, for example, choosing the three most important sites for more detailed analysis. An objective ranking procedure can also be used to analyze statistically the results of an assessment (see section 1B.3.4).

If habitat factors are not easily quantified, then a simple ranking procedure involving a check-list matrix can be used. In the example of table 2F.5, five habitats (A through E) are evaluated by recording the presence or absence of each of six factors, such as the presence of rare or endangered species, habitat area greater than 10 hectares, etc. A "+" is tallied if a factor occurs and "0" is tallied if that factor does not apply to the habitat. (The check-list matrix may also include a "−" tally if the factor, such as pollution, is judged to be negative in its effect of habitat quality). The +'s are then summed (and each − cancels out one +) for each habitat and the habitats are ranked from 1 to 5, based on the sums, with 5 indicating the habitat of highest quality. When equal sums occur, an average rank is used. (For example, in table 2F.5, habitats A and D are each assigned a rank of 2.5, which is the mean of 2 and 3.)

A limitation of this ranking procedure is that the relative size of the factor is not measured, only its presence or absence. Thus, habitat C, for example, cannot be interpreted to be twice as important as D. The likelihood of equal ranks is reduced by selecting more factors to be used in ranking.

Table 2F.4. *Matrix of factors for habitat assessment, indicating the numerical values assigned in Table 2F.3.*

Habitat	Disturbance	Soil	Vegetation Cover	Wildlife	Rare or Endangered Species	Stage of Succession	Plant Community	Total
Beech-maple forest	disturbed (2)	loam (3)	moderate (2)	rabbits (2)	none (0)	climax (3)	forest (3)	15
Cattail marsh	pristine (3)	clay (2)	dense (3)	deer (3)	endangered (3)	subclimax (2)	wetland (3)	19
Oak forest	degraded (1)	sand (1)	moderate (2)	rabbits (2)	rare (2)	subclimax (2)	forest (3)	13
Pasture	degraded (1)	clay (2)	sparse (1)	field mice (1)	none (0)	pioneer plants (1)	field (1)	7
Old field	disturbed (2)	sand (1)	moderate (2)	pheasant (2)	none (0)	old field (1)	field (1)	10

Table 2F.5. *Ranking of habitats by comparing multiple factors.*

Assessment Factor	Habitat				
	A	B	C	D	E
Soil nutrients not limiting	+	0	+	0	+
Undisturbed habitat	−	−	+	+	0
Rare or endangered species occur	+	0	+	0	+
High habitat diversity (> a specified number)	+	0	+	+	0
High vegetation cover (e.g., > 50%)	+	+	+	0	+
Large habitat area (e.g., > 10 ha)	0	+	+	+	+
Total	3	1	6	3	4
Rank	2.5	1	5	2.5	4

Another application of this procedure is the Environmental Impact Assessment Matrix formulated by Leopold et al. (1971). An abbreviated example of this procedure is given in table 2F.6. A matrix is set up with column headings consisting of a comprehensive list of proposed actions or processes expected from a construction or development project, such as a power plant, reservoir, or strip mine. The rows of the matrix are labeled with existing ecological conditions. In each row and column combination it is recorded whether a significant impact is probable and the total number of such probable impacts is obtained. This matrix tabulation is then performed for each of several alternative sites. This matrix can be expanded to incorporate a rating procedure discussed below, where numerical ratings are substituted for the X's of the present example.

Ranking of habitats for assessment has several advantages. Several habitats can be assessed quickly from qualitative reconnaissance surveys and aerial photographs, using procedures in section 2A. Detailed and time-consuming quantitative sampling methods are not required. Rank assessments are useful for scanning sites and for preliminary planning work. These procedures also allow one to make an assessment in a systematic and quantitative manner and to reduce subjective interpretations by setting quantitative, although often arbitrary, criteria.

One common problem with these procedures is that habitat factors are all treated as if they were of equal importance. For example, a wetland is protected by law and therefore its presence may be considered a more important habitat factor than the size of the habitat or the presence of a forest. One may prefer, then, to give extra weight to wetlands (e.g., by recording two X's instead of one in table 2F.6). An alternative is to select two factors pertaining to wetlands, where the second factor may be the average depth of water in the spring being less than a specified amount. It is important that such factors and weighting be chosen and agreed upon prior to conducting the assessment.

3.3. Rating Some assessments may require a more specific quantitative analysis than is accomplished by ranking items. For example, in table 2F.5, habitats C, E, and A were ranked 5, 4, and 2.5, respectively. Suppose, though, that C had 6 rare species, E had 1, and A had 5. On this basis alone, C and A may be relatively closer in value than indicated by their rank. One means of deriving an index that considers relative value is called rating. (Some ecologists also use the term "scoring" to refer to this method.) In this procedure, an arbitrary scale of 1 to 10 is often set up. (Other scales, such as 1 to 5, or 1 to 100, may be used instead.) The scale values or ratings have no units of measurement, and generally have no true value of zero. Ratings are intended for relative comparisons only.

Table 2F.6. *Environmental impact assessment matrix.*

An "X" is assigned to each combination of condition and action for which a significant adverse impact is probable.

		Proposed Actions or Processes That May Cause Adverse Environmental Impacts										
		Excavation	Dredging	Blasting or Drilling	Lumbering or Clearcutting	Draining or Landfilling	Liquid Waste Discharges	Air Emissions	Soiled Waste Disposal	Traffic	Urban or Industrial Growth and Development	Totals
Physical and Chemical	Soil fertility											
	Land form											
	Surface water quality											
	Air quality											
	Floods											
	Erosion											
Community and Ecosystem	Vegetation cover											
	Eutrophication											
	Food chains											
	Habitat diversity											
	Stratification											
	Species diversity											
	Wetlands											
	Mature ecosystems											
Fauna	Game species											
	Endangered species											
	Nongame birds											
	Nongame mammals											
	Fishes											
Flora	Trees											
	Shrubs											
	Grasses and herbs											
	Aquatic plants											
	Totals											

Existing Conditions of the Environment

Table 2F.7. *An example of using criteria to establish an arbitrary rating of a wildlife habitat.*

Rating Score	Timber dbh* (cm)	Canopy Closure (%)	All Tree Size Classes (%)
	Criteria for Rating Tree Size and Canopy Closure		
10	> 22.7	40–69	–
9	> 22.7	10–39	–
8	> 22.7	70–100	–
7	mixed	10–39	< 50
6	mixed	40–69	< 50
5	mixed	70–100	< 50
4	< 5	–	–
3	5–22.7	10–39	–
2	5–22.7	40–69	–
1	5–22.7	70–100	–

Adapted from Flood et al. (1977).
*Diameter of a tree at breast height. (See section 3A.)

On such a scale (e.g., of 1 to 10) each item may be rated relative to a high score (e.g., 10) and a low score (e.g., 1). Objective criteria must be set in order to assign a rating. Table 2F.7 is an example of using the tree diameter breast high (dbh) and percentage of canopy closure to rate a wildlife habitat. Such criteria are based on previous observations of game abundance as related to forest age and cover of the trees in the canopy.

Another rating procedure converts measured absolute or relative values (e.g., density, coverage, chemical concentration) to a scale of 1 to 10. Table 2F.8 is an example of rating 5 habitats on the basis of three factors. As the first factor, the coverage of aquatic plants is determined for each of the 5 sites. As the second factor, the depth of standing water is estimated. The habitat area is determined as the third factor. Criteria must be set for the minimum value of concern (to have a score of 1) and for the maximum value of concern (which will have a score of 10), for in comparative studies it is useful to have scores ranging from 1 to 10. Based on the determined criteria for minimum and maximum value, a rating index may be obtained as follows:

$$IR_i = \frac{(X_{max} - X_i) + 10(X_i - X_{min})}{X_{max} - X_{min}}, \qquad (1)$$

Table 2F.8. *Converting habitat factor measurements to scores on a scale of 1 to 10.*

Habitat Factor	A	B	C	D	E	Total
	Habitat Factor Measurements, by Site					
Plant coverage (m²/100 m²)	87.00	73.00	22.00	14.00	45.00	241.00
Water depth (m)	4.05	1.78	0.86	0.23	2.56	9.48
Area (ha)	24.50	15.20	12.80	18.30	35.70	106.50

Habitat Factor	Minimum	Maximum
	Minimum and Maximum Rating Criteria for Habitat Factors	
Plant coverage (m²/100 m²)	5.00	95.00
Water depth (m)	0.10	5.00
Area (ha)	10.00	50.00

Habitat Factor	A	B	C	D	E	Total
	Rating Index Values for Habitat Factors (from equation 1)					
Coverage	9.20	7.80	2.70	1.90	5.00	26.60
Water depth	8.26	4.09	2.40	1.24	5.52	21.51
Area	4.26	2.17	1.63	2.87	6.78	17.71
Total	21.72	14.06	6.73	6.01	17.30	65.82
Mean	7.24	4.69	2.24	2.00	5.77	

where IR_i is the rating index for habitat i
  X_i is the value of X for the ith habitat
  X_{max} is the maximum value set for X
  X_{min} is the minimum value set for X

For example, if habitat area is the factor to be rated, we may set the minimum area necessary to support a plant community (say, of 15 species) at 10 hectares and declare that a habitat greater than 50 hectares has an insignificant effect on increasing the number of species; if the area of the ith habitat is 24.50 ha, then:

$$IR_i = \frac{(50 - 24.5) + 10(24.5 - 10)}{(50 - 10)}$$
$$= \frac{25.5 + 145}{40}$$
$$= 4.26.$$

Equation 1 can thus be used to determine a relative rating for any measured variable on a scale of 1 to 10.[1] In this way, ratings for different factors measured on different scales can be assessed on a single unitless scale, as in table 2F.8.

3.4. Scoring   Rating of habitat factors on a fixed-range scale is useful for comparing data derived from different variables, especially if any of the variables are not measured on an interval scale or ratio scale. (Ratio scales have a constant interval size and a true zero point—e.g., continuous variables such as height and weight or discrete variables such as counts of items. Interval scales possess a constant interval size but not a true zero point—e.g., the Celsius and Fahrenheit temperature scales.) If all measurements are made on ratio or interval scales, scoring of the data on a relative, unitless scale is useful.

One method of converting data to unitless indices for comparing habitat factors is to determine the value of the variable relative to some standard. A relative index is

$$RI_i = \frac{100X_i}{X_s}, \qquad (3)$$

where X_i is the value of factor X in habitat i and X_s is the standard value of X against which measurements are to be compared.

The standard may be of three kinds: X_s may represent a predetermined criterion of importance, a fixed environmental standard representing some minimum or maximum allowable quantity, or a relative standard such as a total or mean value of X determined during the study. (See unit 3 for examples of relative measures.)

An example of a predetermined criterion of importance might be the minimum habitat area, habitat diversity, or cover that would support a given number of species

or population size. Such criteria could be derived from data in the literature or from an analysis of species area curves or performance curves from the current study (see section 1A.3).

Environmental standards are fixed by law or set by professional organizations, and are obtained from local, state, or federal agencies regulating air, water, or biotic resources. Water quality standards, for example, are available for numerous factors, and they define criteria for drinking water, fishing, swimming, and industrial uses.

A rating index based on an environmental standard can be used as an index of environmental quality. For example, if the recommended minimum dissolved oxygen for a cold-water fishing stream were set by a state agency at 6.0 ppm, and the measured mean value were 8.3 pm at a given site, then

$$RI = 100(8.3/6.0) = 138;$$

thus, an index of water quality > 100 for that factor would be an indication of sufficient oxygen to support healthy fish populations, and a value below 100 would be an index of poor water quality for that factor.

Several factors can be compared using the index of equation 3, and an average environmental quality index can be determined for each site, using the mean of all values of RI for each factor measured. Ott (1978) discusses in more detail the development of environmental indices using this procedure. When determining RI for a given factor, one must be certain that each RI is comparable. For example, some standards represent a maximum allowable value for a pollutant and others a minimum value for a limiting factor. Dissolved oxygen (DO), for example, is a limiting factor, and biochemical oxygen demand (BOD) a pollutant factor (see section 2D). An RI of 120 for BOD would be a pollution index indicating low water quality, and a value of 0.60 would be an indicator of relatively clean water. To compare BOD and DO, simply calculate the inverse of the pollution index:

$$RI_i = \frac{100X_s}{X_i}. \qquad (4)$$

The value of 60 for BOD would then become 167, which could be averaged with the DO value of 138 to express the overall water quality.

A relative standard is useful for comparative studies for which no fixed standard or criterion of importance has been established or is needed. Variables such as relative density, relative frequency, relative coverage, and relative biomass (see unit 3) are examples of a relative standard index. In this procedure X_s is the total value for a given variable, such as density, and the index ranges from 0 to 100% (or from 0 to 1.00). A second type of relative standard is use of the mean of X. In this case RI for a value equal to $\bar{X}$, the mean of X, would be 1.00, and the index can range from 0 to infinity.

---

1. Other rating scales can be used; in general, for a scale of a to b the rating index is computed as

$$IR_i = \frac{a(X_{max} - X_i) + b(X_i - X_{min})}{X_{max} - X_{min}} \qquad (2)$$

Although, in equation 3, $\bar{X}$ can be used in place of X_s, a preferred standardized scoring method is the use of the Z score:

$$Z_i = \frac{X_i - \bar{X}}{s}, \qquad (5)$$

where Z_i is the standard score for the factor in habitat i, X_i is the measured value of X, and $\bar{X}$ and s are the mean and standard deviation, respectively, for all sites for that factor (see section 1B.2).[2] Z scores can range from negative infinity to positive infinity, with a value of 0 when $X_i = \bar{X}$.

The environmental quality indices and pollution indices discussed above are not designed to develop complex predictive ecological models. Rather, they are intended as environmental management tools for the applied ecologist.

4. Selected references

Atkinson, S. F. 1985. Habitat-based methods for biological impact assessment. Environ. Professional 7:265–282.

Cairns, J., Jr. 1980. The recovery process in damaged ecosystems. Ann Arbor Science Publishers, Woburn, Mass.

Daniel, C., and R. Lamaire. 1974. Evaluating effects of water resource developments on wildlife habitat. Wildlife Soc. Bull. 2:114–118.

Flood, B. S., M. E. Sangster, R. D. Sparrowe, and T. S. Baskett. 1977. A handbook for habitat evaluation procedures. U.S. Fish and Wildlife Service Resource Publ. No. 132.778.

Ghiselin, J. 1980. Preparing and evaluating environmental assessments and related documents, 473–487. In S. D. Schemnitz (ed.), Wildlife management techniques. 4th ed. The Wildlife Society, Washington, D.C..

Graber, J. W., and R. R. Graber. 1976. Environmental evaluations using birds and their habitats. Biological Notes No. 97. Illinois Natural History Survey, Urbana, Ill.

Gysel, L. W. and L. J. Lyon. 1980. Habitat analysis and evaluation, 305–347. In S. D. Schemnitz (ed.), Wildlife management techniques. 4th ed. The Wildlife Society, Washington, D.C.

Hammond, K. A., G. Macinko, and W. B. Fairchild (ed.). 1978. Sourcebook on the environment. University of Chicago Press, Chicago.

Heer, J. E., Jr., and D. J. Hagerty. 1977. Environmental assessments and statements. Van Nostrand Reinhold Co., New York.

Leopold, L.B., F. E. Clarke, B. B. Hanshaw, and J. R. Balsley. 1971. A procedure for evaluating environmental impact. U.S. Department of Interior, Geological Survey Circular 645.

Ott, W. R. 1978. Environmental indices: Theory and practices. Ann Arbor Science Publ. Inc., Ann Arbor, Mich.

States, J. B., P. T. Haug, T. G. Schomaker, L. W. Reed, and E. B. Reed. 1978. A system approach to ecological baseline studies. U.S. Department of Interior, Fish and Wildlife Service, Fort Collins, Colo.

U.S. Fish and Wildlife Service. 1980. Habitat evaluation procedures. Division of Ecological Services, U.S. Fish and Wildlife Service, Department of the Interior, Washington, D.C.

Westman, W. E. 1985. Ecology, impact assessment, and environmental planning. John Wiley & Sons, New York.

---

2. The use of the Z score should be limited to variables that can be assumed to have a normal distribution and therefore should not be applied generally to proportions or percentages or to pH values and other variables known to be nonnormal.

introduction: ecological measurements

A number of basic measurements are used in describing populations and communities. Among these are density, frequency, coverage, and biomass. From them other important ecological measures are determined, such as population distribution, species diversity, and productivity. The most critical ecological measurements are introduced below, and methods for obtaining them will be discussed throughout unit 3.

Ecologists call the total count of all individuals in a population or other group of interest a *census*. It is seldom possible to count every individual in a population or other large group, however, but a relatively small part or a sample of the whole can be studied. Then, inferences may be made from this sample about the entire population or community (see section 1A.1). If a sampling method tends to underestimate or overestimate a characteristic of a population or community, it is *biased*. Bias inherent in biotic sampling will be discussed in the appropriate sections of unit 3.

All measurements are to be in metric units; any English measurements encountered should be converted to metric equivalents (see Appendix B).

1. Density

In ecological population studies, numbers of individuals are basic information. *Abundance* (*N*) is the number of individuals in a given area, and *density* (*D*) is that number expressed per unit area or unit volume. For example, a species may have an abundance of 100 individuals in a particular area. If the total area is 2.5 hectares, then the density of this species would be 40 per hectare (40/ha).

While density (often called *absolute density*) is the number of organisms per unit area or unit volume, much of the area may be habitat unsuitable for that species. Therefore, it may be more meaningful to speak of the number per unit of habitable area. Thus in the above example of 100 individuals in 2.5 hectares, if only half the area provides suitable habitat for the species in question, then the species would have an *ecological density* of 80 per hectare (80/ha). (Appendix B give various units of area and their equivalents.)

A problem sometimes encountered in plant sampling is the determination of individuals. Counting trees and many herbaceous plants poses little trouble, but when plants are growing in clumps or are reproducing vegetatively from underground rhizomes the common concept of the individual falters. Then you must count individual shoots or stems; or, if the plants are growing in distinct clumps, count the whole clump and treat it as an individual. The counting of clumps is recommended when measuring aerial cov-

unit 3

biotic sampling methods

erage or basal coverage (see below) for this type of vegetation. In these situations, coverage or biomass has more ecological significance than does density.

In many kinds of faunal sampling, accurate absolute density determinations often are difficult or impossible to obtain. However, if a standardized sampling procedure is used, then at least an *index of density* (*ID*) may be calculated and used for comparative purposes. Such an index might be the number of individuals per unit of habitat or the number per unit effort of sampling, rather than the number per unit area. Sometimes this is called *population intensity*. For example, the number per unit of habitat might be the number of beetles per leaf or the number of parasites per host organism. Density expressed per unit of sampling effort might be the number of grasshoppers per sweep of a net, the number of fish caught per hour, the number of birds seen per kilometer of walking, or the number of mice caught per trap per night.

In comparative studies, one generally wants to know the number of individuals relative to other populations or relative to the same population at other times. As will be seen in section 5A, *relative species density* (*RD*) is important in community studies. Relative species density is the total number of individuals of a species expressed as a proportion (or percentage) of the total number of individuals of all species. If, for example, there are 50 trees in a given area, and 30 of them are sugar maple, then the relative species density of sugar maple is 30/50 = 0.60, or 60%.

Relative population density is the number of individuals of a given species from one location or time expressed as a proportion of the total number of individuals of that species for all locations or times sampled. For example, if one caught 10 locusts with 100 sweeps of a collecting net in July of one year and 70 locusts in the same location and with the identical sweeping effort in September, the relative density for July would be 10/80, or 0.125, and for September it would be 70/80, or 0.875.

2. Frequency

Frequency (*f*) is the number of times a given event occurs. Thus an ecologist might speak of the frequency of measuring water temperatures, or the frequency with which an animal feeds. In many studies, the term frequency indicates the number of samples in which a species occurs. This is expressed as the proportion of the total number of samples taken that contains the species in question. Thus if a species were found in 7 out of 10 samples taken, it would have a frequency of 7/10, or 0.7. This is the same as saying that the probability of finding that species in a sample is 0.7. Since frequency is sensitive to distribution patterns of individuals, it is also useful in describing and testing for such patterns (see section 4C). The *relative frequency* (*Rf*) of a species is the frequency of that species divided by the sum of the frequencies of all species in the community.

3. Biomass

Biomass (*B*) is the weight of the individuals of a population or group of populations, and often is expressed per unit area or volume. For example, we might speak of kilograms of a species per hectare of forest, or of milligrams of a species per liter of pond water. Biomass is useful in visualizing the trophic structure of a community. In terrestrial communities, for instance, species having large biomass often strongly influence the flow of energy and materials through their trophic levels. Where there are large differences in the sizes of species, biomass data often are more useful than density measurements and some ecologists prefer the use of energy over biomass. Special considerations required in the estimation of biomass are discussed in section 6A.

4. Coverage

Coverage (*C*) is the proportion of the ground occupied by a perpendicular projection to the ground from the outline of the aerial parts of the members of a plant species. (This can be visualized as expressing the proportion of ground covered by the species, as the habitat is viewed from above.) As will be seen, coverage is calculated as the area covered by the species divided by the total habitat area; for example, a species' coverage might be 180 m²/ha.

In measuring *foliage cover*, the diameter of the crown of foliage is taken at its densest portion, and the coverage area is determined by assuming a circular outline. *Basal coverage* is generally used in a field or prairie situation and consists of measuring the circumference or the diameter of a clump of grass 2 cm to 3 cm above the ground and calculating the circular area for the foliage. *Basal areas* of trees are determined from trunk circumferences measured 1.5 m ("breast height") above the ground. A direct measurement of foliage coverage is difficult in trees, but the basal area generally is proportional to coverage and hence a useful index of the latter. The *relative coverage* (*RC*) of a species is the proportion of its coverage compared to that of all species in the community. Table 3.1 can be used to convert diameters and circumferences to circular coverage areas.

The degree of cover is sometimes considered as a measure of *dominance* in a community. However, dominance may include additional factors, so the term coverage is preferred.

Table 3.1. *Relation between diameters, circumferences, and circular areas.*

Diameter	Circumference	Area	Diameter	Circumference	Area	Diameter	Circumference	Area
1.00	3.14	0.7854	5.00	15.71	19.6349	9.00	28.27	63.6172
1.10	3.46	0.9503	5.10	16.02	20.4282	9.10	28.59	65.0388
1.20	3.77	1.1310	5.20	16.34	21.2372	9.20	28.90	66.4761
1.30	4.08	1.3273	5.30	16.65	22.0618	9.30	29.22	67.9291
1.40	4.40	1.5394	5.40	16.96	22.9022	9.40	29.53	69.3978
1.50	4.71	1.7671	5.50	17.28	23.7583	9.50	29.85	70.8822
1.60	5.03	2.0106	5.60	17.59	24.6301	9.60	30.16	72.3823
1.70	5.34	2.2698	5.70	17.91	25.5176	9.70	30.47	73.8981
1.80	5.65	2.5447	5.80	18.22	26.4208	9.80	30.79	75.4296
1.90	5.97	2.8353	5.90	18.54	27.3397	9.90	31.10	76.9769
2.00	6.28	3.1416	6.00	18.85	28.2743	10.00	31.42	78.5398
2.10	6.60	3.4636	6.10	19.16	29.2247	10.10	31.73	80.1184
2.20	6.91	3.8013	6.20	19.48	30.1907	10.20	32.04	81.7128
2.30	7.23	4.1548	6.30	19.79	31.1724	10.30	32.36	83.3229
2.40	7.54	4.5239	6.40	20.11	32.1699	10.40	32.67	84.9486
2.50	7.85	4.9087	6.50	20.42	33.1831	10.50	32.99	86.5901
2.60	8.17	5.3093	6.60	20.73	34.2119	10.60	33.30	88.2473
2.70	8.48	5.7256	6.70	21.05	35.2565	10.70	33.62	89.9202
2.80	8.80	6.1575	6.80	21.36	36.3168	10.80	33.93	91.6088
2.90	9.11	6.6052	6.90	21.68	37.3928	10.90	34.24	93.3131
3.00	9.42	7.0686	7.00	21.99	38.4845	11.00	34.56	95.0332
3.10	9.74	7.5477	7.10	22.31	39.5919	11.10	34.87	96.7689
3.20	10.05	8.0425	7.20	22.62	40.7150	11.20	35.19	98.5203
3.30	10.37	8.5530	7.30	22.93	41.8539	11.30	35.50	100.2875
3.40	10.68	9.0792	7.40	23.25	43.0084	11.40	35.81	102.0703
3.50	11.00	9.6211	7.50	23.56	44.1786	11.50	36.13	103.8689
3.60	11.31	10.1788	7.60	23.88	45.3646	11.60	36.44	105.6831
3.70	11.62	10.7521	7.70	24.19	46.5662	11.70	36.76	107.5131
3.80	11.94	11.3411	7.80	24.50	47.7836	11.80	37.07	109.3588
3.90	12.25	11.9459	7.90	24.82	49.0167	11.90	37.38	111.2202
4.00	12.57	12.5664	8.00	25.13	50.2655	12.00	37.70	113.0973
4.10	12.88	13.2025	8.10	25.45	51.5300	12.10	38.01	114.9901
4.20	13.19	13.8544	8.20	25.76	52.8102	12.20	38.33	116.8986
4.30	13.51	14.5220	8.30	26.08	54.1061	12.30	38.64	118.8229
4.40	13.82	15.2053	8.40	26.39	55.4177	12.40	38.96	120.7628
4.50	14.14	15.9043	8.50	26.70	56.7450	12.50	39.27	122.7184
4.60	14.45	16.6190	8.60	27.02	58.0880	12.60	39.58	124.6898
4.70	14.77	17.3494	8.70	27.33	59.4468	12.70	39.90	126.6768
4.80	15.08	18.0956	8.80	27.65	60.8212	12.80	40.21	128.6796
4.90	15.39	18.8574	8.90	27.96	62.2114	12.90	40.53	130.6981

To convert values outside the range of those given above, simply divide or multiply by a power of ten that will result in a value within the range. For example, if one has a diameter of 190 cm, it can be divided by 100 to obtain 1.90 cm, which is on the table. The desired circumference is then the circumference for a diameter of 1.90 cm multiplied by the factor of 100 (namely, 5.97 cm × 100 = 597 cm); and the area is the area for a diameter of 1.90 cm multiplied by the square of 100 (2.8353 cm^2 × 100^2 = 28353 cm^2).

1. Introduction

The plot method is a basic and commonly-used procedure for sampling many types of organisms. A *plot* generally is a rectangle or a square, but circles or other shapes can be used. The term *quadrat* often is used interchangeably with plot, but strictly speaking a quadrat is a square plot.

In plot sampling, one takes a manageable area of known size and identifies, counts, and often measures all individuals within it. In sampling soil or aquatic organisms a volume of the habitat often is sampled and analyzed. This sampling procedure is then repeated (i.e., "replicated") for a number of plots (preferably of the same size) to obtain an adequate representation of the population or community.

The plot method is most widely used for sampling land plants, but may also be used for sampling relatively sessile or slow-moving animals (such as soil fauna), for benthos in aquatic environments, or for animal burrows, nests, and hills, or other animals signs.

Accurate and unbiased application of plot sampling requires random samples (see section 1A.2). If plot sampling is to be used for invertebrates, consult sections 3D or 3E for specific sampling techniques.

2. Procedure

For sampling plants, rectangular plots have been found to yield better results than other shapes; a rectangle with sides in a 1:2 ratio works well. For general class purposes the following plot sizes are recommended: For closely spaced herbaceous vegetation, use a rectangular plot one square meter in area on a scale of 1:2 (i.e., 0.71 $\times$ 1.41 m). For bushes, shrubs, and saplings up to 3 or 4 m tall, use 10-m² plots (2.24 $\times$ 4.47 m). For forest trees over 3 to 4 m high, use 100 m² (7.07 $\times$ 14.14 m) sampling areas. Adjustment of these recommendations can be made by considering species-area curves and performance curves, as described in section 1A.3.

For sampling soil macroinvertebrates or benthic invertebrates in a stream, it is common to use a square plot (quadrat) with an area of 0.1 m² (31.6 $\times$ 31.6 cm) with size of plot adjusted upward when sampling large organisms (see sections 3D.2 and 3E.2.4, respectively). A circular plot is commonly used in pond benthos sampling (see section 3E.2.1).

The location of each plot should be determined either by a grid or other systematic method or by a standard random procedure (see section 1A.2), such as using a randomly selected point as the center of the plot or using a set of random coordinates to define the plot boundaries. These methods of plot selection minimize bias by purposefully ignoring the nature of the vegetation and terrain; that is, one should not include or exclude any plants or animals that are subjectively felt either "good" or

"poor" representations of the community. If different strata in a forest community are being sampled, the three different plot sizes given above can be nested inside each other, the smaller plots lying within the larger ones. Nested plots are also useful for determining the most efficient plot size from species-area curves or performance curves (see section 1A.3).

After the plots have been marked out, identify each species and count the number of individuals of each within each plot. If samples are to be analyzed in the laboratory, separate and label each sample. In plant sampling, also measure foliage coverage, basal coverage, or basal area of each individual, the measurement depending on the type of vegetation sampled (see section 4 of the introduction to unit 3).

3. Data and calculations

For each plot, record the number of individuals and for plants, the area covered for each species, as on data sheet 3A.1, and enter the appropriate data for all samples on the class data sheet 3A.2 as soon as possible. Data sheet 3A.3 may then be used in performing the following calculations:

Density (*D*) is the number of individuals in a unit area:

$$D_i = n_i/A, \qquad (1)$$

where D_i is the density for species *i*, n_i is the total number of individuals counted for species *i*, and *A* is the total area sampled. *Relative species density* (*RD*) is the number of individuals of a given species (n_i) as a proportion of the total number of individuals of all species (Σn):

$$\mathrm{RD}_i = n_i/\Sigma n, \qquad (2)$$

or

$$RD_i = D_i/TD = D_i/\Sigma D, \qquad (3)$$

where *TD* is the density for all species (which is equivalent to ΣD, the sum of the densities of all the species).

Data Sheet 3A.1. *Tabulation of Raw Data for Plant Plot Sampling.*

Date _____ Observers _____

Habitat and Stratum _____

Locality _____

Plot identification number _____ Plot size _____

Plant number	Species:		Species:		Species:		Species:	
	Diameter or circumference (cm)	Area covered (cm²) (a)	Diameter or circumference (cm)	Area covered (cm²) (a)	Diameter or circumference (cm)	Area covered (cm²) (a)	Diameter or circumference (cm)	Area covered (cm²) (a)
1								
2								
3								
4								
5								
6								
7								
8								
9								
10								
11								
12								
13								
Totals	///////////		///////////		///////////		///////////	

The above totals are to be summarized from a series of similar data sheets on data sheet 3A.2.

Data Sheet 3A.2. *Summary of Data from Plot Sampling.*

Date _____ Observers _____

Habitat and Stratum _____

Locality _____

Species (*i*)	Unit*	Plot										Total for species
		1	2	3	4	5	6	7	8	9	10	

* Unit of measurement might be numbers, area covered, or biomass. The data entered in this data sheet are from the totals recorded on data sheet 3A.1. If they are numbers, then n_i will appear in the "total for species" column; if they are areas covered, then the totals column will contain values of a_i; if they are biomasses (see section 6A), then the totals will be B_i.

83

Data Sheet 3A.3. *Class Summary of Data from Plant Plot Sampling.*

Date _____ Observers _____

Habitat and Stratum _____

Locality _____

Total area sampled (A) _____ Total number of plots (k) _____

Species (i)	Number of individuals (n_i)*	Density (D_i)	Relative Density (RD_i)	Present in how many plots? (j_i)*	Frequency (f_i)	Relative Frequency (Rf_i)	Coverage (C_i)*	Relative Coverage (RC_i)	Importance value (IV_i)
Totals	$\Sigma n =$	$\Sigma D =$	$\Sigma RD = 1.0$		$\Sigma f =$	$\Sigma Rf = 1.0$	$\Sigma C =$	$\Sigma RC = 1.0$	

* Data in these columns are from the totals on data sheet 3A.2. Data in the other columns are calculated from them. ($C_i = \Sigma a_i / A$)

84

Frequency (*f*) is the chance of finding a given species within a sample:

$$f_i = j_i/k, \qquad (4)$$

where f_i is the frequency of species, *i*, j_i is the number of samples in which species *i* occurs, and *k* is the total number of samples taken. Frequency is highly dependent on the size and shape of the plots used. If plots are too large, then one is almost certain to find most of the species in a given sample plot. On the other hand, if plots are too small, then the same species will seldom be encountered in more than one plot.

Relative frequency (*Rf*) is the frequency of a given species (f_i) as a proportion of the sum of the frequencies for all species (Σf):

$$Rf_i = f_i/\Sigma f. \qquad (5)$$

Coverage (*C*) is the proportion of the ground occupied by a vertical projection to the ground from the aerial parts of the plant:

$$C_i = a_i/A, \qquad (6)$$

where a_i is the total area covered by species *i* (estimated by basal area, foliage area, or basal coverage), and *A* is the total habitat area sampled. The *relative coverage* (RC_i) for species *i* is the coverage for that species (C_i) expressed as a proportion of the total coverage (*TC*) for all species:

$$RC_i = C_i/TC = C_i/\Sigma C, \qquad (7)$$

where ΣC is the sum of the coverages of all the species.

The sum of the above three relative measures for species *i* is an index called the *importance value* (IV_i):

$$IV_i = RD_i + Rf_i + RC_i. \qquad (8)$$

The value of *IV* may range from 0 to 3.00 (or 300%). Dividing *IV* by 3 results in a figure that ranges from 0 to 1.00 (i.e., 100%), and this is referred to as the *importance percentage*. The importance value, or the importance percentage, gives an overall estimate of the influence or importance of a plant species in the community. Although such an estimate has the advantage of using more than one measure of influence, it has the disadvantage of giving equal weight to each and yielding similar values for different combinations of the three relative values. Also, the term "importance" is confusing since it means different things to different ecologists. It also involves summing of three different yet not independent measures, which is difficult to defend mathematically.

4. Suggested exercises

1. Sample a forest or grassland community using plots.
   a. Determine the density, relative density, frequency, relative frequency, coverage, and relative coverage for the species sampled in given strata.
   b. Calculate importance values for the predominant species; interpret these measures.

2. Sample a benthic or soil habitat for macroinvertebrates using plot sampling with the techniques given in sections 3D or 3E. Determine the density, relative species density, frequency, relative frequency, and/or biomass (section 6A).
3. Compare two similar community types, (e.g., two different forests, two different prairies, two different benthic communities). Analyze the data collected according to one or more of the following considerations:
   a. Species diversity (section 5B)
   b. Community similarity (section 5C)
   c. Species area curve (section 1A.3)
   d. Relative abundance curve (section 5A)
   e. Statistical significance (by Mann-Whitney testing, section 1B.3.4) of difference between density (or coverage) of the major species.
4. Compare plot sampling results to those obtained from the point-quarter (section 3C) or transect (section 3B) methods.
   a. Compare species-sample size curves and performance curves (section 1A.3).
   b. Evaluate the precision of the methods by determining the standard error for the density of a few of the more important species (see sections 1B.2.2 and 1B.2.3). Which method has greater precision? Which has less bias? Which is more efficient in terms of the amount of data collected per time and effort sampling?

5. Selected references

Barbour, M. G., and W. D. Billings (eds.). 1987. North American terrestrial vegetation. Cambridge University Press, New York.

Becker, D. A., and J. J. Crockett. 1973. Evaluation of sampling techniques on tall-grass prairie. J. Range Management 26:61–65.

Bormann, F. H. 1953. The statistical efficiency of sample plot size shape in forest ecology. Ecology 34:474–487.

Cain, S. A., and G. M. DeO. Castro. 1959. Manual of vegetation analysis, Harper & Row, New York.

Daubenmire, R. 1968. Plant communities. Harper & Row, New York.

Greig-Smith, P. 1983. Quantitative plant ecology. 3d ed. University of California Press, Berkeley, Calif.

Kershaw, K. A., and J. H. Looney. 1985. Quantitative and dynamic ecology. 3d ed. Edward Arnold, London.

Mueller-Dombois, D., and H. Ellenberg. 1974. Aims and methods of vegetation ecology. John C. Wiley & Sons, New York.

Oosting, H. J. 1956. The study of plant communities. W. H. Freeman and Co., San Francisco.

Phillips, E. A. 1959. Methods of vegetation study. Holt, Rinehart and Winston, New York.

Pielou, E. C. 1977. Mathematical ecology. John C. Wiley & Sons, New York.

Seber, G. A. F. 1982. The estimation of animal abundance and related parameters. 2d ed. Macmillan Publishing Co., New York.

6. Vascular plant identification manuals   (Applicable to sections 3A through 3C.)

Abrams, L. 1923, 1944, 1951, 1960. Illustrated flora of the Pacific states. 4 volumes. Stanford University Press, Stanford, Calif.

Audubon Society Staff. 1979. The Audubon Society field guide to North American wildflowers. Alfred A. Knopf, New York.

Audubon Society Staff and E. L. Little, Jr. 1980. The Audubon Society field guide to North American trees, eastern edition. Alfred A. Knopf, New York.

Audubon Society Staff and E. L. Little, Jr. 1980. The Audubon Society field guide to North American trees, western edition. Alfred A. Knopf, New York.

Baerg, H. J. 1973. How to know the western trees. Wm. C. Brown Co. Publishers, Dubuque, Iowa.

Benson, L. 1982. The cacti of the United States and Canada. Stanford University Press, Stanford, Calif.

Benson, L., and R. A. Darrow. 1981. Trees and shrubs of the southwestern deserts. 3d ed. University of Arizona Press, Tucson.

Brockman, C. F. 1968. Trees of North America. Western Publishing Co., New York.

Cobb, B. 1977. A field guide to the ferns and their related families: Northwestern and central North America. Houghton Mifflin Co., Boston.

Conard, H. S., rev. by R. L. Redfearn, Jr. 1979. How to know the mosses and liverworts. 2d ed. Wm C. Brown Co. Publishers, Dubuque, Iowa.

Cronquist, A. 1979. How to know the seed plants. Wm. C. Brown Co. Publishers, Dubuque, Iowa.

Cronquist, A. 1980. Vascular flora of the southeastern United States. Volume 1. University of North Carolina Press, Chapel Hill, N.C.

Cuthbert, M. J. 1948. How to know the fall flowers. Wm. C. Brown Co. Publishers, Dubuque, Iowa.

Cuthbert, M. J., and S. Verhoek. 1982. How to know the spring flowers. Wm. C. Brown Co. Publishers, Dubuque, Iowa.

Dawson, E. Y. 1963. How to know the cacti. Wm. C. Brown Co. Publishers, Dubuque, Iowa.

Fassett, N. C. 1957. Manual of aquatic plants. University of Wisconsin Press, Madison, Wis.

Fernald, M. L. 1987. Gray's manual of botany. Dioscorides Press, Portland, Oreg.

Gleason, H. A. 1975. New Britton and Brown illustrated flora of the northeastern United States and adjacent Canada. 3 volumes. Hafner, New York.

Harlow, W. M. 1946. Fruit key and tree twig key to trees and shrubs. Dover Publications, New York.

Harlow, W. M. 1957. Trees of the eastern and central United States and Canada. Dover Publications, New York.

Harrar, E. S., and G. J. Harrar. 1962. Guide to southern trees. Dover Publications, New York.

Hitchcock, A. S. 1971. Manual of the grasses of the United States. Dover Publications, New York.

Hotchkiss, N. 1972. Common marsh, underwater and floating-leaved plants of the United States and Canada. Dover Publications, New York.

Jaques, H. E. 1948. How to know the plant families. 2d ed. Wm. C. Brown Co. Publishers, Dubuque, Iowa.

Knobel, E. 1977. Field guide to the grasses, sedges, and rushes of the United States. Dover Publications, New York.

Little, E. L. 1980. Forest trees of the United States and how to identify them. Dover Publications, New York.

McMinn, H. E., and E. Maino. 1946. An illustrated manual of Pacific Coast trees. 2d ed. University of California Press, Berkeley, Calif.

Mickel, J. T. 1979. How to know the ferns and fern allies. Wm. C. Brown Co. Publishers, Dubuque, Iowa.

Miller, H. A., and H. E. Jacques. 1978. How to know the trees. Wm. C. Brown Co. Publishers, Dubuque, Iowa.

Muenscher, W. C. 1944. Aquatic plants of the United States. Cornell University Press, Ithaca, N.Y.

Muenscher, W. C. 1950. Keys to woody plants. 7th ed. CSD, Sundburg, Mass.

Niehause, T. F., and C. L. Ripper. In press. A field guide to Pacific states wildflowers. Houghton Mifflin Co., Boston.

Peterson, R. T., and M. McKenney. 1974. A field guide to the wildflowers of northeastern and north central North America. Houghton Mifflin Co., Boston.

Petrides, G. A. 1973. A field guide to trees and shrubs. Houghton Mifflin Co., Boston.

Petrides, G. A. 1988. A field guide to eastern trees. Houghton Mifflin Co., Boston.

Pohl, R. W. 1978. How to know the grasses. 3d ed. Wm. C. Brown Co. Publishers, Dubuque, Iowa.

Prescott, G. W. 1980. How to know the aquatic plants. 3d ed. Wm. C. Brown Co. Publishers, Dubuque, Iowa.

Reid, G. K. 1967. Pond life. Western Publishing Co., New York.

Runkel, S. T., and D. M. Roosa. 1989. Wildflowers of the tallgrass prairie. The Upper Midwest. Iowa State University Press, Ames, Iowa.

Rydberg, P. A. 1965. Flora of the prairies and plains of central North America. Hafner, New York.

Sargent, C. S. 1922. Manual of the trees of North America. 2 volumes. Dover Publications, New York.

Small, J. K. 1972. Manual of the southeastern flora. 2 volumes. Hafner, New York.

Venning, F. D. 1984. Wildflowers of North America. Western Publishing Co., New York.

Verhock, S., and M. J. Cuthbert. 1982. How to know the spring flowers. 2d ed. Wm. C. Brown Co., Dubuque, Iowa.

Vines, R. A. 1960. Trees, shrubs and woody vines of the Southwest. University of Texas Press, Austin, Tex.

Wilkinson, R. E. 1979. How to know the weeds. Wm. C. Brown Co. Publishers, Dubuque, Iowa.

Winchester, A. M., and H. E. Jacques. 1981. How to know the living things. 2d ed. Wm. C. Brown Co., Dubuque, Iowa.

Zim, H. S., and A. C. Martin. 1956. Trees. Golden Press, New York.

1. Introduction

In some types of vegetation, the use of plots (section 3A) may be impractical and prohibitively time-consuming. Also the point-quarter method (section 3C) is often difficult to apply due to nonrandom distribution of the sampled individuals. Transects are useful in these instances and are especially advantageous and efficient in studies of contiguous stages in ecological succession or of communities at transition zones. Three major types of transect are introduced here.

A *belt transect* is a long strip of terrain in which all organisms are counted and measured. Knowing the width and length of the transect, one may use the computational procedures of plot sampling (section 3A), considering the belt transect to be a very long rectangular plot. In addition, the belt may be divided into intervals representing zones to be studied, and each interval may be treated as a plot.

Another transect method, used mainly by plant ecologists, is the *line-intercept* technique. Data are tabulated on the basis of plants lying on a straight line cutting across the community under study. Because an area is not being sampled, only density indices and relative estimates of density can be calculated. Line-intercept transects have been widely used in grassland community studies, as true estimates of absolute density either cannot be made or are difficult to intepret because of the problem of distinguishing between individual plants. In cases where relative estimates are sufficient, line-intercept transects may efficiently obtain them.

A third transect procedure, used widely by terrestrial vertebrate ecologists, is the *strip census,* or *line transect.* (Plant ecologists use the term line transect synonymously with line-intercept, although the two are different sampling techniques.) A strip census involves walking a line established through an area and recording individuals observed from that line. The data recorded are a population index rather than an absolute measure of density (see the introduction to unit 3). For example, the number of individuals observed per unit distance traveled, numbers per unit of time spent in observation, or numbers caught per trap per unit time would be such indices. Examples of using line transects in vertebrate population studies include road kill censuses, bird counts, and small mammal trapping (see section 3H). This method can be quantified to yield density estimates useful in studies where the animals are highly mobile, yet often difficult to see until flushed. To do so, one walks a line through an area and records the distance to each animal seen. Mathematical methods have been developed to estimate density in this fashion, as explained in section 3H.2.

3b

transect sampling

2. Procedure

If the objective of transect sampling is to determine species composition in a given habitat, as it was in section 3A, then the directional orientation of the transect should be determined by connecting two randomly selected points in the community to be studied (see section 1A.2 for a discussion of randomness). If, however, the specific desire is to study a community transition or some ecological gradient then the transect length should be oriented along that transition or gradient. Several replicate transects should be used in the same study area.

If a transect is used to sample mobile animals, such as by a strip census, the act of laying out and marking the transect may disturb and disperse the objects of the census. Therefore, first establish the transect on a map, afterward mark it in the field, and wait a while before censusing to allow the animals to resume normal activities.

If a transect is divided into contiguous segments, then the data for the several segments may be used to compute frequency by recording the presence or absence of a species in each interval. This procedure is explained below for the line-intercept method, with which it is most commonly used.

For line-intercept sampling, extend a wire, cord, or measuring tape to mark the line between two points (identified by stakes, flags, or marked vegetation). The line may be 10, 25, 50, or 100 meters long, with longer transects useful for more widely spaced organisms. Mark off 1-, 5-, or 10-meter intervals on the line, using larger intervals for communities consisting of widely spaced individuals. Each interval will be treated as a separate unit of the transect.

Begin counting at one end of the line, and record data for each interval. In very dense vegetation (as some grasslands), count only those plants physically intercepted by

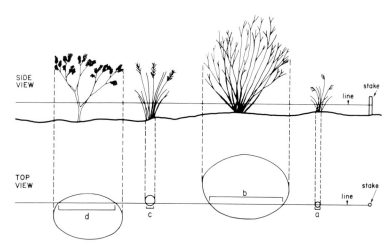

Figure 3B.1. The intercept length (brackets) is that portion of a line intercepted by a plant (or clump of plants, as the basal intercept length for plants a and c) or by a perpendicular projection of the foliage to the line (as the aerial intercept length for plants b and d).

the tape measure. Otherwise, count all plants that are intercepted within a 1-cm strip of the line. Include also those plants whose aerial foliage overlies the transect. In grassland communities, individual plants are difficult to distinguish and count. Though you could count individual stems, stalks, or shoots in such cases, it is more common to count clumps.

Coverage data collected from sampling plants by the line-intercept method differ from those obtained from plots or belt transects. In the latter sampling (section 3A), coverage is estimated from areas covered, but in line-intercept sampling, the measurement of *intercept length* (or "intercept distance") is used to estimate coverage. This length is that portion of the transect length intercepted by the plant measured at or near the base of the plant or clump of plants or by a perpendicular projection of its foliage intercepted by the line (figure 3B.1).

For each plant counted, measure the intercept length and record the values on the raw data sheet (data sheet 3B.1). Then summarize all such data on data sheet 3B.2. Where several strata exist each stratum may be surveyed separately. Sampling only one component of the community, such as grasses, forbs, or shrubs is often desired and may be done to simplify a class experience.

3. Data and calculations

After the summarized data from data sheet 3B.1 have been recorded for each transect interval on data sheet 3B.2, the following quantities should be determined using data sheet 3B.3. If belt transects are used, then apply the equations for plot sampling in section 3A.3, and treat each transect interval as an individual plot.

For a given species, *i*, the *linear density index* (ID_i) is calculated as:

$$ID_i = n_i/L, \tag{1}$$

where n_i is the total number of individuals of species i collected, and L is the total length of all transects sampled. The species' *relative density* (RD_i) is:

$$RD_i = n_i/\Sigma n, \tag{2}$$

where Σn is the total number of individuals counted for all species, or:

$$RD_i = ID_i/\Sigma ID, \tag{3}$$

where ΣID is the sum of the density indices for all species.

The *linear coverage index* (IC_i) for this species is:

$$IC_i = l_i/L, \tag{4}$$

where l_i is the sum of the intercept lengths for species i (i.e., the total length of the transects intercepted by the species). And the *relative coverage* of species i (RC_i) is:

$$RC_i = l_i/\Sigma l, \tag{5}$$

where Σl is the sum of the intercept lengths for all species, or:

$$RC_i = IC_i/\Sigma IC, \tag{6}$$

where ΣIC is the sum of the values of the linear coverage index indices for all species. Note that because basal and aerial coverage distances of various individual plants may overlap, the sum of the intercept lengths (Σl) may be larger than the transect length (L).

The *frequency* of species i is defined as:

$$f_i = j_i/k, \tag{7}$$

Data Sheet 3B.1. *Tabulation of Raw Data from Line-Intercept Plant Sampling*

Date _____ Observers _____

Habitat and stratum _____

Location _____

Transect (or interval) identification _____

Length of transect (or interval) _____

Plant Number	Species: Intercept length (l)	Species: Intercept length (l)	Species: Intercept length (l)	Species: Intercept length (l)	Species: Intercept length (l)
1					
2					
3					
4					
5					
6					
7					
8					
9					
10					
11					
12					
13					
Totals					

The above totals are to be summarized from a series of similar data sheets on data sheet 3B.2.

Data Sheet 3B.2. *Summary of Data from Line-Intercept Plant Sampling*

Date _____ Observers _____

Habitat and stratum _____

Locality _____

Species (*i*)	Unit*	Intercept interval										Total for species
		1	2	3	4	5	6	7	8	9	10	

* Unit of measurement might be numbers or intercept length. The data entered in this data sheet are from the totals recorded on data sheet 3B.1. If they are numbers, then n_i will appear in the "total for species" column; if they are intercept lengths, then the totals column will contain values of l_i.

90

Data Sheet 3B.3. *Class Summary of Data from Line-Intercept Plant Sampling*

Date _____ Observers _____

Habitat and stratum _____

Locality _____

Total transect length (L) _____

Total number of transect intervals _____

Species (i)	Number of individuals (n_i)	Linear density index (ID_i)	Relative density (RD_i)	Present in how many transect intervals? (j_i)*	Frequency (f_i)	Relative frequency (Rf_i)	Intercept length (l_i)*	Linear coverage index (IC_i)	Relative coverage (RC_i)	Importance value (IV_i)
Totals	$\Sigma n =$	$\Sigma ID =$	$\Sigma RD = 1.0$		$\Sigma f =$	$\Sigma Rf = 1.0$	$\Sigma l =$	$\Sigma IC =$	$\Sigma RC = 1.0$	

* Data collected in these columns are from the totals on data sheet 3B.2. Data in the other columns are calculated from them.

where j_i is the number of line-intercept intervals containing species i, and k is the total number of intervals on the transects. The *relative frequency* of species i (Rf_i) is:

$$Rf_i = f_i/\Sigma f, \qquad (8)$$

where Σf is the sum of the frequencies of all species. As discussed in section 3A.3, the *importance value* (IV_i) of species i is:

$$IV_i = RI_i + RC_i + Rf_i. \qquad (9)$$

In the line-intercept method, the probability of being sampled is dependent on the size of the plant. A large rare plant is more likely to be detected than a small rare plant. Large, dense species will appear more frequently than small, dense species. The pattern of distribution can also affect the estimates of frequency. The problems of interval length and number of transects desired are similar to the problems of plot size and number discussed in section 1A. Performance curves can be made, and one may graph species-interval curves as well as species-area curves (see section 1A.3).

4. Suggested exercises

1. Sample a forest or grassland community using the intercept technique.
   a. Determine the density index, relative density, frequency, relative frequency, coverage index, and relative coverage for the species sampled.
   b. Calculate importance values for the predominant species; interpret the meaning of these measures.
2. Compare two similar community types (such as two different forests, two different prairies). Analyze the data collected according to one or more of the following considerations:
   a. Species diversity (section 5B)
   b. Community similarity (section 5C)
   c. Species-interval length curve (section 1A.3)
   d. Relative abundance curve (section 5A)
   e. Statistical significance (by Mann-Whitney testing, section 1B.3.4) of difference between density (or coverage) index of the major species.
3. Compare transect sampling results to those obtained from the plot (section 3A) or point-quarter (section 3C) methods.
   a. Compare species-sample size curves and performance curves (section 1A.3).
   b. Evaluate the precision of the methods by determining the standard error for the density of a few of the more important species (see section 1B.2.2 and 1B.2.3). Which method has greater precision? Which has less bias? Which is more efficient in terms of the amount of data collected per time and effort sampling?

5. Selected references

Barbour, M. G., and W. D. Billings (eds.). 1988. North American terrestrial vegetation. Cambridge University Press, New York.

Becker, D. A., and J. J. Crockett. 1973. Evaluation of sampling techniques on tall-grass prairie. J. Range Manage. 26:61–65.

Cain, S. A., and G. M. DeO. Castro. 1959. Manual of vegetation analysis. Harper & Row, New York.

Canfield, R. 1941. Application of the line interception method in sampling range vegetation. J. Forestry 39:338–394.

Lindsey, A. A. 1955. Testing the line-strip method against full tallies in diverse forest types. Ecology 36:485–494.

Mikol, S. A. 1980. Field guidelines for using transects to sample nongame bird populations. Office of Biological Services, Fish and Wildlife Service, U.S. Department of the Interior, Washington, FWS/OBS-80/58.

Parker, K. W., and D. A. Savage. 1944. Reliability of the line-interception method in measuring vegetation of the southern great plains. J. Amer. Soc. Agron. 36:97–110.

Phillips, E. A. 1959. Methods of vegetation study. Holt, Rinehart, and Winston, New York.

6. Vascular plant identification manuals   See section 3A.6.

1. Introduction

point-quarter sampling

Plot, or quadrat, methods of sampling (section 3A) are often very laborious and time-consuming, and their results are dependent on the size, shape, and number of the plots used. Not surprisingly, other methods have been advanced to reduce these problems without sacrificing the information or accuracy obtainable from sampling plots. Plotless methods are useful for plants or sessile animals and have the advantage of not having to demarcate sampling areas of a certain size or shape. The most popular plotless sampling method is the *point-quarter*, or *quadrant*, method. ("Quadrant" is not to be confused with the "quadrat" of section 3A.)

The accuracy of the point-quarter method is sensitive to departures from a random distribution of individuals, especially if only small numbers of individuals are counted. Thus the method should not be used for populations with either highly aggregated or uniformly-spaced individuals (see section 4C). An aggregated, or clumped, population will give an underestimate of density, while a population with a highly regular pattern tends to give an overestimate. This and related criticisms have been discussed by Catana (1963), Cottam and Curtis (1956), Pileou (1977), and Risser and Zeldler (1968). The point-quarter technique should be checked against the plot method before using it in a given area, especially if clumping or if uniformity of distribution occurs. (Plot methods can handle nonrandom populations by increasing the number and size of the plots.) Distribution pattern may be less important when using a modification of the point-quarter method called "wandering-quarter" sampling, described below. (Little information is available as to the efficiency and reliability of this method compared to other plotless procedures.) The point-quarter method is superior to other plotless techniques, however, such as the so-called nearest-neighbor, closest-individual, and random pairs procedures (Mueller-Dombois and Ellenberg, 1974).

2. Procedure

Select a number of randomly-determined points (as described in section 1A.2) and mark them with flags. These points may be set randomly through the entire stand or randomly along a line transect running though it. Each point represents the center of four compass directions (N, S, E, W), which divide the sampling site into four quarters, or quadrants (figure 3C.1). In each quadrant measure the distance from the center point to the center of the nearest individual, regardless of species. Only one plant per quadrant is measured so that a total of four plants are recorded for each point sampled. Identify the plant and record its area covered as described in section 4 of the introduction of unit 3. If plants are widely and/or non-randomly spaced, then the same plant is liable to be counted more than once, and the point-quarter method should not be used in such situations.

The center of the rooted stem or the center of the crown of a clump of stems should be used when measuring the point-to-plant distance. If two plants are fairly close, be sure to measure the distance of both and record the smaller distance from the center point. Do not depend on visual perception of the distance to judge the closest plant, for there is a tendency to judge the larger of two plants as the closer. Record all measurements on the raw data sheet (data sheet 3C.1). This procedure may be modified to determine the density of single species by measuring the point-to-plant distances in each quadrant for that species only. But whereas the plants of all species may be distributed randomly, the members of a single species very often are not, and the point-quarter method will then yield inaccurate results due to this nonrandomness.

A variation of point-quarter sampling, the wandering-quarter method, appears to be somewhat independent of the pattern of distribution. A line transect is set up as described in section 3B, and a starting point near the beginning of the transect is selected at random. With the aid of a compass, set up one quadrant (90° angle) with the

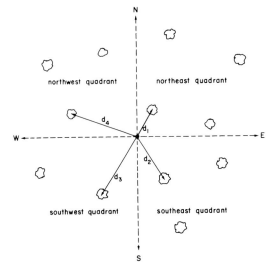

Figure 3C.1. In point-quarter sampling, determine the shortest point-to-point distance (d_i) in each of four quadrants.

Data Sheet 3C.1.   *Tabulation of Raw Data from Point-Quarter Plant Sampling.*

Date _____ Observers _____

Habitat and Stratum _____

Locality _____

Number of points sampled _____

Point number	Quadrant number	Species	Diameter or circumference (cm)	Area covered (cm²) (a)	Point-to-plant distance* (m)
1	1				
1	2				
1	3				
1	4				
2	1				
2	2				
2	3				
2	4				
3	1				
3	2				
3	3				
3	4				
4	1				
4	2				
4	3				
4	4				

The above data are to be summarized from a series of similar data sheets on data sheet 3C.2.

* If only relative measures of density, frequency, and coverage are desired, then omit this column.

Data Sheet 3C.2. *Summary of Data from Point-Quarter Plant Sampling.*

Date _____ Observers _____

Habitat and Stratum _____

Locality _____

Species (*i*)	Unit*	Point										Total for species
		1	2	3	4	5	6	7	8	9	10	

* Unit of measurement might be numbers, area covered, or biomass. The data entered in this data sheet are obtained from data sheet 3C.1. If they are numbers, then n_i will appear in "total for species" column; if they are areas covered, then the totals column will contain values of a_i; if they are biomass (see section 6A), then the totals will be B_i.

Data Sheet 3C.3. *Class Summary of Data from Point-Quarter Plant Sampling.*

Date _____ Observers _____

Habitat and stratum _____

Locality _____

Total number of points (*k*) _____ Total of point-to-plant distances (Σd_i) _____

Species (*i*)	Number of individuals (n_i)*	Relative density (RD_i)	Density (D_i)	Number of points with species (j_i)*	Frequency (f_i)	Relative frequency (Rf_i)	Area covered from all quadrants (a_i)*	Coverage (C_i)	Relative coverage (RC_i)	Importance value (IV_i)
Totals		$\Sigma RD = 1.0$	$\Sigma D =$		$\Sigma f =$	$\Sigma Rf = 1.0$	$\Sigma a =$	$\Sigma C =$	$\Sigma RC = 1.0$	

* Data in these columns are from the totals on data sheet 3C.2. Data in the other columns are calculated from them.

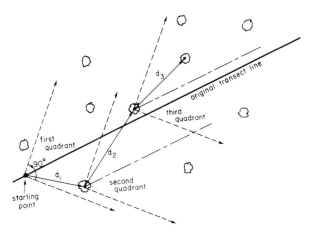

Figure 3C.2. In wandering-quarter sampling, determine the nearest plant in a quadrant bisected by a transect, and the plant becomes the apex of a similar quadrant bisected by a line parallel to the transect. Each value of d_i is the distance from the quadrant apex to the nearest plant in that quadrant.

transect line bisecting the angle as in figure 3C.2. Then measure point-to-plant distance of the nearest plant in that quadrant, identify the plant, and estimate the coverage. This plant then serves as the apex of a new quadrant whose angle is bisected by a line running parallel to the transect. Determine in the manner above the nearest point-to-plant distance in this quadrant. Repeat this procedure until you reach the end of the transect.

3. Data and calculations

Calculations for the point-quarter and the wandering-quarter methods are the same (except that frequency calculations are not applicable in the latter procedure). The mean density per unit area is estimated as follows:

Sum all point-to-plant distances taken for all species and compute the mean:

$$\bar{d} = \Sigma d_i / \Sigma n, \qquad (1)$$

where $\bar{d}$ is the mean point-to-plant distance, d_i is the point-to-plant distance for individual number i (so Σd_i is the sum of the point-to-plant distances for all individuals), and Σn is the total number of individuals measured. Theoretically and empirically, the mean area in which a single plant occurs is equal to the mean distance squared (Cottam and Curtis, 1956; Morisita, 1954). This can be visualized as one individual in a square area in which the side of the square is equal to the mean point-to-plant distance. Then,

$$\bar{A} = \bar{d}^2, \qquad (2)$$

where $\bar{A}$ is the mean area per plant. $\bar{A}$ is the inverse of the *total density* (TD) computed in plot sampling (section 3A.3), where TD is the total number of individuals of all species per unit area ($TD = \Sigma D$, the sum of the individual densities of the several species):

$$TD = u/\bar{A}, \qquad (3)$$

where u represents the number of area units to be used in expressing density. If the mean area, $\bar{A}$, is in terms of square meters, and the desired number of area units is one square meter, then $u = 1$ and $TD = 1/\bar{A}$, the units of $\bar{A}$ are square meters, and the units for TD are numbers per square meter. If the units of $\bar{A}$ are square meters, and it is desired to compute density on the basis of 100 square meters, then u is 100 and $TD = 100/\bar{A}$ will be an expression of numbers per 100 square meters. If a hectare were the desired unit area, then u would be 10,000 (as there are 10,000 square meters in a hectare), and the resultant total density would be numbers per hectare. (See Appendix B for area units of measurement and their equivalents.)

As mentioned in section 3C.1, the point-quarter method overestimates density if the members of the population are uniformly distributed (such as in regular plantings as orchards or crops). In such a case, equations 2 and 3 may be used by considering $\bar{d}$ to be the mean distance between plants (Mueller-Dombois and Ellenberg, 1974).

The *relative density* (RD) for each species is calculated as:

$$RD_i = n_i / \Sigma n, \qquad (4)$$

where n_i is the number of individuals of species i counted, Σn is the total number of individuals of all species counted, and RD_i is the relative density of species i. The *absolute density* (D) for species i is:

$$D_i = (n_i / \Sigma n)(u / \bar{A}), \qquad (5)$$

or,

$$D_i = (RD_i)(TD) = (RD_i)(\Sigma D), \qquad (6)$$

which is simply the relative density times the total density (and is nothing more than a rearrangement of equation 3 in section 3A).

Frequency for a given species is estimated in a similar manner as in plot sampling:

$$f_i = j_i / k, \qquad (7)$$

where f_i is the frequency of species i, j_i is the number of sampling points at which species i was counted, and k is the total number of points sampled. *Relative frequency* for species i (Rf_i) is estimated as:

$$Rf_i = f_i / \Sigma f, \qquad (8)$$

where Σf is the total of the frequencies for all species.

Coverage for species i (C_i) is estimated from the sum of the areas covered for that species and the species density:

$$C_i = (a_i)(D_i) / n_i, \qquad (9)$$

where a_i is the sum of the foliage coverages, basal areas, or basal coverages, for species i, D_i is the density of species

i (see equation 5 or 6), and n_i is the total number of individuals sampled of that species. *Relative coverage* for species i is:

$$RC_i = C_i/\Sigma C, \qquad (10)$$

where ΣC is the total coverage or basal areas for all species.

If only relative measures are desired, then point-to-plant distances need not be determined. Relative density (equation 4) and relative frequency (equation 8) are calculated simply from the counts and identification of species in the quadrants at the sampling points. Calculation of relative coverage (equation 10) additionally requires measurements of coverage of each individual counted.

As described in section 3A.3, the *importance value* of species i is:

$$IV_i = RD_i + Rf_i + RC_i. \qquad (11)$$

4. Suggested exercises

1. Using the data collected by point-quarter sampling:
   a. make species-sample curves (section 1A.3), where a sample is the data for a point;
   b. make performance curves (section 1A.3).
2. Compare the point-quarter method to plot sampling (section 3A) and the line-intercept method (section 3B).
3. Employ plot sampling (section 3A) and the point-quarter and wandering-quarter methods in stands of vegetation known to be random, uniform, and aggregated; compare the results from the different procedures.

5. Selected references

Barbour, M. G., and W. D. Billings (eds.). 1988. North American terrestrial vegetation. Cambridge University Press, New York.

Barbour, M. G., J. H. Burk, and W. D. Pitts. 1980. Terrestrial plant ecology. Benjamin/Cummings Publishing Co., Menlo Park, Calif.

Catana, H. J. 1963. The wandering quarter method of estimating population density. Ecology, 44:349–360.

Cottam, G., and J. T. Curtis. 1956. The use of distance measures in phytosociological sampling. Ecology 37:451–460.

Cottam, G., J. T. Curtis, and A. J. Catana, Jr. 1957. Some sampling characteristics of a species of aggregated populations. Ecology, 38:610–622.

Dix, R. L., 1961. An application of the point-centered quarter method to the sampling of range land vegetation. J. Range Management 14:63–69.

Greig-Smith, P. 1983. Quantitative plant ecology. 3d ed. University of California Press, Berkeley, Calif.

Morisita, M. 1954. Estimation of population density by spacing method. Mem. Fac. Sci. Kyushu Univ. E., 1:187–197.

Mueller-Dombois, D. and H. Ellenberg. 1974. Aims and methods of vegetation ecology. John Wiley & Sons, New York.

Pileou, E. C. 1977. Mathematical Ecology. John Wiley & Sons, New York.

Risser, P. G. and P. H. Zeldler. 1968. An evaluation of the grassland quarter method. Ecology 49:1006–1009.

6. Vascular plant identification manuals   See section 3A.6.

terrestrial invertebrate sampling

1. Introduction

Sections 3A, 3B, and 3C deal with sampling methods especially suitable for plants and sessile or slow-moving animals. Sections 3D, 3E, 3F, 3G, and 3H discuss methods applicable to sampling a variety of animals, including highly mobile forms. Data collected by these methods may then be used for analysis of populations (unit 4), communities (unit 5), and productivity (unit 6).

Some methods, such as capture-recapture and removal methods (sections 3F and 3G), intentionally concentrate on single species, but techniques in this section are suitable for sampling aggregations of animal species. Some of these (e.g., sweep nets, bait traps, and drop-boards) are difficult to quantify for studies of animal communities, for they are typically biased toward collecting certain species and the area sampled is difficult to determine. Indeed, all methods are limited in that some species will be missed, and the relative densities may be biased toward or against one or more species. But even if some of these procedures contain inherent biases, they can be standardized for comparative studies. Thus they can provide some useful information about community patterns and relative population changes.

General principles of sampling are discussed in section 1A.3. Also, if an animal sampling procedure employs a plot or a transect, portions of sections 3A and 3B, respectively, should be consulted. In animal sampling, the organisms collected are commonly examined in the laboratory. Therefore, take special care in preserving, preparing, and labeling each sample. Labels should contain the date, names of collectors, locality, and some sample description (such as an identification number corresponding to field notes). Use good quality bond and ink for labeling. (Standard notebook paper disintegrates, and most fountain pen and ball point pen inks run or fade when placed in preservatives.) Careless labeling can render a sample useless.

The format of Data Sheet 3A.2 provides for a convenient summary for animal community studies.

2. Ground macroinvertebrates

Because of the small size of ground invertebrates, use small sampling plots (0.1 or 0.2 m²). The data are then treated as discussed in plot sampling (section 3A). The area of the plot may be outlined by stakes, or very conveniently by a metal or wood frame. Generally, remove the vegetation from within the plot, but carefully do this by cutting rather than uprooting or tearing so that animals from the vegetation are not shaken to the ground. Dig up the ground, including both litter and soil, remove it from the plot down to a depth of 10 cm, and place this sample in a tray. Then separate the animals from the soil, and either identify and count them or preserve them for laboratory analysis at a later time. This method is the terrestrial analogue of the Surber stream sampling technique described in section 3E.2.4.

For subsequent laboratory examination, place each plot's contents into a separate plastic bag, label it, and seal it. A more thorough analysis of the soil fauna is possible by placing the soil and litter in a pan of warm water (45–50°C) and stirring gently. Most of the animals will then float to the surface. This procedure is especially useful with frozen ground.

The data may be recorded as coming from either the surface stratum or the subterranean stratum, or one may collectively refer to sampling the ground stratum. The data are then expressed as the number of individuals of a given taxon per m² (less commonly, for subterranean animals, as numbers per m³ of soil). Bear in mind that smaller invertebrates are more easily missed than larger ones, and more sedentary animals such as worms are favored over quick-moving species such as sowbugs. Also, contamination of the litter with foliage invertebrates can occur during careless sampling.

3. Ground microinvertebrates

For collecting small soil and litter arthropods, the Berlese-Tullgren funnel is commonly used (figure 3D.1). Here the soil sample of a known surface area and depth is inverted in a large smooth funnel, and an incandescent light bulb is placed near the top of the funnel. Take these samples with a soil corer or any other method for obtaining a standard area and depth of litter and soil, as described in section 2D. The light and heat, and the subsequent drying of the soil, force the animals downward toward the funnel stem, through which they fall into a container of 80% ethanol or other preservative. The soil being sampled should be wrapped in cheesecloth to prevent substrate from falling down into the preservative. The data are then treated as they would be for a plot sample (section 3A).

The Berlese method is limited, however, because it favors those species that do not desiccate easily, as well as those species that move through the soil unimpeded by the

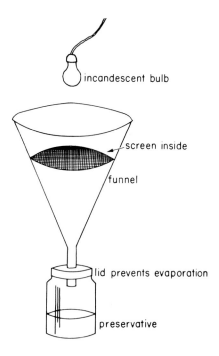

Figure 3D.1. The Berlese-Tullgren funnel may be constructed easily and inexpensively.

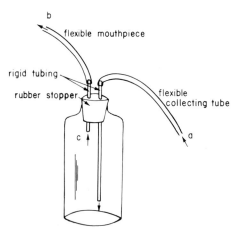

Figure 3D.2. With a simple aspiration bottle, small arthropods can be drawn into the bottle at point a by applying suction at point b with a rubber bulb or by mouth. (The fastidious user of mouth suction will desire to place a piece of cheesecloth over the end of the tube at point c.)

of the total ground surface. This method is the terrestrial analogue of employing artificial substrates to collect benthic organisms in an aquatic habitat (section 3E.2.5).

5. Pit trap

Another simple technique commonly used to obtain relative estimates of some invertebrate populations is the *pit trap*. Though it can also sample small mammals and amphibians, too many traps would be needed for a reasonably sized sample of vertebrates. An individual trap consists of a large can or jar (those of 1- or 2-liter capacity work well) set in the ground with its rim at the level of the soil surface. Many animals traversing the area fall into the can and cannot escape up the smooth sides. You may bait the traps with such items as oatmeal, peanut butter, freshly cut fruit, or raw meat. However, an unbaited trap with a few centimeters of water in the bottom to drown the caught animals can collect a surprising variety of invertebrates. A variety of spiders and beetles generally dominate a pit trap sample.

The traps may be set randomly, along a transect, or in a grid. An estimate of the area from which the animals were attracted is difficult, so the abundance estimates can only be population indices. Express the data as numbers per unit effort of sampling. This is done by calculating the number of individuals caught per trap per number of sampling days. If trapping is done over a period of days using a grid, then one may attempt a capture-recapture or a removal study as described in sections 3F and 3G.

6. Foliage invertebrates

Most procedures available for studies of invertebrates that live above the ground—predominantly insects and spiders—are difficult to quantify. They usually result in population indices rather than measures of absolute density.

drying process. Thus it may miss many insect larvae and other soft-bodied invertebrates. Sampling microscopic soil animals, such as rotifers and nematodes, requires special techniques not described here (see Southwood, 1978).

4. Cryptozoans

One procedure for sampling ground-dwelling cryptozoans ("hidden animals") is by use of *drop-boards*. A series of wooden boards, each $0.1\,m^2$, is laid out on the ground over a given area. These may be arranged randomly, along a transect, or in a grid. Those species of animals that prefer dark, cool, and moist microhabitats will collect under these boards. Although this procedure samples a particular component of the ground fauna, it involves only a few easily identified species for the beginning student to learn. Set out the drop-boards a few days to a week before collecting the specimens to allow representatives of a larger number of species to find the boards.

When sampling, lift each board carefully but quickly. It is advisable to mark the area of the board before lifting it to ensure gathering animals only from the area covered by the board. Work fast, since most of the animals are negatively phototactic and will quickly seek hiding places. Collect small animals rapidly by sucking them up in a glass jar fitted with a rubber bulb (figure 3D.2). Identify each species and record the number of individuals for each board. The data can be expressed as the number of animals per square meter, but the value represents the number of animals per unit of drop-board area. Therefore, it is a measure relative to sampling effort and is a density index; it does not provide true estimates of population densities

The following procedures will provide the student effective sampling of numerous species.

6.1. Herb invertebrates   Ideally, sample foliage invertebrates can be sampled by enclosing a volume of foliage within a tent-type structure or a bottomless cylinder (such as an inverted garbage can) and spraying the inside of the enclosure with an anesthetic or an insecticide. All the animals captured within are then identified, counted and/or weighed, and the data treated as in plot sampling (section 3A). This method, though quantitative and quite thorough in sampling the invertebrates present, is tedious and often impractical (particularly in dense vegetation).

A very common method of sampling is to sweep herbs with a sturdy net. The sweeping must be standardized as much as possible: The length of each step, the rate of walking, the arc of the sweep, the level of the arc, the speed of sweeping, and the number of sweeps are all variables that can affect the results. No two sweeps should traverse the same foliage. The sweep net method has been so widely used by ecologists that much literature exists regarding the standardization of this method for quantitative studies (see Carpenter, 1936; Shelford, 1951; Southwood, 1978), and in spite of its crudeness, it can give reliable and repeatable results. For example, an investigator might find that fifty sweeps of a foliage net for a given habitat collects approximately the same number of individuals as obtained per square meter using the plot techniques described above.

To remove the collected animals from the sweep net, first close the bag of the net by hand gripping it. Then carefully spray or pour ether, ethyl acetate, chloroform, or some other anesthetic agent on the closed part of the bag. After a few moments, the animals are immobilized or killed, and the bag may be opened and emptied into a tray. The animals may be identified and tabulated in the field or placed in jars or vials with 70% ethanol for later enumeration in the laboratory. Record the data as a density index representing the number of individuals per ten sweeps. If the sweeping has been standardized using a plot, then the data may be expressed as the number per square meter or hectare. Sweep net sampling may also be used to collect data for capture-recapture or removal estimates of population size (sections 3F and 3G).

Be aware of the sources of bias in sweep net sampling. For instance, there is a tendency to sample most thoroughly those arthropods that either do not fall off the vegetation or do not fly away as the net approaches. Thus many dipterans (flies, gnats, and mosquitoes) may be missed. The efficiency of the sampling also depends on variables other than the sweeping technique: type of habitat, species sampled (which depends on daily and seasonal cycles of activity), vertical distribution of animals, weather, and size and shape of the net.

6.2. Shrub and tree invertebrates   The shrub stratum may be sampled by the sweep net method described above, although the standardization of the procedure may be more difficult. Also, some shrub habitats may be too dense or thorny to move through with a sweep net. Alternatively, shrubs can be covered with cheesecloth and sprayed with anesthetic, the fallen invertebrates then being collected in trays of alcohol on the ground. Unfortunately, not all small animals release their grasp under such treatment.

Another method for capturing tree and shrub invertebrates involves the spreading of a white cloth, large trays, or pans beneath the vegetation to be sampled and beating the branches or stems with poles. Smaller trees can also be shaken vigorously by a few sturdy students (holding the collecting cloth above the heads of the shakers). Alcohol-filled pans are useful in killing captured animals. Surprisingly, most of the animals (insects and spiders) will drop to the ground, even insects capable of flight. However, many diptera (flies, mosquitoes, and gnats) and hymenoptera (bees and wasps) do fly away and thus are missed by this procedure.

7. Suggested exercises

1. Using some of the techniques given above, sample invertebrates in different forest, grassland, and/or field communities. Compare the relative abundance and diversity (sections 5A and 5B) in the different areas.
2. Compare the densities and taxonomic compositions of invertebrates found in different strata in a grassland or forest community. Relate the different forms to their functions or niches in the community. Propose as well as possible a food web for the community studied.
3. Compare the efficiency and accuracy of sweep net sampling with the tent or cylinder method described above.
4. Using some of the techniques given above, sample a chosen community at different times or for various weather conditions. Evaluate the results and relate them to possible sampling biases and to the community structure.
5. Examine the reliability of your sampling by means of species-sample size curves and performance curves (section 1A.3) using any of the sampling techniques given above.

8. Selected references

Andrews, W. 1972. A guide to the study of soil ecology. Prentice-Hall, Englewood Cliffs, N.J.

Carpenter, J. R. 1936. Quantitative community studies of land animals. J. Animal Ecol. 5:231–245.

Davis, D. E., and R. L. Winstead. 1980. Estimating the numbers of wildlife populations, 221–245. In S. D. Schemnitz (ed.), Wildlife management techniques manual. Wildlife Society, Washington, D.C.

Flowerdew, J. P. 1976. Ecological methods. Mamm. Rev. 6:123–160.

Jackson, R. M. 1966. Life in the soil. Arnold, London.

Morris, R. F. 1960. Sampling insect populations. Annu. Rev. Entomol. 5:243–264.

Phillipson, J. 1971. Methods of study in quantitative soil ecology. IBP handbook no. 18. Blackwell Scientific Publishers, Oxford, England.

Shelford, V. E. 1951. Fluctuation of forest animal populations in east-central Illinois. Ecol. Monogr. 21:183–214.

Southwood, T. R. E. 1978. Ecological methods 2d ed. Chapman and Hall, London.

9. Terrestrial and aquatic animal identification manuals
(Applicable to sections 3D through 3H.)

Arnett, R. H., Jr., N. M. Downie, and H. E. Jaques. 1980. How to know the beetles. 2d ed. Wm. C. Brown Co. Publishers, Dubuque, Iowa.

Audubon Society Staff. 1983. The Audubon Society field guide to North American fishes, whales, and dolphins. Alfred A. Knopf, New York.

Audubon Society Staff and F. W. King. 1979. The Audubon Society field guide to North American reptiles and amphibians. Alfred A. Knopf, New York.

Audubon Society Staff and L. Milne. 1980. The Audubon Society field guide to North Amerian insects and spiders. Alfred A. Knopf, New York.

Audubon Society Staff and R. M. Pyle. 1981. The Audubon Society field guide to North American butterflies. Alfred A. Knopf, New York.

Audubon Society Staff and M. D. Udvardy. 1977. The Audubon Society field guide to North American birds: Western region. Alfred A. Knopf, New York.

Audubon Society Staff and J. O. Whitaker, Jr. 1980. The Audubon Society field guide to North American mammals. Alfred A. Knopf, New York.

Ballinger, R. E., and J. D. Lynch. 1983. How to know amphibians and reptiles. Wm. C. Brown Co., Dubuque, Iowa.

Blair, W. F., A. P. Blair, P. Brodkorb, F. R. Cagle, and G. A. Moore. 1968. Vertebrates of the United States. McGraw-Hill Book Co., New York.

Bland, R. G., and H. E. Jaques. 1978. How to know the insects. Wm. C. Brown Co. Publishers, Dubuque, Iowa.

Booth, E. S. 1982. How to know the mammals. Wm. C. Brown Co. Publishers, Dubuque, Iowa.

Borror, D. J., D. M. DeLong, and C. A. Triplehorn. 1981. An introduction to the study of insects. Saunders College Publishing, Philadelphia.

Borror, D. J., and R. E. White. 1970. Field guide to the insects of America north of Mexico. 5th ed. Houghton Mifflin Co., Boston.

Burch, J. B. 1962. How to know the eastern land snails. Wm. C. Brown Co. Publishers, Dubuque, Iowa.

Burt, W. H. 1972. Mammals of the Great Lakes region. University of Michigan Press, Ann Arbor.

Burt, W. H. 1976. A field guide to the mammals. 3d ed. Houghton Mifflin Co., Boston.

Chu, H. F. 1949. How to know the immature insects. Wm. C. Brown Co. Publishers, Dubuque, Iowa.

Collins, H. H., Jr. (assembler). 1981. Complete field guide to American wildlife. (eastern ed. and western ed.) Harper & Row, New York.

Conant, R. 1975. A field guide to the reptiles and amphibians of eastern and central North America. 2d ed. Houghton Mifflin Co., Boston.

Covell, C. V., Jr. 1984. Field guide to the moths of eastern North America. Houghton Mifflin Co., Boston.

Eddy, S., A. C. Hodson, J. C. Underhill, W. D. Schmid, and D. E. Gilbertson. 1982. Taxonomic keys to the common animals of the north central states, exclusive of the parasitic worms, terrestrial insects, and birds. Burgess Publishing Co., Minneapolis.

Eddy, S., and J. C. Underhill. 1978. How to know the freshwater fishes. Wm. C. Brown Co. Publishers, Dubuque, Iowa.

Edmondson, W. T. (ed.). 1959. Freshwater biology. John Wiley & Sons, New York.

Ehrlich, P. R. and A. E. Ehrlich. 1961. How to know the butterflies. Wm. C. Brown Company Publishers, Dubuque, Iowa.

Farrand, J., Jr. (ed.). 1983. The Audubon Society master guide to birding. Volume 1, Loons to sandpipers; Volume 2, Gulls to dippers; Volume 3, Old World warblers to Old World sparrows. Alfred A. Knopf, New York.

Fitzpatrick, J. F., Jr. 1982. How to know the freshwater crustacea. Wm. C. Brown Co. Publishers, Dubuque, Iowa.

Hall, E. P. 1981. Mammals of North America. Ronald Press Co., New York.

Hamilton W. J., Jr., and J. O. Whitaker, Jr. 1979. Mammals of the eastern United States. 2d ed. Cornell University Press, Ithaca, N.Y.

Harrison, C. 1978. A field guide to the nests, eggs, and nestlings of North American birds. Collins, New York.

Hubbs, C., and K. F. Lagler. 1964. Fishes of the Great Lakes region. University of Michigan Press, Ann Arbor, Mich.

Jahn, T. L., E. C. Bovee, and F. F. Jahn. 1979. How to know the Protozoa. 2d ed. Wm. C. Brown Co. Publishers, Dubuque, Iowa.

Kaston, B. J. 1978. How to know the spiders. Wm. C. Brown Co. Publishers, Dubuque, Iowa.

Klots, A. B. 1977. Field guide to the butterflies of North America, east of the Great Plains. Houghton Mifflin Co., Boston.

Kozloff, E. N. 1974. Seashore life of Puget Sound, the Strait of Georgia, and the San Juan Archipelago. University of Washington Press, Seattle.

Kudo, R. R. 1977. Protozoology. 5th ed. Charles C. Thomas Co., Springfield, Ill.

Lehmkuhl, D. M. 1979. How to know the aquatic insects. Wm. C. Brown Co. Publishers, Dubuque, Iowa.

Levi, H. W., L. R. Levi, and H. S. Zim. 1969. A guide to spiders and their kin. Golden Press, New York.

McCaffery, W. P. 1981. Aquatic entomology. The fishermen's and ecologists' illustrated guide to insects and their relatives. Science Books International, Boston.

McDaniel, B. 1979. How to know the mites and ticks. Wm. C. Brown Co. Publishers, Dubuque, Iowa.

Merritt, R. W., and K. W. Cummins (eds.). 1984. An introduction to the aquatic insects of North America. 2d ed. Kendall/Hunt Publishing Company, Dubuque, Iowa.

Morris, P. A. 1973. Field guide to Atlantic coast shells. Houghton Mifflin Co., Boston.

Murie, O. J. 1975. A field guide to animal tracks. Houghton Mifflin Co., Boston.

Pennak, R. 1978. Freshwater invertebrates of the United States. 2d ed. John Wiley & Sons, New York.

Perlmutter, A. 1961. Guide to marine fishes. New York University Press, New York.

Peterson, R. T. 1972. A field guide to western birds. Houghton Mifflin Co., Boston.

Peterson, R. T. 1988. A field guide to eastern birds. A field guide to birds east of the Rockies. Houghton Mifflin, Boston.

Ransom, J. E. (assembler). 1981. Complete field guide to North American wildlife: Western edition. Harper & Row Publishers, New York.

Reid, G. K. 1967. Pond life. Western Publishing Co., New York.

Robbins, C. S., B. Brunn, and H. S. Zim. 1983. Birds of North America. 2d ed. Golden Press, New York.

Scott, W. B., and E. J. Crossman. 1973. Freshwater fishes of Canada. Bull. Fish. Res. Bd. Can. 184.

Slater, J. A., and R. M. Baranowski. 1978. How to know the true bugs (Hemiptera: Heteroptera). Wm. C. Brown Co. Publishers, Dubuque, Iowa.

Smith, DeB. L. 1977. A guide to the marine coastal plankton and marine invertebrate larvae. Kendall/Hunt Publishing Company, Dubuque, Iowa.

Smith, H. M. 1978. A guide to field identification of amphibians of North America. Golden Press, New York.

Stebbins, R. C. 1966. A field guide to western reptiles and amphibians. 2d ed. Houghton Mifflin Co., Boston.

Swain, R. B. 1948. The insect guide. Doubleday and Co., Garden City, N.Y.

Tarjan, A. C., R. P. Esser, and S. L. Chang. 1977. An illustrated key to nematodes found in fresh water J. Water Pollut. Contr. Fed. 49:2318–2337.

3e

aquatic sampling

1. Introduction

Aquatic habitats are markedly stratified. Organisms at the water's surface are called *neuston;* those at or in the bottom substrate are the *benthos;* and those in between are either *nekton* or *plankton,* depending on whether they are strong swimmers or relatively free floating, respectively. Organisms on submerged substrates above the bottom are called *periphyton* or *aufwuchs,* though some ecologists unfortunately define these terms differently.

Techniques and equipment for quantitatively collecting aquatic organisms depend on the stratum. In aquatic environments, accurate estimates of absolute density are difficult to obtain, and variation in size, habits, and activity contribute to bias in such measures. One commonly estimates relative density, or a density index based on a unit of habitat or unit of sampling effort (see the introduction to unit 3). The techniques given below may be adapted to the plot or transect sampling considerations of sections 3A and 3B.

Most of the sampling techniques discussed here are examples of *grab sampling.* In grab sampling one removes a small portion of the habitat (substrate or water) at one time. On the other hand, traps, seines, plate samplers, and townets sample the organisms without removing any of the habitat. Grab sampling collects a measurable quantity of the habitat, thus enabling us to calculate absolute densities. However, without highly specialized equipment this type of sampling is limited to shorelines, shallow water, or the uppermost portions of a body of water. Also, grab samples cannot always be easily quantified. This is particularly true for many types of benthic habitats, because uniform areas or volumes of substrate are difficult to collect. (For example, one square meter of habitat sampled may have five square meters of inhabitable surface area of rocks or gravel, allowing only the estimation of absolute density, rather than ecological density.) Also, grab samples do not always collect every desired organism. The process usually misses fast-moving organisms such as crayfish and fish. Delicate organisms may be destroyed by the sampling process, and individuals may be missed during sorting. In addition, sample preparation of grab samples is often time-consuming.

2. Benthos sampling

2.1. Wilding Sampler An ideal benthos collection device—the Wilding, or stovepipe, sampler—consists of a topless and bottomless cylinder with a known cross-sectional area (0.1 to 0.2 square meters is common, and a sturdy metal garbage can with the bottom cut off will work well). Furthermore, it may also be used for sampling plankton and slow-moving nekton. Push the sampler into the substrate with the bottom of the cylinder firmly planted and the open top above water. Carefully remove nekton and plankton using a large aquarium dipnet or by carefully dipping out the water within the cylinder. If you also wish to analyze the nekton and plankton components of the community, pour this water through a set of sieves and save the accumulated material. After removing most of the water, scoop out the top 10 centimeters of substrate and place into a bucket. Separate the substrate from the organisms by washing it through a series of graded screens and sorting out the specimens using fine-pointed forceps. The saline flotation method, described in section 3E.4.2, is also useful.

This type of sampling is a form of plot sampling (see section 3A). While sampling the benthos in this manner is fairly accurate in shallow ponds and lake littoral zones, it is impractical in deep water and on rocky substrates; it is also time-consuming.

Density of each benthic species is expressed as the number of individuals per square meter. This is calculated by dividing the number of organisms in the sample by the area of the sampler (equation 1 in section 3A). Express plankton or nekton density as numbers per liter or per cubic meter.

2.2. Scoop Net Long-handled nets are useful in scooping up soft substrate, but they may be unsuitable for rocky-bottomed areas. Instead, nets mounted on a sturdy triangular frame are easier to manipulate. Place the collected material in a sieve or a series of graded sieves to wash out the silt, mud, and sand particles, leaving behind the vegetation, large detritus, and animal life.

Quantifying net catches is difficult. However, a careful collector can express the catch as numbers per unit of substrate by measuring the weight or the volume of the substrate actually scooped. If only an index of abundance is desired, then express the data as numbers per scoop as long as you carefully obtain the same size scoop each time and avoid scooping the same site more than once. Standardization involves holding the net at arm's length and pulling it along the substrate toward yourself. The sampling is analogous to taking short belt transects (see section 3B). The area sampled can be estimated by

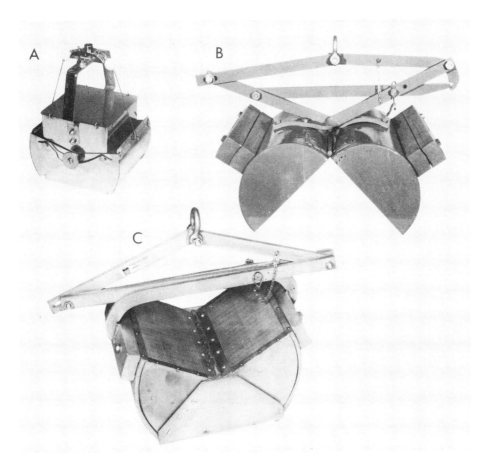

Figure 3E.1. Common benthic grab samplers are the (A) Ekman, (B) Peterson, and (C) Ponar devices, which are lowered to the bottom with their jaws locked agape. The jaws of the Ekman are unlocked by dropping a "messenger" (a weight) down the line after the sampler has been lowered to the bottom; a spring then closes them as the sampler is raised. The jaws of the other two devices are unlocked when the sampler strikes the bottom, and the weight of the jaws causes them to close as the sampler is raised. (Photographs courtesy of the Wildlife Supply Company, Saginaw, Michigan.)

multiplying the width of the net opening by the length of the sample taken. The density then is the number of individuals per square meter.

Realize that a scoop net is likely to miss capturing many organisms present in an area and may collect some organisms swept from upstream. For accuracy, determine the number of scoops necessary to collect the number of animals that would have been collected in 0.1 m² using the more effective cylinder sampler discussed above. This calibration is similar to that applicable to the terrestrial sweep-net technique (section 3D.6.1).

2.3. Dredges and Other Benthic Grab Samplers   Dredges of various sorts may be dragged along the bottom, scraping the surface or digging into the substrate (depending on their design). By knowing the width of the dredge opening and the distance dragged, you can express the resultant catch as numbers per unit volume or unit area of sub-

strate. Dredges are typically inefficient collecting devices, however, as they often skip over the bottom surface and are likely to damage many of the invertebrates present. As a result, this method of collection generally results in an underestimate of invertebrate abundance.

Several types of bottom grab samplers are in use, some of which are unfortunately called dredges, an inaccurate term because these devices are not dragged along the bottom, as are true dredges. Grab samplers are lowered from a boat or a bridge, and, if heavy, are winch-raised to the surface. The grab sampling device generally consists of an arrangement of jaws designed to remain open until the sampler reaches the bottom, whereupon they snap shut, scooping up the substrate as the sampler is raised to the water surface. The animals may be recovered from the trapped material by means of a sieve or series of sieves. Some commonly used bottom samplers are shown in figure 3E.1. Common sizes of the lightweight Ekman sampler opening are from 15 to 30 cm, collecting from a few

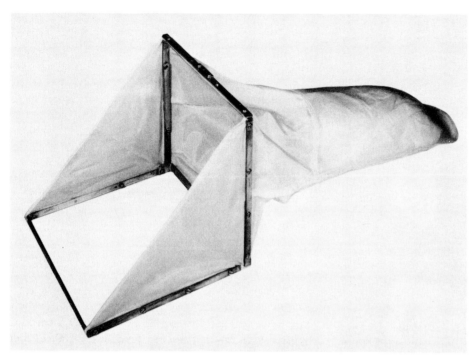

Figure 3E.2. The Surber net is designed for stream benthic sampling. (Photograph courtesy of the Wildlife Supply Company, Saginaw, Michigan.)

Figure 3E.3. A Hester-Dendy type sampler provides artificial substrate to which benthos and periphyton attach. (Photograph courtesy of the Wildlife Supply Company, Saginaw, Michigan.)

hundred cm² of substrate; the heavier Ponar device (20–30 kg) grabs substrate from an area of about 530 cm²; and the Peterson grab sampler, which weighs about 20–40 kg, grabs about 930 cm². The choice of device depends to a large extent on the hardness of the substrate, for heavy grab samplers (such as the Peterson) may be used in rocky or clayey as well as in muddy or sandy habitats. Because the area described by the open jaws is known, the data can be expressed as numbers per unit area. However, as the volume of the sample is a variable difficult to control, the number per unit area can be a biased estimate of numbers per unit volume. Direct measurement of the weight or volume of the sample may be a better measure of the amount of habitat sampled.

2.4. Stream Bottom Samplers   In shallow streams the Surber net (sometimes called a Surber swift-water net) is very practical. This device consists of a square metal frame describing a known area (typically 0.1 m²) (figure 3E.2). This frame is placed on the stream bottom such that a net trails out from it, swept downstream by the current. Rocks and gravel are then dislodged by hand, and the substrate is otherwise manually disturbed within the frame. The animals within this area are swept into the trailing net by the current. Rocks (particularly their undersurfaces) must be examined very closely, for many invertebrates cling to them. Do not put the rocks and gravel in the net as they may crush the animals or damage the net. However, rocks and gravel may be placed into a pan and examined for organisms later on shore, or in the laboratory using the flotation method in section 3E.4.2. The data thus collected may be reported as numbers per unit area. This method is the aquatic analogue of the terrestrial invertebrate collecting technique using a frame to outline a plot of known area (see sections 3D.2 and 3A).

Some bias may result from collection of drift from upstream, particularly if the substrate has been disturbed by other members of the collecting party. Remember also that more conspicuous animals are often collected in favor of the smaller and more sessile species, and by ignoring those firmly attached to rocks and gravel you will underestimate population density and species diversity. Also, a non-uniform depth sample will result in additional error in sampling.

The Hess bottom sampler is a hybrid between the Surber sampler and the Wilding sampler. It consists of a sturdy metal cylinder that can be pushed into the substrate, with a net extending from the side of the cylinder and ending in a collecting jar. If the substrate is too hard to be penetrated by the sampler, a foam rubber ring may be placed on the bottom of the cylinder to prevent organisms from escaping.

2.5. Artificial Substrates   Another method of benthos sampling makes use of artificial surfaces onto which the benthic fauna attach themselves (figure 3E.3). Such sampling does not permit estimation of the actual densities on the natural substrate, but it is still very useful and efficient for comparative ecological studies. Artificial substrates may be glass microscope slides for algae and periphyton, or larger plates of Plexiglas or board for benthic invertebrates. Wire mesh baskets containing rocks or concrete balls of known surface area also have been used. The use of artificial substrates is analogous to drop-board sampling of cryptozoans in a terrestrial habitat (section 3D.4). As with drop-boards, artificial substrates favor the attraction and colonization of certain species and discourage others. Therefore, the naturally-occurring benthic subcommunity is not actually represented, but only those organisms that favor the offered substrate conditions.

The Hester-Dendy-type sampler (Hester and Dendy, 1962; Mathers and Martin, 1969) is a convenient invertebrate plate sampler. It may be constructed inexpensively as a series of separated hardboard plates (figure 3E.3), which may be suspended in the water or bolted to a heavy block of cement or steel plate. Figure 3E.3A shows a sampler with circular plates, each about 7.5 cm in diameter; each square plate in figure 3E.3B is about 8 cm on a side. The total surface area of each sampler is about 0.12 m². The plate sampler is placed in the water for a period of time (three to five weeks works well) to permit colonization by aquatic organisms. Afterward, place a can over the sampler and invert the sampler within the can. Then remove the can with the sampler from the water, thus preventing loss of the invertebrates to the current as the sampler is taken from the habitat.

Factors that affect results from artificial substrate sampling include siltation of the substrate, variation in physical factors (e.g., current and temperature) in the stream between different samplers, and differences in the depth at which the samplers are placed. For monitoring water quality these samplers provide a standard replication of samples that is difficult to find in nature. Thus for conducting controlled field experiments these samplers have important advantages despite their inherent biases toward certain species.

3. Nekton and plankton sampling

3.1. Traps and Pumps   Plankton traps are commonly used to collect algae and zooplankton. With such traps sampling can be performed at specifically desired depths. This type of grab sampling is essentially a three-dimensional plot whose size is the known volume of water. This technique can also be used to standardize other sampling methods, such as townets. Unfortunately, the sample must be concentrated so enough organisms can be examined; this is a potential source of error. Often a large number of samples must be taken and pooled when the density of the organisms is very low. Though tedious and time-consuming, this type of sampling is considered by many aquatic ecologists to be the most precise and least biased method for sampling plankton. Even so, some error stems from inhibition of flow through the sampler and dispersion of organisms by the sampler as it moves through the water.

The Kemmerer sampler (Figure 2D.1) may be lowered to a desired depth. Then, a metal weight (called a *messenger*) is dropped along the supporting line to close the sampling chamber, thus enclosing a known volume of water to be brought to the surface to be filtered through plankton netting. The Schindler-Patalas trap consists of a cube-shaped chamber (12 and 30 liters are common sizes), with top and bottom doors that open as the trap is lowered through the water. When the device is raised, the doors close tightly and the chamber's contents empty into an attached plankton net. Plankton sampling may also employ pumps that pass large volumes of water through filtering sieves. However, these methods typically are inconvenient for class studies and will not be discussed here.

3.2. Dip Net   Dip nets generally have a circular opening and are fastened to a long pole. Limited to sampling shallow waters, this type of net is swept through the water either from the shore or while the collector is wading. As in the benthic sampling discussed above, this method is difficult to quantify. The speed of the net movement and the distance sampled should be kept constant. The data may be expressed as the number per sweep, or standardized as in the bottom sampling discussed above. However, if the diameter of the net and the distance of its movement are known, then absolute density may be estimated. The distance the net is swept multiplied by the area of the net opening approximates the volume of water sampled. Problems of standardization against a Wilding or Kemmerer sampler are the same as those discussed for benthic scoop nets.

3.3. Townet   These nets consist of a mesh bag on a metal frame fastened to a towline (see figure 3E.4). A collecting chamber, often simply a jar or vial, is at the end of the net. The mesh size of the net determines the size of plankton collected.

Townets may be cast from shore and pulled in, towed behind a boat, or lowered from a boat or bridge and pulled upward through the water. If the rate of towing is kept constant and slow during each sampling period, the density of organisms may be expressed as the number per cubic meter of water. The distance towed multiplied by the area of the net opening approximates the volume of water sampled. However, because of water turbulence, this calculation generally overestimates the actual volume sampled. If the tow is too fast, then water turbulence at the net opening results in less water flowing through the net. Too lengthy a towing period will clog the net with organisms and silt, thus reducing the actual volume of water sampled. On the other hand, if towing is too slow, many organisms will escape capture. Also, the depth of a townet sample is difficult to control, for at different speeds of tow

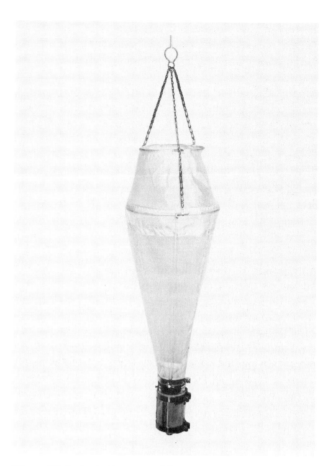

Figure 3E.4. A plankton net with collecting chamber. Many designs of such nets are available; that pictured is the Wisconsin-style plankton net. (Photograph courtesy of the Wildlife Supply Company, Saginaw, Michigan.)

the net may be at different depths. To correct for these biases, standardize the technique against a grab-type water sampler.

3.4. Seine   Seines are large nets that can be either hauled through the water along the bottom of a stream or littoral zone of a lake or pond, or held stationary while fish or crayfish are driven into them. Bag seines are generally more efficient than standard minnow seines but are harder to haul through the water. Absolute densities of samples obtained by seining are difficult to quantify and standardize and are subject to many biases. For example, stronger swimmers escape capture easily. You need a good deal of skill and care to obtain an adequate sample of fish or crayfish by this method. Two procedures are described here.

In shallow, swift streams stretch a 6-m (20-ft) bag seine across the stream. A person at each end of the net holds it upright so that the current fills out the bag. Be sure the weighted bottom of the seine is on the bottom of the stream. Additional persons form a line about 20 to 30 meters upstream and move noisily downstream, kicking up the bottom, splashing the water, and forcing the fish downstream into the net. Immediately on reaching the net, the splashers rapidly raise the weighted edge out of the water so that no fish escape. The net and captured fish are then carried ashore; the catch placed in buckets, sorted, and counted. The fish commonly are also weighed and their body lengths measured. The sample may be preserved in 10% formalin for later identification and enumeration in the laboratory. Scales of a particular species may also be collected for age analysis (section 4A). The area sampled is approximated by the distance the group traveled multiplied by the distance between the two ends of the seine.

A second method of sampling involves two people hauling the seine through the water. This technique takes both muscle and practice so that both haulers move rapidly and at the same speed while keeping a constant distance apart. Mark off a distance, perhaps 20 to 30 meters, to drag the seine. In a stream, move upstream against the current. Long hauls are not recommended because fish are lost as the net operators tire and slow down. When the end of the haul is near, pull the seine rapidly toward shore, where its weighted edge can be raised out of the water (preferably with the aid of other persons). Always drag the weighted bottom edge of the net along the bottom to prevent fish from swimming under it. The area sampled is computed as the length of the haul times the distance between the ends of the seine during the sampling.

4. Sorting and preparation of samples

Once a sample is taken, separate the organisms from the water, sediment, litter, or vegetation, and when necessary, preserve for sorting, identification, and tabulation at a later date. This is often the most tedious step in the collection of ecological data and may introduce serious bias if not done conscientiously and carefully. The type of separation used depends on the stratum and type of organisms sampled, the time available, and the accuracy needed.

4.1. Sieves One of the standard items of equipment for sorting aquatic samples is a series of graded screens for washing and separating. The screens are stacked atop each other, with the coarsest mesh on the top and the finest on the bottom. Place the sample on top of the screens and pour large volumes of clean water through the screens. If pond water is used, first pour it through a fine mesh screen to remove organisms that would bias the collections. After washing, inspect carefully the large pieces of debris left behind and remove all organisms by using a pair of fine-pointed forceps. Inspect to see that each screen is cleaned of all organisms. Set upper and lower size limits on the animals to be studied by examining material from only the desired screens in the series.

4.2. Flotation This procedure is especially useful for separating benthic animals from their substrate; it involves subjecting both to a solution having a specific gravity between that of the organisms and that of the sediments. Place a fresh substrate sample in a tray of water containing a high concentration of salt or sugar. Concentrations (by weight) necessary to achieve solutions within the commonly used specific gravity range of 1.10 to 1.20 are as follows: sodium chloride, 14–26%; or calcium chloride, 12–22%; or magnesium sulphate, 10–18%; or sucrose, 24–44%. The organisms (and organic detritus) will float to the surface. Agitate the material gently by hand to free the animals from the substrate; adding a few drops of 10% chloral hydrate solution per liter of salt or sugar solution helps relax the organisms' grasp. Then scoop up the specimens with a fine-mesh aquarium dip net and save. Rocks and debris should be examined and carefully cleaned of any animals left behind. Unfortunately, members of some species may be distorted by the osmotic effects of salt solutions, a situation that may cause difficulty in taxonomic identification. The flotation procedure works better if the organisms are not subjected to preservative (formalin or alcohol) before placing them in the flotation solution (Flannagan, 1973).

4.3. Settling This technique is useful for concentrating plankton. After obtaining a water sample, add a few drops of 10% chloral hydrate solution per liter of sample to kill and relax the animals. Then add 100 ml of undiluted formalin as a preservative. (Addition of formaldehyde prior to the chloral hydrate will often cause the animals to become contracted or distorted.) Adding 5 drops of Biebrich Scarlet-Eosin B stain will render the animals more easily identified and counted.[1] If the sample contains debris or large invertebrates, remove them by passing the water through a coarse screen.

Gently mix the water sample and place 0.5–1 liter of it in a glass cylinder. If algae are to be examined, add 5–10 drops of Lugol's iodine solution to the sample and stir gently.[2] After 24 hours, carefully siphon or decant off 90% of the clear water, leaving in the cylinder the settled cloudy material containing the plankton; this gives a concentration factor of 10. Microscopically examine a portion of this concentrated sample.

4.4. Preservation of Samples After animals are separated from the water, vegetation, or substrate, they can be preserved until later analysis. Place the organisms from

1. A combination of Biebrich Scarlet and Eosin B makes a stain that colors animal tissue but not algae (Williams, 1974). Prepare a solution of each of the two stains by dissolving 1 gram in 1 liter of water; then combine equal amounts of each of the two solutions and add to the water sample to stain the zooplankton.
2. Lugol's iodine solution darkly stains starch in algal cells. Prepare by dissolving 1 gram iodine and 2 grams potassium iodide in 300 ml water. Killing, fixing, and staining may be done simultaneously by using a solution of 10 grams iodine, 20 grams potassium iodide, and 20 milliliters glacial acetic acid in 200 ml water.

110   Biotic Sampling Methods

each sample in a separate jar or vial with a preservative such as 70% ethyl or 40% isopropyl alcohol or 5% formalin.[3] Alcohol is less irritating to use than formalin but should be used only for short-term storage unless the animals are fixed first in formalin. For zooplankton, add 5 to 10 drops of Biebrich Scarlet-Eosin B solution (see section 4.3) to stain the animals so that they are more easily identified and enumerated. For algae, add 5 to 10 drops of Lugol's iodine solution (see section 4.3).

Place a label in each jar or vial, and seal tightly. Each label should contain the date, location, sample number, type of sample, and your name. Waterproof black ink is recommended for labeling, but a pencil will do for a temporary label. Do not use ordinary fountain pen or ball point pen inks, as these dissolve or fade away. Use a high quality bond paper for the label since standard notebook paper easily disintegrates.

4.5. Analysis of Samples   For macroinvertebrates, sort one sample at a time in a dish or tray. Separate the individuals into groups of similar forms. This does not require any particular taxonomic experience; simply place different-looking organisms into different groups. After forming the groups, check them to make sure that all organisms within each group look alike. Then, with the aid of a taxonomic key and/or a reference collection, identify each group. Some taxa may be keyed to an order while others may be keyed to a family or lower taxonomic category. Some useful identification guides are in sections 3D.9 and 3E.6. Count the number of individuals in each of the identified taxa and record the data. For community studies, the format of data sheet 3A.3 offers a useful and convenient data summary procedure. Save the organisms and pool them with other samples from the same habitat if determination of biomass (section 6A) is needed.

For plankton analysis, take subsamples of the prepared sample for microscopic examination. Exercise care in taking subsamples so that they represent the entire sample. A Sedgwick-Rafter counting cell is commonly used. This device is 50 mm long and 1 mm deep, and holds 1 milliliter of the water sample. Take care in filling the cell to avoid air bubbles. Because the cell is 1 mm deep, counting is difficult under high-power magnification; many organisms not in focus will be missed. Examine strips of the counting cell under the microscope instead of laboriously counting the contents of the entire cell. This procedure is a transect sample of a 1 ml subsample from a sample collected and prepared from the field. Count all the organisms in two to four such 50-mm strips. The volume sampled (v, in mm³) is:

$$v = (k)(d)(w)(L), \qquad (1)$$

where k is the number of strips, d is the depth of the cell (in mm), w is the width of the microscope field (in mm), and L is the length of the cell (in mm). The width of the microscopic field is calibrated by an ocular micrometer or Whipple grid and a stage micrometer. The number of organisms of a given species in 1 ml is:

$$N = (n)(V/v)(c), \qquad (2)$$

where n is the number of individuals counted for that species, V is the volume of the sample on the slide (1 ml, or 1000 mm³), v is the volume of the strips counted (in mm³), and c is the concentration factor. For example, if a sample was concentrated from 1000 ml to 100 ml, 1 ml of this concentrated sample was counted, the diameter of the microscopic field was found to be 0.6 mm, and 8 organisms of a particular species were counted in 5 strips of the counting cell, then:

$$V = 1 \text{ ml} = 1000 \text{ mm}^3,$$
$$v = (5)(1 \text{ mm})(0.6 \text{ mm})(50 \text{ mm}) = 150 \text{ mm}^3,$$

and,

$$c = 1000/100 = 10.$$

Therefore,

$$N = (8)(1000 \text{ mm}^3/150 \text{ mm}^3)(10)$$
$$= 533,$$

and we can express this as a density, D, of 533/ml.

One may also count plankton using an ordinary microscope slide and coverslip. However, use a very small subsample, generally 0.1 to 0.5 ml. Thus you would need more replicate slide counts than through use of a Sedgwick-Rafter slide. If the entire volume of sample is covered exactly by the coverslip, then estimate the depth of the sample between the coverslip and slide by dividing the area of the coverslip into the volume of the sample.

5. Suggested exercises

1. Using some of the techniques given above, sample animals in different stream and/or pond communities. Compare the relative abundance and diversity (sections 5A and 5B) in the different areas.
2. Compare the densities and taxonomic compositions of invertebrates in different portions of a pond community. Relate the different forms to their functions or niches in the community. Propose, as well as possible, a food web for the community studied.
3. Compare the efficiency and accuracy of scoop net sampling with Wilding cylinder sampling.
4. Using some of the techniques given above, sample a chosen community at different times or for various weather conditions. Evaluate the results and relate them to possible sampling biases and the community structure.
5. Examine the reliability of your sampling by means of species-sample size curves and performance curves (section 1A.3), using any of the sampling techniques given above.

3. Commercially, formalin is available as a 40% aqueous solution of formaldehyde. Therefore, a 5% formalin solution is obtained by combining 7 parts water and 1 part 40% formalin.

6. Selected references

Beak, T. W., T. C. Griffing, and A. G. Appleby. 1972. Use of artificial substrate samplers to assess water pollution, 227–241. In J. C. Cairns, Jr. and K. L. Dickson (eds.), Biological methods for the assessment of water quality. American Society for Testing and Materials, Philadelphia.

Brinkhurst, R. O. 1975. The benthos of lakes. St. Martin's Press, New York.

Cairns, J., Jr. 1982. Artificial substrates. Ann Arbor Science Publishers, Woburn, Mass.

Cummins, K. W. 1962. An evaluation of some techniques for the collection and analysis of benthic samples with special emphasis on lotic waters. Amer. Midland Natur. 67:477–503.

Davis, C. C. 1955. Marine and fresh-water plankton. Michigan State University Press, Lansing, Mich.

Edmondson, W. T. (ed.). 1959. Freshwater biology. John Wiley & Sons, New York.

Edmondson, W. T., and G. G. Winberg (eds.). 1971. A manual on methods for the assessment of secondary productivity in fresh waters. IBP handbook no. 17. Blackwell Scientific Publishers, Oxford, England.

Fassett, N. C. 1957. Manual of aquatic plants. University of Wisconsin Press, Madison.

Flannagan, J. F. 1970. Efficiencies of various grabs and corers in sampling freshwater benthos. J. Fish. Res. Bd. Can. 27:1691–1700.

Flannagan, J. F. 1973. Sorting benthos using flotation media. Technical report no. 354. Fisheries Research Board of Canada, Winnipeg.

George, E. A. 1976. A guide to algal keys (except seaweeds). Brit. Phycol. J. 11:49–55.

Hester, F. E. and J. S. Dendy. 1962. A multiple-plate sampler for aquatic macroinvertebrates. Trans. Amer. Fisheries Soc. 91:420–421.

Hutchinson, G. E. 1967. A treatise on limnology. Volume 2. Introduction to lake biology and the limnoplankton. John Wiley & Sons, New York.

Kudo, R. R. 1977. Protozoology. 5th ed. Charles C. Thomas, Springfield, Ill.

Lind, O. T. 1985. Handbook of common methods in limnology. 2d ed. Kendall/Hunt Publishing Co., Dubuque, Iowa.

Mason, W. T., C. I. Weber, P. A. Lewis, and E. C. Julian. 1973. Factors affecting the performance of basket and multiplate macroinvertebrate samplers. Freshwater Biol. 3:409–436.

Mathers, C. K., and T. Martin. 1969. A new multiple-plate sampler for collecting macroinvertebrates of the stream biocies. Trans. Illinois State Acad. Sci. 62:331–333.

Merritt, R. W., and K. W. Cummings (eds.). 1983. An introduction to the aquatic insects of North America. 2d ed. Kendall/Hunt Publishing Co., Dubuque, Iowa.

Needham, J. G., and P. R. Needham. 1962. A guide to the study of freshwater biology. 5th ed. Holden-Day, San Francisco.

Omori, M., and T. Ikedo. 1984. Methods in marine zooplankton ecology. John Wiley & Sons, New York.

Pennak, R. W. 1978. Freshwater invertebrates of the United States. John Wiley & Sons, New York.

Prescott, G. W. 1978. How to know the freshwater algae. Wm. C. Brown Co. Publishers, Dubuque, Iowa.

Vinyard, W. C. 1979. Diatoms of North America. Mad River Press, Eureka, Calif.

Welch, P. S. 1948. Limnological methods. McGraw-Hill Book Co., New York.

Wetzel, R. G. 1983. Limnology. 2d ed. Saunders College Publishing, Philadelphia.

Wetzel, R. G., and G. E. Likens. 1979. Limnological analysis. W. B. Saunders Co., Philadelphia.

Williams, G. E., III. 1974. New techniques to facilitate hand-picking macrobenthos. Trans. Amer. Microscop. Soc. 93:220–226.

Winberg, G. G. (ed.). 1971. Methods for the estimation of production of aquatic animals. Academic Press, New York.

Woelkerling, W. J., R. R. Kowal, and S. B. Gough. 1976. Sedgwick-Rafter cell counts: A procedural analysis. Hydrobiologia 48:95–107.

7. Animal identification manuals   See section 3D.9.

3f

capture-recapture sampling

1. Introduction

None of the numerous techniques available for estimating the size of animal populations is foolproof, and none can apply equally well to all populations. Sections 3D and 3E introduce some population sampling methods that can yield reasonably good size estimates only for small and relatively immotile animals. This exercise presents a popular method useful for estimating the population size of a single species of highly mobile animal, such as most vertebrates. It is called the *capture-recapture,* or *mark-recapture,* technique. In honor of some early contributors to its development, fishery biologists refer to the basic procedure as Peterson's method, ornithologists and mammalogists call it Lincoln's method, and diplomatic ecologists refer to it as the Lincoln-Peterson method.[1] A large body of literature has grown around capture-recapture methodology (e.g., see the references in section 5), and only a basic introduction is given here.

2. Lincoln-Peterson method

A number of individuals from a population of interest are captured, marked by some identifiable means, and released within a short period of time (e.g., a day). At a later date (perhaps after a week or two), a second sample of individuals is taken from the population. Some of the individuals in this second sample may be identified as being members of the first sample because they were previously marked. Obviously, if the population is large, the marked individuals will have become "diluted" within it, and only very few would be expected to appear in the second sample. But if the population is relatively small, then more previously marked animals would be in the second sample. Indeed, if certain assumptions about sampling and the animals' distribution are correct, then the proportion of

marked individuals in the second sample is the same as that in the entire population, and the total population may be estimated as follows:

Assume the total population size to be estimated contains N individuals. From this population take a sample of M individuals, mark these animals, and return them to the population. At a later time, take a second sample of n individuals from the population; this sample contains R recaptured animals (i.e., individuals captured and marked in the first sampling). Then the population size, N, may be estimated by the following considerations:

$$\frac{M}{N} = \frac{R}{n} ; \tag{1}$$

$$\frac{N}{M} = \frac{n}{R} . \tag{2}$$

Equation 1 says that the proportion of marked individuals in the entire population is equal to the proportion of marked animals in a random sample taken from that population. Equivalently, equation 2 says that the ratio of the total population to the number of animals marked on the first date is equal to the ratio of the number caught on the second date to the number that were recaptured on the second date. By rearrangement of equation 2:

$$N = \frac{Mn}{R} , \tag{3}$$

So equation 3 estimates the population size, N.

The theory behind this method of population size estimation is exemplified by laboratory exercises using inanimate objects, as indicated in the following example: Suppose you take 200 white balls out of a pot having an unknown number of white balls, paint them black, return them to the pot, and mix all balls in the pot thoroughly. If you then take 250 balls from the pot and find 50 of them to be black, then $M = 200$, $n = 250$, $R = 50$, and the unknown total number of balls (N) could be estimated using equation 3:

$$N = Mn/R = (200)(250)/50 = 1000.$$

Note that if someone came along after you replaced the marked balls in the pot, and removed 100 balls *at random,* you would still have the same ratio of white to black balls in the pot, and therefore you would still estimate the original number of balls. This situation is analogous to random mortality or random emigration in a population.

It has been shown, however, that equation 3 overestimates the actual population size in the long run (i.e., it is biased), but this bias can be reduced by computing

$$N = \frac{M(n + 1)}{R + 1} \tag{4}$$

1. Interestingly, Peterson's use of marking and recapturing in the 1890s was not for the purpose of estimating population size, and Lincoln's 1930 reference is preceded by 13 years by a published description of the method by Dahl (LeCren, 1965), so assignment of priority of this technique has been misleading for many decades.

(Bailey, 1951, 1952), a formula that assumes n/N to be a very small fraction (Seber, 1982:61).[2] For the present example,

$$N = (200)(250 + 1)/(50 + 1) = 984.$$

The following assumptions must be met to validly use this capture-recapture procedure:

1. All individuals in the population must have an equal and independent chance of being captured. That is, the two samples taken from the population must be random samples.
2. There is no change in the ratio of marked to unmarked animals. During the time from initial capture to recapture, there must be no significant additions of unmarked animals to the population through births or immigration, and population losses from death and emigration must remove the same proportion of marked and unmarked animals. The estimation procedure will work if mortality or emigration occurs randomly for marked and unmarked animals, for then the ratio n/R will be unaltered. If there are additions of new individuals to the population but no mortality then N will be an estimate of the population size at the time the second sample is taken. If there is both mortality and recruitment, then this method will overestimate the size of the population at the time of either of the two samples.
3. Marked individuals distribute themselves homogeneously with respect to unmarked ones so that unmarked animals have the same opportunity for capture in the second sample as do marked ones. That is, there must be a random distribution of marked individuals throughout the entire population, and marking an animal must not affect the subsequent likelihood of that animal's being recaptured. One must be careful not to alter the catchability of an animal by the acts of capturing and marking it; this can happen if the catching or marking technique causes significant changes in the animal's behavior or vigor.

These assumptions require a good deal of knowledge of the natural history of the species under study. When applying this technique, you should know the following:

1. Reproductive history of the population. Are young being added to the population? Is the catchability of animals changing during the period of measurement due to reproduction-induced changes in behavior?
2. Mortality pattern of the species. Is the population undergoing a decline? Remember that the population may experience mortality without affecting the population estimate, as long as mortality affects the marked and unmarked alike.

3. Effects of marking on the behavior and physiology of the animal. Is the animal's movement or behavior altered? Is the probability of mortality changed?
4. Seasonal patterns of activity and movement. Do not use this method during hibernation or migration seasons.
5. Biases in the capture of the animals. Do different individuals, sexes, or ages avoid capture or become prone to capture? Is the animal highly mobile or relatively sessile?

In addition, of course, one must use a marking technique enduring enough so the marks will last from the time of marking until the time of recapture.

As with all population estimates made from samples, there is an uncertainty caused by the error associated with examining a sample rather than the entire population. A measure of this error that expresses the uncertainty of a capture-recapture population estimate is the *standard error*, computed as:

$$SE = \sqrt{\frac{M^2(n + 1)(n - R)}{(R + 1)^2(R + 2)}}. \tag{5}$$

For our example above,

$$SE = \sqrt{\frac{200^2(250 + 1)(250 - 50)}{(50 + 1)^2(50 + 2)}} = 121.8.$$

The precision with which the capture-recapture technique estimates population size is inversely dependent on the number of marked animals recaptured. Thus, attempt to obtain a reasonably large R, by making n large.

Approximate confidence intervals (see section 1B.2.4) for mark-recapture estimates may be approximated from the standard error. A $1 - \alpha$ confidence interval may be computed as

$$N \pm (t)(SE), \tag{6}$$

where t is Student's t (table 1B.1) for DF $= \infty$ (i.e., a 95% confidence interval calculation would use $t = 1.96$; and one would use $t = 2.58$ for a 99% confidence interval). For our above example, the 95% confidence interval could be calculated as:

$$984 \pm (1.96)(121.8) = 984 \pm 238.$$

Thus we could say with 95% confidence that the true population size is between 746 and 1222 animals.

An approximate test for the difference between two population sizes estimated by capture-recapture is:

$$t = \frac{N_1 - N_2}{\sqrt{(SE)_1^2 + (SE)_2^2}}, \tag{7}$$

where the subscripts 1 and 2 refer to the two populations being compared (see section 1B.3.1 for principles of t-testing). For this test use DF $= \infty$.

Many modifications have been proposed to correct for some of the limitations of the Lincoln-Peterson technique.

2. A modification of this procedure is to keep collecting animals for the second sample until a predetermined number of recaptures are obtained. This calls for different computations (Caughley, 1977:143).

Discussions of a large number of them are in Andrewartha (1961), Caughley (1977), Seber (1982, 1986), and Southwood (1978). These newer methods often use theory and sampling procedures based on multiple markings and/or multiple recaptures. While such procedures may in some circumstances yield more accurate estimates than the Lincoln-Peterson procedure, they typically require more time and effort (and they may have more restrictive assumptions).

Some of these methods are applicable if death, migration, or emigration alters the ratio of marked to unmarked animals.

3. Application

Capture-recapture methods may be applied to a variety of animals. Field sampling may be done with sweep nets for larger and slower arthropods (section 3D.6.1), or dip nets for benthic sampling of aquatic invertebrates (section 3E.2.2). Collect clams and other relatively sessile animals in a small body of water by hand, employing random plots or transects as described in sections 3A and 3B. Capture fish or crayfish with seines (section 3E.3.4). Small mammals such as field mice may be live-trapped using a grid of 100 traps spaced 10 to 20 meters apart. Be sure to use the same sampling effort on all sampling days. In the laboratory, populations such as tenebrionid beetles, goldfish, or other convenient animals, as well as inanimate objects, such as beans, corks, or marbles, may be used to demonstrate the principles of capture-recapture. Inanimate objects may also be subjected to experiments to test the importance of the assumptions given above and to investigate the effect of sample sizes on the standard error of the population estimate.

Identify, mark, record, and release animals as soon after capture as possible. Mark animals with an exoskeleton or shell with rapid-drying weatherproof paint; dyes, or more permanent tags or bands, may be applied to vertebrates. Fish may be marked by clipping off a portion of the dorsal, caudal, or anal fin. Amphibians or mammals may be toe-clipped. Consider the permanence of the marking procedure. Markings on exoskeletons or feathers will be lost if molting occurs between samplings; also, some types of tags can physically wear off.

The proper elapsed time between samplings depends largely on how long it takes for members of the first sample to distribute themselves randomly within the entire population. A week or two should suffice for most animals, but slow-moving forms may take longer.

Data in addition to numbers (e.g., weight, length, age) may also be collected from these samples for use in section 4A or in other studies. To obtain an estimate of density (D), an estimate of the area (A) sampled must be had in addition to the estimate of population size (N):

$$D = N/A. \qquad (8)$$

4. Suggested exercises

1. Vary the technique of sampling and analyze the results for possible sources of error. Find the technique resulting in the least bias favoring certain age classes, sexes, marked, or unmarked animals.
2. Using a laboratory population, systematically violate each of the three assumptions underlying the Lincoln-Peterson method. Examine the data to see the amount of bias generated by failing to satisfy these assumptions.
3. Compare capture-recapture to removal methods (section 3G) in terms of bias and sampling random sampling error (SE).
4. Using capture-recapture techniques, estimate the population density of a species in each of two or more areas displaying different environmental conditions. Evaluate the results.

5. Selected references

Andrewartha, H. G. 1961. An introduction to the study of animal populations. University of Chicago Press, Chicago.

Bailey, N. T. J. 1951. On estimating the size of mobile populations from capture-recapture data. Biometrika 38:293–306.

Bailey, N. T. J. 1952. Improvements in the interpretation of recapture data. J. Animal Ecol. 21:120–127.

Begon, M. 1979. Investigating animal abundance: Capture-recapture for biologists. University Park Press, Baltimore.

Blower, J. G., L. M. Cook, and J. A. Bishop. 1981. Estimating the size of animal populations. George Allen & Unwin, London.

Caughley, G. 1977. Analysis of vertebrate populations. John Wiley & Sons, New York.

Cormack, R. M. 1968. The statistics of capture-recapture. Oceanogr. Mar. Biol. Annu. Rev. 6:455–506.

Cormack, R. M. 1979. Models for capture-recapture, 217–255. In R. M. Cormack, G. P. Patil, and D. S. Robson (eds.), Sampling biological populations. International Co-operative Publishing House, Fairland, Md.

Davis, D. E., and R. L. Winstead. 1980. Estimating the numbers of wildlife populations, 221–245. In S. D. Schemnitz (ed.), Wildlife management techniques manual. Wildlife Society, Washington, D.C.

LeCren, E. D. 1965. A note on the history of mark-recapture population estimates. J. Animal Ecol. 34:453–454.

Nichols, J. D., B. R. Noon, S. L. Stokes, and J. E. Hines. 1981. Remarks on the use of mark-recapture methodology in estimating avian population size, 121–136. In C. J. Ralph and J. M. Scott, Estimating numbers of terrestrial birds. Studies in avian biology no. 6. Cooper Ornithological Society, Allen Press, Lawrence, Kans.

Otis, D. L., K. P. Burnham, G. C. White, and D. R. Anderson. 1978. Statistical inference from capture data on closed animal populations. Wildlife Monogr. 62.

Pielou, E. C. 1974. Population and community ecology. Gordon and Breach, New York.

Seber, G. A. F. 1982. The estimation of animal abundance and related parameters. Macmillan Publishing Co., New York.

Seber, G. A. F. 1986. A review of estimating animal abundance. Biometrics 42:267–292.

Southwood, T. R. E. 1978. Ecological methods. Chapman and Hall, London.

Watt, K. 1967. Ecology and resource management. McGraw-Hill Book Co., New York.

White, G. C., D. R. Anderson, K. P. Burnham, and D. L. Otis. 1982. Capture-recapture and removal methods for sampling closed populations. Los Alamos National Laboratory, Los Alamos, N. Mex.

1. Introduction

As seen in section 3F, capture-recapture methods allow you to estimate population size with a minimum of harm to the population. However, these methods are sometimes difficult to use on natural populations because the underlying assumptions cannot be realized. If these assumptions are difficult to meet, and if the population being studied can be sampled without replacement of captured individuals, so-called removal methods can be used. We shall consider only two of the simplest removal methods to convey some of the basic principles behind this approach. In practice, more elaborate sampling procedures and theory can obtain more precise and reliable results. References in section 8 discuss removal methods of animal sampling more extensively.

2. Theoretical considerations

Removal methods of population estimation involve successive trapping of members of the population. The basis of the methods is the expectation that the number caught and removed from the population at a given time of trapping will be greater than the number caught at a later trapping. That is, as one reduces the population size, the size of the catch will decrease. Necessary assumptions of removal methods are:

1. Each individual in the population has an equal and independent chance of being captured. That is, the sampling must be random.
2. Except for the effects of the trapping, the population is not increasing or decreasing in size.
3. The probability of capturing an individual is the same for each period of sampling.

Violation of any of these assumptions will result in inaccurate population estimates.

The first assumption relates to the collecting technique. Capture should not favor one sex, one age class, or one individual over another. Each sample taken should represent a random collection from the entire population. For example, if some members of the population consistently avoid capture, then the size of the population will be underestimated.

In capture-recapture theory, births or immigration may not occur during the study, although random mortality and emigration are acceptable. When removal methods are used, a population may have births and immigration, but (according to assumption 2 above) they must balance with normal deaths and emigration so that population size remains constant. An increase in the population size between sampling periods would give an overestimate of population size.

According to the third assumption, if the chance of capturing animals in the population changes from the first sampling period to the second sampling period, then the

removal sampling

population estimate will be biased. Therefore, the sampling effort must be the same for each sampling period. If the probability of capture increases between sampling periods, then the population would be overestimated. In capture-recapture sampling, the probability of capture need not be constant; what is required instead is that the ratio of marked to unmarked individuals remains the same.

Trap-shyness or trap-proneness can affect the probability of capture. If trap-prone animals are more likely caught the first sampling period, the probability of their capture will be high. Therefore, during the second sampling period, the probability of capture will be lower, and fewer animals will be caught due to a higher proportion of trap-shy animals. A variety of other factors, such as bait acceptance, weather conditions, and differential activity of ages and sexes affect the probability of capture. Thus one must assure that sampling and environmental conditions are as identical as possible during all sampling periods and that the sampling effort remains constant.

Finally, it should be realized that the error in estimating population size by these methods is smallest when large proportions of the population have been sampled. If only a very small proportion of the total population has been captured, then the confidence in the population size estimate is very low.

3. Regression method

One useful method for graphical estimation of population size, N, is based on successive removals of animals from the population; such procedures have developed since the early part of this century. Presented here is the method of Hayne (1949), which is a modification of Leslie's method (Leslie and Davis, 1939). In this procedure, one obtains a series of collections, capturing animals at different times and removing them from the population. The amount of collecting effort must be the same each time. So, for example, one may tabulate the number of fish caught in an eight-hour period, or the numbers of mammals trapped in a twenty-four-hour period, on each of several dates. The

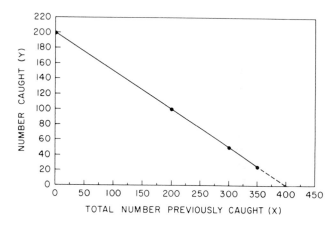

Figure 3G.1. The number of captures each sampling period as a function of the accumulated prior catch. The data are from table 3G.1.

Table 3G.1. *Hypothetical numbers of animals caught in four successive sampling periods. These data are plotted in figure 3G.1.*

Sampling period (i)	1	2	3	4
Number of animals caught				
($Y_i = n_i$)	200	100	50	25
Accumulated prior catch (X_i)	0	200	300	350

numbers of animals caught are then plotted against the total numbers previously caught, as shown in figure 3G.1, for the data of table 3G.1.

In this example, 200 animals were caught, with a given sampling effort, in sample 1; 100 were caught with the same sampling effort in sample 2; and so on. A total of 0 animals were accumulated prior to sampling period 1; 200 animals were accumulated before sample 2 was taken; $200 + 100 = 300$ animals were removed from the population prior to the third sample; and so on. If the probability of capture remains constant, the points on the graph should fall along a straight line. In this example, the probability of capture is 0.50 (50% of the remaining population is removed with each sampling). If this line is extrapolated to the horizontal axis (the dashed line in figure 3G.1), the total accumulated catch of 400 would then be the total original population size. This point of extrapolation represents the theoretical condition of a total census when all animals have been removed and counted.

For the extrapolation to be dependable, however, one must count a very large proportion of the population and obtain enough samples to draw a reliable line through the several data points. The line may be drawn by eye if the points are obviously along a straight line (although the linearity of points in figure 3G.1 is unlikely to be encountered with real data). But in general, a regression line should be calculated as described in section 1B.6.1, in which case the computed slope of the line indicates the average proportion of the population removed with each sampling.

See section 1B.6.1 to determine the statistics a and b in the equation:

$$Y_i = a + bX_i, \qquad (1)$$

where X_i is the accumulated catch at period i, and Y_i is the number caught at period i ($Y_i = n_i$, the number caught in sample i). The slope, b, of the regression line will be a negative value, for Y_i decreases as X_i increases. Once a and b have been calculated by regression analysis, one can

calculate N, which is the value of X_i when $Y_i = 0$. That is, by substituting in equation 1,

$$0 = a + bN, \qquad (2)$$

and it follows that:

$$N = -a/b \qquad (3)$$

(which is equivalent to obtaining N by equation 31 in section 1B). For example, if $a = 200$, and $b = -0.50$, as in figure 3G.1, then,

$$N = -(200)/(-0.50) = 400.$$

A confidence interval for the population size may be determined by equation 32 in section 1B.

4. The Moran-Zippin method

This procedure for estimating population size (Moran, 1951; Zippin, 1956, 1958) requires fewer sampling periods than Hayne's extrapolation method, although the two population estimates should be the same or very similar. The basis of the method is as follows.

Let N be the population size, n_1 be the number of animals caught and removed during the first sampling period, and n_2 be the number caught and removed on the second sampling period. The proportion of the original population captured in the first sampling period is then, n_1/N, and after the first group of n_1 animals is removed from the population, $N - n_1$ animals remain. The proportion of this remaining number of animals captured in the second sample is $n_2/(N - n_1)$. If we can assume that the two trappings caught the same proportion, p, of animals (which is the same as saying that p is the probability of an animal being captured), then,

$$p = n_1/N \qquad (4)$$

and

$$p = n_2/(N - n_1). \qquad (5)$$

Therefore,

$$\frac{n_1}{N} = \frac{n_2}{N - n_1}. \qquad (6)$$

Solving this equation for N, the population size, we find that

$$N = \frac{n_1^2}{n_1 - n_2}. \qquad (7)$$

Applying this equation for estimating N to the first two samples in table 3G.1, we compute:

$$N = (200)^2/(200 - 100) = 40000/100 = 400,$$

the same answer obtained using Hayne's regression method.

The standard error of this population estimate is

$$SE = \frac{(n_1)(n_2)\sqrt{n_1 + n_2}}{(n_1 - n_2)^2} \qquad (8)$$

(Seber, 1982:318). For the above example,

$$SE = \frac{(200)(100)\sqrt{300}}{(200 - 100)^2} = \frac{346410}{10000} = 34.6.$$

An approximate confidence interval for our estimate of N may be calculated as:

$$N \pm (t)(SE), \qquad (9)$$

where t is Student's t (see table 1B.1) for $DF = \infty$. (Thus for a 95% confidence interval, use $t = 1.96$, and for a 99% confidence interval use $t = 2.58$.) Therefore, we may conclude, with 95% confidence, that the size of the population we sampled is:

$$400 \pm (1.96)(34.6) = 400 \pm 68;$$

that is, the true population size is estimated to be between 332 and 468.

Population estimates by the Moran-Zippin method may be based on more than two samples, but the arithmetic becomes quite involved. While using data from three successive removal samples improves accuracy over estimates made using two samples, little is gained by using more than three samples. See Seber (1982), Southwood (1978), and Zippin (1958) for the computations needed.

5. Further considerations

Often one does not want to remove animals from a study area, to prevent affecting behavior or vigor of the population. You may use these removal methods by marking each captured animal and releasing such animals back into the population but not counting any marked animals subsequently caught. If 30 animals are captured, marked, and released during the first trapping, and 20 unmarked and 10 marked individuals are caught during the second, then using Zippin's equation, $N = (30)^2/(30 - 20) = 90$. Using the Lincoln-Peterson method (section 3F.2), we find $N = (30)(30)/10 = 90$. Although the estimates of N are both identical, their standard errors are 42 and 19, respectively. In general, populations that can be sampled by either method are sampled more reliably and efficiently using capture-recapture procedures (Zippin, 1958).

6. Procedure

The following exercise may be done in the laboratory or in the field. Field collecting and sampling techniques as outlined in sections 3D and 3E will depend on the sampled species. In nature, a population of small fish or crayfish can be seined in a small pond on successive days (section 3E.3.4), successive sweep net samples of large insects may be used (section 3D.6.1), or one may snaptrap field mice (section 3D.7). In the laboratory, sample a variety of animals, such as flour beetles, meal worms, or aquarium fish. Having an estimate of the population size, N, and the area, A, sampled, the density of the animals can be expressed as:

$$D = N/A. \qquad (10)$$

In the laboratory, one can set up a variety of sampling schemes, combining plots, belt transects, capture-recapture, and removal sampling. These methods can then be compared with each other. To examine the amount of error and critically evaluate the sampling technique, sampling should be repeated six to ten times, with each repeat being a replicate sample. Standardization of the chosen sampling method (e.g., net or trap) is necessary to permit an exact repeat of the procedure each time a sample is taken. Inconsistent replicate sampling almost always results in a biased population estimate.

The Moran-Zippin procedure described here requires two periods of sampling. The animals captured in the first sampling (n_1) are not returned to the habitat (unless they are marked so that they will not be counted again if recaptured in the second sample). After this first sampling, the remaining members of the population are allowed time to redistribute themselves in the habitat. In some laboratory population studies, this process may be facilitated by gentle but thorough mixing of the habitat medium. The second sample is obtained by the same method as the first; the size of this sample is n_2. Then the population size, N, and its standard error may be computed, and absolute density may be expressed as numbers per unit area or numbers per unit volume.

To obtain a graphical estimate of population size, use four or five sampling periods (three periods at a minimum) to lower the standard error for the population estimate. On graph paper, plot the number of animals caught during each sampling period as a function of the accumulated catch obtained during prior sampling periods (figure 3G.1). Draw a straight line through these points to give the best fit for those points and read the extrapolated estimate of N. If a precise fit to these points is desired, then use linear regression analysis, as described in section 3G.3.

7. Suggested exercises

1. Compare the results of the two removal methods. Which method is likely to have a greater amount of error and why? If possible, obtain a census of the entire population, or obtain an estimate from plot sampling. Compare the results to the removal method estimates.
2. Compare removal sampling with capture-recapture sampling (section 3F). Perform the sample collecting simultaneously for both methods by treating marked animals as removed. Which method is more efficient in terms of both time expenditure and sampling error?
3. Using a laboratory population, systematically violate the assumptions of removal sampling. Examine the data and evaluate the errors generated.
4. Using a removal method of sampling, compare the population density of a given species under two or more times of year or other environmental conditions.

8. Selected references

Blower, J. G., L. M. Cook, and J. A. Bishop. 1981. Estimating the size of animal populations. George Allen & Unwin, London.

Caughley, G. 1977. Analysis of vertebrate populations. John Wiley & Sons, New York.

Davis, D. E., and R. L. Winstead. 1980. Estimating the numbers of wildlife populations, 221–245. In S. D. Schemnitz (ed.), Wildlife management techniques manual. 4th ed. Wildlife Society, Washington, D.C.

Hayne, D. W. 1949. Two methods for estimating animal populations. J. Mammal. 30:399–411.

Leslie, P. H., and D. H. S. Davis. 1939. An attempt to determine the absolute number of rats on a given area. J. Animal Ecol. 8:94–113.

Menhinick. E. F. 1963. Estimation of transect population density in herbaceous vegetation with emphasis on removal sampling. Ecology 44:617–621.

Moran, P. A. P. 1951. A mathematical theory of animal trapping. Biometrika 38:307–311.

Pielou, E. C. 1974. Population and community ecology. Gordon and Breach, New York.

Seber, G. A. F. 1982. The estimation of animal abundance and related parameters. Macmillan Publishing Co., New York.

Seber, G. A. F. 1986. A review of estimating animal abundance. Biometrics 42:267–292.

Smith, R. L. 1980. Ecology and field biology. 3d ed. Harper & Row, New York.

Southwood, T. R. E. 1978. Ecological methods. Chapman and Hall, London.

White, G. C., D. R. Anderson, K. P. Burnham, and D. L. Otis. 1982. Capture-recapture and removal methods for sampling closed populations. Los Alamos National Laboratory, Los Alamos, N. Mex.

Zippin, C. 1956. An evaluation of the removal method of estimating animal populations. Biometrics 12:163–189.

Zippin, C. 1958. The removal method of population estimation. J. Wildlife Manage. 32:325–339.

3h

terrestrial vertebrate sampling

1. Introduction

Many procedures have been used to sample terrestrial vertebrate populations. Two of these, mark-recapture and removal methods, have been presented earlier in sections 3F and 3G, respectively. In the discussions below we provide an introduction to several additional methods, which have particular utility in the sampling of wildlife (i.e., mammal and bird) populations.

As no single procedure is capable of sampling all wildlife species, a variety of direct and indirect methods are used to collect data on populations within a community. Generally it is impractical to conduct an accurate census of most wildlife populations, owing to limitations of cost and personnel, as well as to the concealing coloration and avoidance behavior of many species; therefore, the relative abundance and standardized indices of abundance for a given species are measured more commonly than the absolute density.

Each of the approaches discussed below has serious biases if caution is not exercised when collecting and interpreting the data. But alternative methods such as capture-recapture and removal sampling (sections 3F and 3G) have their own sets of biases, are relatively time-consuming and costly, and often would not improve the acquisition of information needed for most wildlife studies. The use of trapping indices, strip census data, and indirect sign indices may yield informative data with much less effort than attempts to estimate absolute densities. If factors such as the duration of observation, distance traveled, and persons performing the sampling are constant for each sampling period and method used, then standardized, useful data can be obtained for comparative purposes.

Section 3B introduced concepts and procedures involving transect sampling. We shall expand on one method that employs transects, the *strip census,* also known among wildlife ecologists as the *line transect.* Relative abundances of species, or density indices, typically are sufficient to monitor seasonal changes in animal populations and to characterize faunal communities. Many wildlife sampling procedures are called "censuses," but these, like those below, are not true censuses, since all the animals in an area typically are not counted.

Strip census indices are efficient in that they collect a large amount of data quickly and with relatively small effort. Moreover, they are sufficiently accurate for many comparative studies if the precautions discussed below are taken. Data may be standardized by walking a measured distance in a fixed time period (e.g., 1 kilometer in one hour), with abundances expressed as the number of animals of a species seen per kilometer or unit area.

2. Line transects

The *line transect,* or *strip census,* method of population estimation involves counting animals seen by an observer traversing a transect line. It is a particularly useful technique when animals are difficult to see and must be "flushed" to be counted. We shall here describe the procedure where a straight-line transect is marked through an area and all animals flushed or otherwise encountered are counted, and the distance from the observer to each animal or group of animals is measured. This method assumes that there is either a random distribution of individuals over the area sampled or that the transect line is located randomly in the area, that all members (e.g., of both sexes and all ages) are equally likely to be flushed, that the sighting of one animal does not influence sightings of other animals, and that no animal is counted more than once. This procedure can be applied to medium-sized animals (e.g., rabbit, grouse, quail, pheasant) in a variety of habitats. Even animals that ordinarily are concealed can be counted, if they are flushed out by the observer. This method gives fairly reliable information on population trends and provides indices of population density.

More than a dozen methods have been suggested for estimating density from strip censuses. Some utilize the angles the sightings make with transect; others use the perpendicular distances of the flushed animals from the transect; and some employ the radial distance, the distances of the flushed animals from the observer. An early example of the latter type of procedure is the so-called *King method:*

$$D = \frac{10^4 n^2}{2L\Sigma d_i},\qquad (1)$$

where D = the population density (in numbers per hectare)

n = the number of animals sighted

L = the length of the transect (in meters)

d_i = the distance from the observer to the ith animal sighted, measured (in meters) to the point where the animal was at the time it was flushed

10^4 = factor for converting m² to ha

Hayne (1949) suggested the following calculation as an improvement over King's formula, as the latter tends to be biased, underestimating population density:

$$D = \frac{10^4 \Sigma(1/d_i)}{2L}. \tag{2}$$

See Gates (1979), Kovner and Patil (1974), and Seber (1982) for other estimation formulas that have been proposed. Many of these have the potential disadvantage of assuming a particular underlying mathematical distribution, which Hayne's estimation procedure does not.

Strip censuses are known for their ease of performance in some situations, not for their accuracy and precision. Estimates of variance and computation of confidence intervals are difficult, as specific theoretical models or mathematical distributions typically must be assumed; but confidence intervals as an expression of precision of the population estimate may be expressed readily if replicate transects are sampled (see section 2.1).

Strip census counts are made in each study area along marked straight transect lines. A reasonable transect length might be 500 to 1000 m, which may vary according to terrain, vegetation, species sampled, and density of wildlife. Sampling generally is conducted during the early morning and/or evening to include one-half hour before sunrise and one-half hour after sunset. All birds or mammals sighted and flushed (without duplication) are counted along the route. The distance to the animal observed is recorded with an optical range finder or measuring tape after the animal is flushed. Estimating the flushing distance to the nearest 2 meters generally is sufficient. Anderson et al. (1979) recommend that at least forty animals should be sighted, preferably sixty to eighty.

In using strip census data, analyses should include the number of individuals of a given species by habitat type. If possible, individuals sighted should be categorized as juveniles, subadults, or adults. Additional observations should include bird and mammal usage of each habitat sampled, and a complete analysis of the habitat (see sections 2A and 5A).

A number of factors other than mathematical and theoretical limitations restrict the accuracy and precision of strip censuses. Sources of inaccuracy and bias include: (1) skill of observer, (2) conspicuousness of animals, (3) weather conditions, (4) species activity related to time of day or season, (5) duplicate counts of individuals driven ahead by flushing, (6) variation of the screening effect of habitat, and (7) distance from the observed animal (Emlen, 1971). Training in observing animals and their behavior helps compensate for the first two problems. Observations must be made during periods of greatest activity and best weather conditions, for standardization of problems 2 through 4. Biases will be minimized and standardized by using one observer, or equally well-trained observers, for all identification and counting. This is often impractical, however, and one must realize that there will be variability in data collecting among observers. By recording accurate information on habitats, screening efficiency, and weather conditions possible biases related to these factors can be identified and taken into consideration in the interpretation of the data. None of these problems can be completely overcome, however; but by taking these precautions one can obtain quantitative measures of population abundance useful for comparative purposes.

2.1. Replicate Transects   Replicate transects may be employed to assess the precision of the population estimates, but each replicate transect line should be long enough that at least twenty-five animals are flushed along it (Burnham et al., 1980). Separate transects must be used, not simply different segments of the same transect line, or the same transect line may be used at different times. Replicate transects do not have to be of the same length, but they must not intersect. An estimate of population density for k transects is

$$D = \frac{\Sigma L_i D_i}{\Sigma L_i}, \tag{3}$$

where D_i is the density estimate from the ith transect (using equation 1 or 2), L_i is the length of the transect i, and Σ indicates the sum over all k transects. The standard error of that overall estimate, D, is

$$SE = \sqrt{\frac{\Sigma L_i(D_i - D)^2}{\Sigma L_i(k - 1)}}, \tag{4}$$

and a $1 - \alpha$ confidence interval may be computed as

$$D \pm (t)(SE), \tag{5}$$

where t is Student's t (table 1B.1) for $k - 1$ degrees of freedom.

3. Roadside counts

Roadside counts have long been a standard method for observing wildlife population trends (Davis and Winstead, 1980; Howell, 1951; Klein, 1965). The observers travel by motor vehicle along roads and trails, while the sighted number of individuals of the species being "censused" is tallied and related to the number of kilometers traveled. Large areas can be covered quickly and easily by using only two persons and a vehicle. A fairly reliable population index may be obtained with this method, but several factors, such as time of day, weather, condition of roadside cover, availability of food, and season, affect the numbers of animals observed. Many of these problems can be minimized by adjusting the time of the census relative to the animal's activity, and avoiding adverse weather conditions. This technique can be used in deserts, forests, grasslands, wetlands, and agricultural areas, as long as they are adjacent to roads. It has been used to "census" medium-size to large mammals such as rabbits, woodchucks, and deer as well as various species of birds. Although this method provides a reliable index of population changes from year to year or between geographical areas with similar habitats, it is limited for seasonal comparisons because of serious sampling biases caused by changes in food supply and cover.

Sampling should occur over at least three days during different weeks of a month. Observations should be made during the species' peak times of activity (commonly during dawn or dusk). The duration of the observation period and the speed of travel should be standardized and kept as constant as possible. Binoculars should be used for observations and identification, while at least two persons should be used to survey optimally the route traveled. For some nocturnal animals, nighttime observations are made using spotlights. Road kills may also be tallied.

At least three "road transects" in each study area should be established and driven each observation period. A 5-m transect traveled twice during an observation period is a reasonable effort but the length of the survey route may vary according to terrain, habitat, and available roads. Data for roadside counts are expressed as the number of animals seen per kilometer traveled. Observations should be standardized so that each road transect is sampled the same time of day for the same duration (for example, two hours, beginning one hour before sunrise or sunset). If different lengths of time are used for the counts, then the results should be standardized by dividing by the time period of observation and expressing the results as numbers per kilometer per hour.

Road surveys may be modified for raptorial birds and large mammals. The survey can be conducted along the route with several five- to ten-minute observation stops. The observer will identify all animals seen or heard during each stop. This modification incorporates characteristics of point surveys, discussed in section 3H.5.

Road surveys of animal populations offer the advantage of traversing large areas quickly and easily. Their usefulness can be limited, however, by numerous factors, such as the animals' activity, the influence of environmental variables, and the condition of roadside cover (Davis and Winstead, 1980; Hewitt, 1967; Howell, 1951). Problems associated with road surveys are similar to those described for strip censuses, although bias from duplicate observations may be minimized in road surveys by adjusting the length of the route and speed traveled.

4. Point surveys

Observation points can be established along roads, edges of ponds or marshes, atop high ground, or at other locations suitable for viewing the habitat. For a period of fifteen to sixty minutes at each observation point, the observer records all sightings of birds or mammals at that site. An index of abundance of each species can be expressed as the number of animals seen per hour of observation.

This method is most applicable to animals moving through (or over) the site (e.g., migrating birds), because individuals residing at the site might well be counted more than once.

5. Small mammal sampling

Vertebrate collecting usually requires special time-consuming skills, but novices can effectively snap-trap small mammals such as rodents and shrews. Standard household snap traps or Museum Special Snap Traps (Animal Trap Co. of America, Lititz, PA) can be used. Traps may be set out in a grid, with the total area covered by the grid considered a plot (see section 3A), or along a transect (section 3B). If grid sampling is done, then one can use section 3G to estimate the population size by removal methods.

Typically, at least one hundred trap stations are required to obtain an adequate number of small mammals, but if population densities are high, then fewer stations may suffice. Each station represents a point along a transect or in a grid. At each station place at least two or three traps, so that if one trap is filled or sprung, that station has remaining traps to sample the population. The stations are set at constant intervals of 10 to 20 meters, the specific distance between them depending on the type of habitat and probable density of animals. Place a small amount of bait, about the size of a pencil eraser, securely on the bait pedal before setting each trap. A mixture of peanut butter and oatmeal is a very good bait for a variety of mammals. Then set the traps and sprinkle a few pieces of oatmeal near them as an added attractant. A more effective sample of small mammals is possible by prebaiting the area for two nights prior to setting the traps. This involves either baiting the traps but not setting them or randomly scattering bits of the bait throughout the habitat.

Traps should be checked twice daily, early in the morning and just before dark. Reset any sprung traps or traps with bait stolen, and replace any lost traps. Place each caught animal in an individual plastic bag, label and seal it. Freeze specimens if they cannot be examined the same day as collection. A density index may be expressed as the number of animals per functional trap per number of nights sampled (e.g., number per one hundred trap-nights), or per unit area or transect length sampled. A functional trap is one that can or has already caught an animal. Thus subtract from the total number of traps set the number of sprung traps, lost traps, and traps with stolen bait.

Such trapping of mammals may be used for a removal study (section 3G), biomass study (section 6A), or age structure analysis (section 4A). If live-traps are used, then capture-recapture studies may be performed (section 3F). Measurements of weight or length of the animals can be used to construct size classes related to age. The reproductive data for population growth may be determined in some species by examining dissected females to ascertain the number of embryos and placental scars.

6. Pellet counts

Counts of fecal material such as pellets are suited to environments where preservation of pellet groups is optimal, such as semiarid regions (Davis and Winstead, 1980). This technique involves removing all pellet groups from plots, then estimating from subsequent observations on those plots the number of groups per hectare to compare animal use of areas between sampling periods. Ten to fifteen 100-m² plots 7.07 by 14.14 m can be used, although narrower and longer plots have been suggested by Van Etten and Bennett (1965). These plots should be checked once every three to seven days; periods between samplings should not be so long that feces will decompose or be destroyed by weather or insects. A random selection of plots should be located in the study area and the number of pellet groups in each plot tallied and summed.

The number of pellet groups per unit area is, then, an index of density (proportional to the true population density) and is determined as

$$ID = n/A, \qquad (6)$$

where n is the sum of pellet groups counted over all plots and A is the total area sampled (i.e., the sum of the areas of all the plots).

This method may be used in those habitats and climates where pellet groups will be preserved between sampling periods (i.e., dry weather, no snow cover, little or no dung beetle activity). After counting pellets, one must be assured that they will not be counted on successive sampling periods (so they might be removed by the observer if they will not disappear by natural processes). Defecation rates for the species studied are closely estimated if it is desired to convert pellet counts to numbers of animals, and the fecal material must be identifiable. This technique employs a relatively small plot size easily sampled by one observer, but accurate results require several periods of observation.

7. Track counts

Track counts, especially after rain or snow, can be used in identifying species and relative abundances of larger animals in a survey area; they may be used in habitats where tracks are retained under such conditions (e.g., mud flats, dirt roads, soft soil, and snow). Because many mammals, such as raccoon, skunk, opossum, and deer, are nocturnal and secretive, tracks may be the first indication of their presence in an area. Species may be identified with the help of Murie (1975). Use of animal tracks to estimate population size requires determination of the relationship between animal numbers, spatial distribution, and abundance of tracks for the animal being "censused"; the

abundance of the tracks without such quantitative calibration gives an index of density.

After a fresh snow or rain, which can eliminate previous tracks in an area, individuals indicated by recent tracks entering or leaving the study area can be counted. Caution must be exercised in using these data, as one animal may cross a transect or plot several times. Therefore, these counts should be considered an index of activity rather than a measure of abundance. This index may be expressed as the number of individual track sightings observed per kilometer of transect traveled. If plots are laid out, then the density index can be expressed as the number per hectare sampled. Such data should be compiled from several census transects or plots.

8. Suggested exercises

1. Using the flushing method for strip-censusing, determine the population density index for a selected species at two locations. Evaluate the adequacy of the sample size, using performance curves (section 1A.3). Consider the effect of possible biases in the data and what could be done to control them.
2. Conduct a road transect survey or a point survey of all sighted mammals or birds. Determine the density indices and relative abundances of the species sighted.
3. Conduct a point survey at an optimal location for a species of migrating bird or resident waterfowl. If data have been collected at these locations and at similar times of the year, compare your population indices to past abundances of the species studied.
4. Conduct a census of animal feces or tracks in a selected area. Use either transect or plot sampling (depending on the species and the substrate bearing the tracks). Compare the reliability and efficiency of one or both of these indices to the effectiveness of one or more of the visual sighting methods described above.
5. Using small snaptraps or live traps, lay out several trap transects in selected habitats. Determine and compare the density index and relative density of all small mammals caught in each of the habitats sampled.

9. Selected references

Anderson, D. R., J. L. Laake, B. R. Crain, and K. P. Burnham. 1979. Guidelines for line transect sampling of biological populations. J. Wildlife Manage. 43:70–78.

Burnham, K. P., D. R. Anderson, and J. L. Laake, 1980. Estimation of density from line transect sampling of biological populations. Wildlife Monogr. 44:1–202.

Davis, D. E. (ed.). 1982. CRC handbook of census methods for terrestrial vertebrates. CRC Press, Boca Raton, Fla.

Davis, D. E., and R. L. Winstead. 1980. Estimating the numbers of wildlife populations, 221–245. In S. D. Schemnitz (ed.), Wildlife management techniques manual. 4th ed. The Wildlife Society, Washington, D.C.

Eberhardt, L. L. 1978. Transect methods for population studies. J. Wildlife Manage. 42:1–31.

Eberhardt, L. L., and R. C. Van Etten. 1956. Evaluation of the pellet group count as a deer census method. J. Wildlife Manage. 20:70–74.

Emlen, J. T. 1971. Population densities of birds derived from transect counts. Auk 88:323–342.

Emlen, J. T. 1977. Estimating breeding season bird densities from transect counts. Auk 94:455–468.

Gates, C. E. 1979. Line transect and related issues, 71–154. In R. M. Cormack, G. P. Patil, and D. S. Robson (eds.), Sampling biological populations. International Co-operative Publishing House, Fairland, Md.

Hayne, D. W. 1949. An examination of the strip census method for estimating animal populations. J. Wildlife Manage. 13:145–157.

Hewitt, O. H. 1967. A road-count index to breeding populations of red-winged blackbirds. J. Wildlife Manage. 31:39–49.

Howell, J. C. 1951. The roadside census as a method of measuring bird populations. Auk 68:334–357.

Klein, P. D. 1965. Factors influencing roadside counts of cottontails. J. Wildlife Manage. 29:665–671.

Kovner, J. L., and S. A. Patil. 1974. Properties of estimators of wildlife population density for the line transect method. Biometrics 30:225–230.

Linhart, S. B., and F. B. Knowlton. 1975. Determining the relative abundance of coyotes by scent station lines. Wildlife Soc. Bull. 3:119–124.

Murie, O. J. 1975. Field Guide to Animal Tracks. Houghton Mifflin Co, Boston.

Otis, D. L., K. P. Burnham, G. C. White, and D. R. Anderson. 1978. Statistical inferences for capture data on closed animal populations. Wildlife Monogr. No. 62.

Robinette, W. L., C. M. Loveless, and D. A. Jones. 1974. Field tests of strip census methods. J. Wildlife Manage. 38:81–96.

Seber, G. A. F. 1982. The estimation of animal abundance and related parameters. 2d ed. Macmillan Publishing Co, New York.

Seber, G. A. F. 1986. A review of estimating animal abundance. Biometrics 42:267–292.

Tilghman, N. G., and D. H. Rusch. 1981. Comparisons of line-transect methods for estimating breeding bird densities in deciduous woodlots, 202–208. In C. J. Ralph and J. M. Scott (eds.), Estimating numbers of terrestrial birds. Studies in avian biology no. 6. Cooper Ornithological Society, Allen Press, Lawrence, Kans.

Van Etten, R. C., and C. L. Bennett. 1965. Some sources of error in using pellet-group counts for censusing deer. J. Wildlife Manage. 29:723–729.

Yapp, W. B. 1956. The theory of line transects. Bird Study 3:93–104.

10. Animal identification manuals   See section 3D.19.

introduction

The *population* is a basic organizational unit in biology. It is an assemblage of individuals of the same species inhabiting a given area. Unfortunately, a few writers have lapsed carelessly into using the term population to mean an aggregation of several species in an area, a situation where "community" or "subcommunity" would be the appropriate designation (see unit 5).

A population has unique structural and functional characteristics not found in the individuals it comprises. In this unit we shall investigate some of these characteristics. One, *population structure,* will be discussed in terms of age distribution (section 4A) and spatial distribution (section 4C). Another, *population dynamics,* will be treated in terms of population growth (section 4B) and survivorship (section 4A).

unit 4

analysis of populations

125

1. Introduction

Populations, whether animal or plant, vary in their proportions of young and old individuals. Time units such as weeks, months, or years can describe ages. Or individuals can be assigned to qualitative age classes, such as nestling, juvenile, subadult, and adult, or egg, larva, pupa, and adult. The proportions of individuals belonging to the various age groups are collectively referred to as the *age structure* or *age distribution* of the population.

Three different procedures may be used for obtaining the age structure of a population. The vertical approach follows a particular *cohort*. A cohort is a group of individuals born within the same time interval. Thus by knowing the age of cohort members, you can follow their survival until all have died. The horizontal approach uses data on all ages within a given population at one time; that is, all cohorts in the population are examined at the same time. In the latter method, one assumes a stable age structure and that the birth and death rates remain constant. A third approach involves knowing the age at death for members of a population. Such data are commonly obtained for game species.

Knowledge of age structure is important, for the age distribution of a population affects its growth and dynamics. From a knowledge of age structure, a table of age-specific mortality, survivorship, and life expectancy can be constructed—a *life table* (section 4). In addition, population growth rates may be estimated from data on births per female in the population.

2. Procedures

Age class data may be collected by randomly sampling a field or laboratory population, or from a preserved (but randomly collected) sample from a population. Methods for aging vertebrates include counting growth rings on horns of mammals or on scales of fishes, examining tooth development and wear, and observing pelage or plumage changes. Weight or length classes may be tabulated for many species of fishes and invertebrates. Growth rings are also used for aging woody plants, as well as certain invertebrates, such as clams and snails. Various environmental variables often affect the measured or observed characters. Therefore, such characters as tooth wear, molt, weight, and length should be standardized to known ages for a given location and time. Insects are often aged by their developmental stage or by molts within stages of metamorphosis. Details on methods of determining age are given in Caughley (1977), Glock (1955), Lagler (1956), Larson and Taylor (1980), Momot (1967), Smith (1980), and Southwood (1978).

Record the number of individuals of each age. Generally, the numbers are recorded for *age classes,* or intervals, rather than exact ages (e.g., those individuals between three and four years old, those between four and five years

4a

age structure and survivorship

old, and so on). The investigator should also record measurements (e.g., length and/or weight of the body or of some part of the body) other than the characters used to indicate age of the organism, as these often provide additional information on population growth and vigor.

3. Age pyramids

The number of individuals in each age class may be plotted as a horizontal histogram forming a "pyramid." Age is placed on the vertical axis and the number, or the proportion, of individuals in each age class is plotted so that a symmetrical "pyramidal" graph results. Often the age class bars are graphed so that data for males appear on one side of the pyramid and females on the other (figure 4A.1). *Age pyramids* such as these usefully compare populations from different sites, or the same population at different times of the year or from year to year.

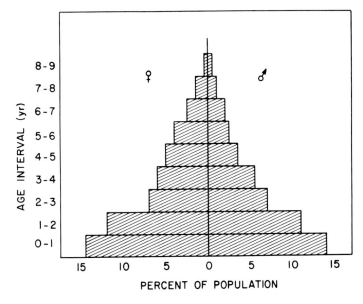

Figure 4A.1. An age pyramid.

Table 4A.1. *A life table. The data in the* x *and* L$_x$ *columns were obtained from a population of animals. Then, all other columns of data were derived from them, as described in section 4A.4.*

age	cohort (age interval) x	number in cohort L_x	number living at start l_x	number dying during x d_x	probability of dying during x q_x	probability of surviving interval x s_x	animal-years left to live T_x	life expectancy e_x
0–1 yr	0	33	46	26	0.57	0.43	63	1.37 yr
1–2 yr	1	16	20	8	0.40	0.60	30	1.50 yr
2–3 yr	2	9	12	6	0.50	0.50	14	1.17 yr
3–4 yr	3	4	6	4	0.67	0.33	5	0.83 yr
4–5 yr	4	1	2	2	1.00	0.00	1	0.50 yr
5–6 yr	5	0	0					

Age structure is dependent on many factors, such as longevity, rate of population increase, mortality, and environmental influences. In general, a growing population will show an increased proportion of young individuals, a stable one will show no increase or decrease in the relative numbers in each age class, and a declining one will show an increase in the proportions of the population in the older age classes and a decrease in membership in the younger age classes.

The size of individuals is usually correlated with the age structure of the population. For added information, plot the length, weight, or other size measurement against size class. In an overcrowded population, one may find stunting of individuals (section 4D.2.2) as well as reduced reproduction and a high proportion of older individuals.

4. Life tables

In a *life table* (table 4A.1), various statistics are compiled for each age class, or *cohort* (designated x). Data are commonly collected as numbers of individuals in each age class. L_x is the number of individuals in age class x. It is assumed that L_x is the number alive at the middle of age class x (for example, in table 4A.1, 33 individuals are assumed to be 0.5 years old, even though the true ages of the 33 might range between 0 and 1 year old).

We designate l_x as the number of individuals alive at the beginning of age class x. Thus, L_x may be defined as:

$$L_x = (l_x + l_{x+1})/2 \qquad (1)$$

(i.e., L_x is the number alive at the midpoint of age class x), and :

$$l_x = 2L_x - l_{x+1}. \qquad (2)$$

For example, in table 4A.1,

$$L_0 = 33$$
$$L_1 = 16$$
$$L_2 = 9$$
$$L_3 = 4$$
$$L_4 = 1$$
$$L_5 = 0$$

Since $L_5 = 0$, we can set $l_5 = 0$. Then, by applying equation 2:

$$l_5 = 0$$
$$l_4 = 2(1) - 0 = 2$$
$$l_3 = 2(4) - 2 = 8 - 2 = 6$$
$$l_2 = 2(9) - 6 = 18 - 6 = 12$$
$$l_1 = 2(16) - 12 = 32 - 12 = 20$$
$$l_0 = 2(33) - 20 = 66 - 20 = 46$$

The number of individuals in the population that die during interval x is:

$$d_x = l_x - l_{x+1}. \qquad (3)$$

Therefore,

$$d_0 = 46 - 20 = 26$$
$$d_1 = 20 - 12 = 8$$
$$d_2 = 12 - 6 = 6$$
$$d_3 = 6 - 2 = 4$$
$$d_4 = 2 - 0 = 2$$

Note that the sum of the d_x values must equal l_0; in our example:

$$\Sigma d_x = l_0 = 46.$$

The age-specific *mortality rate* is the proportion of individuals at the start of an age interval who die during that age interval:

$$q_x = d_x/l_x, \qquad (4)$$

also expressed as the probability of an individual dying during that interval. For our data, the age-specific mortality rates are:

$$q_0 = 26/46 = 0.57$$
$$q_1 = 8/20 = 0.40$$
$$q_2 = 6/12 = 0.50$$
$$q_3 = 4/6 = 0.67$$
$$q_4 = 2/2 = 1.00$$

The age-specific *survival rate* for age interval x is the proportion of individuals alive at the start of the interval surviving (i.e., not dying) during that interval (or in other words, the probability of surviving that age interval):

$$s_x = 1 - q_x. \qquad (5)$$

For the present data,

$$s_0 = 1.00 - 0.57 = 0.43$$
$$s_1 = 1.00 - 0.40 = 0.60$$
$$s_2 = 1.00 - 0.50 = 0.50$$
$$s_3 = 1.00 - 0.67 = 0.33$$
$$s_4 = 1.00 - 1.00 = 0.00$$

We can calculate age-specific *life expectancy* (commonly done for human populations) as follows. Let us define T_x as the number of time units left for all individuals to live from age x onward; this is obtained by summing L_x values as follows:

$$T_x = \sum_{i=x}^{last} L_i, \qquad (6)$$

or,

$$T_x = L_x + T_{x+1}, \qquad (7)$$

and expressing T_x as time units; so:

$$T_4 = L_4 = 1 \text{ yr}$$
$$T_3 = 4 + 1 = 5 \text{ yr}$$
$$\text{or } T_3 = L_3 + T_4 = 4 + 1 = 5 \text{ yr}$$
$$T_2 = 9 + 4 + 1 = 14 \text{ yr}$$
$$\text{or } T_2 = L_2 + T_3 = 9 + 5 = 14 \text{ yr}$$
$$T_1 = 16 + 9 + 4 + 1 = 30 \text{ yr}$$
$$\text{or } T_1 = L_1 + T_2 = 16 + 14 = 30 \text{ yr}$$
$$T_0 = 33 + 16 + 9 + 4 + 1 = 63 \text{ yr}$$
$$\text{or } T_0 = L_0 + T_1 = 33 + 30 = 63 \text{ yr}$$

Then, the life expectancy for an individual of age x is:

$$e_x = T_x/l_x; \qquad (8)$$

so:

$$e_0 = 63 \text{ animal-yr}/46 \text{ animals} = 1.37 \text{ yr}$$
$$e_1 = 30/20 = 1.50 \text{ yr}$$
$$e_2 = 14/12 = 1.17 \text{ yr}$$
$$e_3 = 5/6 = 0.83 \text{ yr}$$
$$e_4 = 1/2 = 0.50 \text{ yr}$$

Life expectancy represents the average additional length of time that an individual will live, once it has reached age x.

To compare different populations, the numbers dying (d_x) or surviving (l_x) are often expressed as numbers per 100 or per 1000 individuals entering the population at age zero; that is, l_0 is set to 100 or 1000, and all other life table entries are expressed relative to this value. For the example in our table, we have $l_0 = 46$ and $l_1 = 20$. If we set $l_0 = 100$, then we would have an l_1 value of $(20/46)(100) = 43.5$. In other words, for every 100 individuals born into the population, 43.5 survive to age 1. If we begin with $l_0 = 100$, then of course $\Sigma d_x = 100$.

In some studies, one collects data that are either l_x or d_x, rather than L_x. In such cases, simple rearrangements of equations 1 through 3 will allow for the computation of all of the above life table statistics.

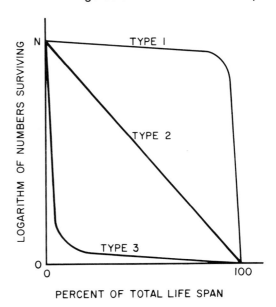

Figure 4A.2. The three types of survivorship curves. N is the number of organisms (on a logarithmic scale) born into a population at a given time. The curves show how the membership of this group decreases with time.

5. Survivorship curves

Various types of graphs may be constructed from life table data, including mortality rate curves, life expectancy curves, and survivorship curves. Most widely used among ecologists, the *survivorship curve* is prepared by plotting (usually on semilogarithmic graph paper) the logarithm of the number of survivors against age. For comparative purposes use l_x data based on 100 or 1000. From this graphical presentation, three basic types of curves are recognizable (as described by Pearl, 1928). (See figure 4A.2). In the type 1 survivorship curve, there is a high survival rate of the young and a low survival rate after a particular old age. In the type 2 curve a constant rate of mortality occurs at all ages (a constant percentage of population decreases each time period). The type 3 curve shows a high juvenile mortality and a relatively low rate of mortality thereafter. Most populations exhibit survivorship curves intermediate between types 1 and 2 or between types 2 and 3. Some adult birds approach type 2, and modern humans approach type 1. Including data from very early life stages (even eggs) tends to introduce type 3 curve characteristics.

6. Suggested exercises

1. Record the dates of birth and death from grave markers in a cemetery. Construct a life table from these data. Plot a survivorship curve. Construct an age pyramid.
2. Sample a population of fish or crayfish using seining (section 3E.3.4). Estimate age by counting annual rings on fish scales (Lagler, 1956) or by measuring carapace length on crayfish (Momot, 1967). Construct an age pyramid, life table, and survivorship curve.

7. Selected references

Andrewartha, H. G., and L. C. Birch, 1954. The distribution and abundance of animals. University of Chicago Press, Chicago.

Begon, M., and M. Mortimer. 1986. Population ecology: A unified study of animals and plants. 2d ed. Sinauer Associates, Sunderland, Mass.

Caughley, G. 1977. Analysis of vertebrate populations. John Wiley & Sons, New York.

Deevey, E. S., Jr. 1947. Life tables for natural populations of animals. Quart. Rev. Biol. 22:283–314.

Downing, R. L. 1980. Vital statistics of animal populations, 247–267. In S. D. Schemnitz (ed.), Wildlife management techniques manual. 4th ed. Wildlife Society, Washington, D.C.

Glock, W. S. 1955. Tree growth. II. Growth rings and climate. Bot. Rev. 21:73–188.

Hutchings, M. J. 1986. Plant population biology, 377–435. In P. D. Moore and S. B. Chapman (eds.), Methods in plant ecology. 2d ed. Blackwell Scientific Publications, Oxford, England.

Lagler, K. F. 1956. Freshwater fisheries biology. Wm. C. Brown Co. Publishers, Dubuque, Iowa.

Larson, J. S. and R. D. Taylor. 1980. Criteria of age and sex, 143–202. In S. D. Schemnitz (ed.), Wildlife management techniques manual. 4th ed. Wildlife Society, Washington, D.C.

Momot, W. T. 1967. Population dynamics and productivity of the crayfish, *Orconectes virilis,* in a marl lake. Amer. Midland Natur. 78:55–81.

Pearl, R. 1928. The rate of living. Alfred A. Knopf, New York.

Pielou, E. C. 1974. Population and community ecology. Gordon and Breach, New York.

Pielou, E. C. 1977. Mathematical ecology. John Wiley & Sons, New York.

Poole, R. W. 1974. An introduction to quantitative ecology. McGraw-Hill Book Co., New York.

Silverton, J. W. 1987. Introduction to plant population ecology. 2d ed. John Wiley & Sons, New York.

Smith, R. L. 1980. Ecology and field biology. 3d ed. Harper & Row, New York.

Southwood, T. R. E. 1978. Ecological methods. Chapman and Hall, London.

Vandermeer, J. 1981. Elementary mathematical ecology. John Wiley & Sons, New York.

1. Introduction

Populations are biological units that may grow larger, grow smaller or stay the same size over time depending on the combined effects of the rates of *natality* (i.e., birth), *mortality* (i.e., death), *immigration* (i.e., physical movement into the population), and *emigration* (i.e., physical movement out of the population). We shall here consider populations in which there is no immigration or emigration.

A rate of birth, death, or population change may be readily expressed as a percent, which is the number per 100 members of the population, over a unit of time. For example, we could speak of a birth rate of 30% per year (meaning that, on the average, 30 individuals are born each year for every 100 members of the population) and a death rate of 10% per year (meaning that on the average 10 individuals die each year for every 100 members of the population), which would result in a population growth rate of 20% per year (meaning that the population increases in size by 20 individuals each year for every 100 members of the population).

2. Potential rate of population growth

Consider a protozoan that reproduces by dividing every twenty-four hours, and assume that it is living under environmental conditions that do not limit its population growth. If we begin observing one such animal we would see that after one day there would be two animals, after two days there would be four animals, after three days there would be eight animals, and so on. The rate of increase in such a case would be 200% per day, and the rate of increase would be referred to as *geometric*. If N_t is the size (or density) of this population at a given time, t, and N_{t+1} is the population size (or density) one day later, then the rate of population increase can be expressed as

$$R = \frac{N_{t+1}}{N_t} \qquad (1)$$

which in this example would be 2 (or 200%).

If N_0 is the population size (or density) of some starting time for our observations, then

$$N_t = N_0 R^t \qquad (2)$$

would be the size (or density) of the population after t units of time (days, in our example). For example, if the starting population size is $N_0 = 5$, and the rate of increase is a constant 200% per day, then the following geometrically increasing population sizes would occur:

after 2 days: $N_2 = (5)(2)^2 = 20$
after 5 days: $N_5 = (5)(2)^5 = 160$
after 10 days: $N_{10} = (5)(2)^{10} = 5,120$
after 20 days: $N_{20} = (5)(2)^{20} = 5,242,880$

and so on.

4b

population growth

In reality, even if the *average* rate of increase were 200% per day, all members of the population would not divide at exactly the same time every twenty-four hours. The population growth would appear continuous instead of stepwise. In a continuously growing population that has no limitations on its growth, the growth rate of the population (also called the *intrinsic rate of increase* or the *biotic potential*) is

$$r = dN/Ndt, \qquad (3)$$

where dN denotes the change in population size from a size of N, and dt is an infinitesimally small time interval over which the change takes place. The number of individuals added to a population of size N over the time interval dt is

$$dN/dt = rN, \qquad (4)$$

and if we start with a population of size N_0, then the size of the population at time t will be

$$N_t = N_0 e^{rt}, \qquad (5)$$

where e is the base of natural logarithms (i.e., 2.71828). Figure 4B.1. shows equation 5 graphically for some combinations of r and N_0. While r is an important ecological concept, it is seldom realized in nature, for populations can not grow without limit (except for relatively short periods of time). Methods for estimating r are discussed in section 2.2

2.1. Population Doubling Time   If we ask how long it takes for a particular population to double in size, we can change equation 5 to:

$$N_t/N_0 = e^{rt}; \qquad (6)$$

and set $N_t/N_0 = 2$. Then,

$$e^{rt} = 2$$
$$rt = \ln 2$$
$$\text{and } t = 0.693/r \qquad (7)$$

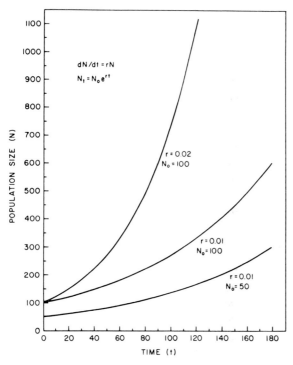

Figure 4B.1. Exponential population growth. The size of a population after t units of time have elapsed is N_t, which is an exponential function of the initial population size (N_o) and the elapsed time (t).

is the *doubling time* for the population.[1] For example, the doubling time for a population (e.g., many human populations) that increases at the rate of 2% per year is:

$$t = \frac{0.693}{0.02/\text{yr}} = 35 \text{ yr.}$$

2.2. Growth in an Unlimited Environment Population growth may be measured under conditions of unlimited space and food in either of two ways:

1. Growth rates may be measured during the very early phase of growth when space, food, and interactions between individuals have not yet begun to affect natality and mortality.

2. A large proportion of the population is periodically removed, thus maintaining the population at a low enough level so that natality and mortality are not affected by crowding.

In an unlimited environment, exponential growth is described by equation 5 (figure 4B.1), and r can be estimated from the population densities measured at different times. Equation 8 is the logarithmic equivalent of equation 5, using natural logarithms.

$$\ln N_t = \ln N_0 + rt \qquad (8)$$

1. The notation "ln" is used to indicate natural logarithms (logarithms to the base e), while "log" denotes common logarithms (logarithms to the base 10); $\ln X = 2.3026 \log X$ (for example, $\log 2 = 0.301$ and $\ln 2 = 0.693$).

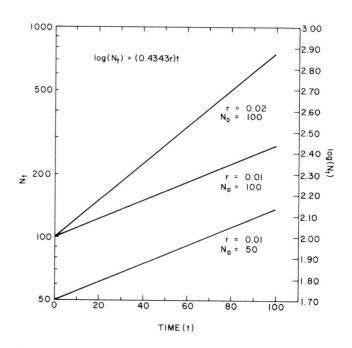

Figure 4B.2. A semilogarithmic plot of exponential population growth, for the three populations in Figure 4B.1.

and equation 9 is the equivalent using common logarithms:

$$\log N_t = \log N_0 + 0.4343rt \qquad (9)$$

If the logarithms of population sizes are plotted against time we have what is termed a semilogarithmic graph. This is shown in Figure 4B.2, using common logarithms. Graph paper showing the logarithmic scale on the left axis is commercially available. The three lines in this figure represent the same three populations that are shown in Figure 4B.1. You can see that a semilogarithmic plot of exponential growth results in a straight line. Note also that populations with the same r will appear as parallel lines.

If population size (N) is observed at each of several times (t), we may subject log N and t to the regression computations of section 1B.5.1 to determine the slope of the line (b). As the slope, b, is 0.4343r in equation 9,

$$r = b/0.4343. \qquad (10)$$

Population densities (N_o and N_t) may be estimated using appropriate sampling procedures (see section 1A and unit 3). Population growth of a variety of species can be studied easily in the laboratory (e.g., yeasts, bacteria, protozoa such as *Paramecium*, aquatic crustaceans such as *Daphnia*, or insects such as *Drosophila*, *Tenebrio*, or *Tribolium*). For best results, sampling intervals should be about one generation in length.

The value of r may also be estimated using the following approximation of equation 3:

$$r = \Delta N/N\Delta t, \qquad (11)$$

where ΔN ("delta N") is a change in population size, Δt ("delta t") is the time interval over which N changed, N

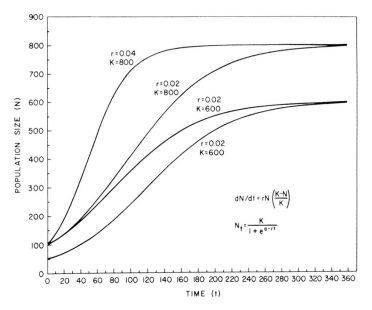

Figure 4B.3. Logistic population growth. The top three curves have $N_o = 100$; the bottom curve has $N_o = 50$.

is the population size at the start of this time interval, and Δt is very short. In fact, a series of values calculated from equation 11 can usefully determine the time at which environmental factors or interactions between individuals affect a population's growth rate.

2.3. Growth in a Limited Environment Population growth in an unlimited environment often cannot be followed for substantial lengths of time. If the environment is limited, population growth will deviate increasingly from exponential as population size increases.

An experimental situation may be set up where a population is allowed to grow in a finite space. Nothing is added to nor subtracted from the environment by the experimenter, except an excessive supply of food is provided to prevent the population from starving before reaching its maximum density. The population size is then measured at several times.

Many species will then exhibit an S-shaped, or *sigmoid*, growth curve. A particular type of sigmoid curve, the *logistic* curve, is predicted by population theory[2]:

$$dN/dt = rN\left(\frac{K - N}{K}\right) \qquad (12)$$

where K is the *carrying capacity* of the habitat, the maximum possible density of a given species that can be sustained there. Four logistic curves are presented in figure 4B.3. When the population is very small, $(K - N)/K$ is approximately equal to 1, so that equation 12 approaches equation 4, and population growth is nearly exponential. When N is close to K, however, $(K - N)/K$ approaches

2. This important relationship was independently formulated by Verhulst (1839, see Harper, 1981:2) and Pearl and Reed (1920) and further discussed by Gause (1934).

zero, meaning that the population growth approaches zero.

The value of K may be determined by allowing the population density to reach its maximum. Typically, however, a population does not maintain a constant maximum N, but exhibits some fluctuation in population size around the carrying capacity. Thus one should determine the mean of several estimates of K over a period of time. An estimate of K through use of equation 15 is explained below and shown in figure 4B.4.

To estimate r, we may use an estimate of K and population size data collected from the rapidly increasing portion of the growth curve (since the early and late phases of growth are generally complicated by various environmental and life history phenomena). Equation 12 can be rewritten as:

$$r = \frac{dN}{Ndt}\left(\frac{K}{K - N}\right), \qquad (13)$$

which can be approximated by:

$$r = \frac{\Delta N}{N\Delta t}\left(\frac{K}{K - N}\right). \qquad (14)$$

You may use the latter equation to calculate r for several values of ΔN and Δt, and then to determine a mean value of r.

The value of r may also be estimated graphically, using equation 15 (derived from equation 14):

$$\frac{\Delta N}{N\Delta t} = r - (r/K)N. \qquad (15)$$

As shown in figure 4B.4, $\Delta N/N\Delta t$ may be plotted on the ordinate (vertical axis) and N plotted on the abscissa (horizontal axis). N is the beginning of the interval ΔN. At each observed density, N, plot the observed rate of increase, $\Delta N/N\Delta t$, and draw a straight line through the

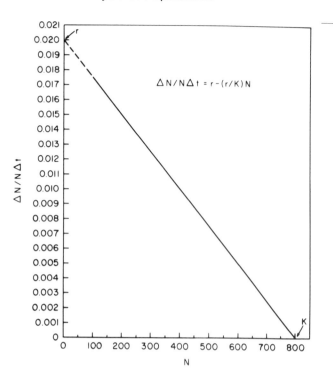

Figure 4B.4. A graphical determination of r and K in logistic population growth. This is the population growth shown in figure 4B.3 for $r = 0.02$, $K = 800$, and $N_o = 100$.

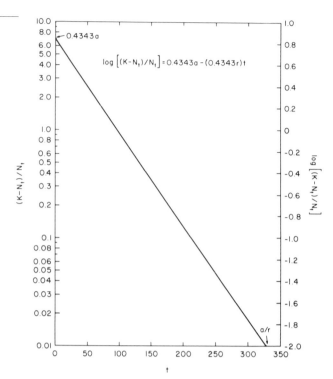

Figure 4B.5. A graphical determination of r in logistic population growth. This is the population growth shown in figure 4B.3 for $r = 0.02$, $K = 800$, and $N_o = 100$.

points. (For maximum accuracy, use the regression techniques of section 1B.6.1.) This line will intercept the abscissa at K and intercept the ordinate at r; thus the graph estimates both r and K for the observed population. The slope of the line is r/K. The accuracy of this determination is greatest if values of ΔN are determined for very small values of Δt. (If Δt is not very small, then a better value of N to plot on the abscissa is the mean of $N_t + N_{t+1}$.)

Another approach to estimating r uses the integrated form of Equation 12:

$$N_t = \frac{K}{1 + [(K - N_0)/N_0]e^{-rt}} \qquad (16)$$

where e is the base of natural logarithms (2.71828). If we define

$$a = \ln \frac{K - N_0}{N_0}, \qquad (17)$$

then, from equation 16:

$$\ln \frac{K - N_t}{N_t} = a - rt. \qquad (18)$$

Like equation 15, this equation can be plotted as a straight line; r can thus be estimated graphically and without requiring a very small Δt. Values of $\ln [(K - N_t)/N_t]$ may be plotted as a function of t. Or we may write equation 18 as:

$$\log \frac{K - N_t}{N_t} = a/2.3026 - (r/2.3026)t \qquad (19)$$

and plot $(K - N_t)/N_t$ against t on semilogarithmic graph paper (figure 4B.5). The absolute value of the slope of the line would then be $r/2.3026$ ($= 0.4343r$), and the line would intercept the ordinate at the value of $a/2.3026$ ($= 0.4343a$). For the example of figure 4B.5, r would be estimated as followed,

Select two points on the ordinate that are far apart, such as 7.0 and 0.01.

Take the logarithm of their ratio: $\log (7.0/0.01) = \log 700 = 2.8451$.

Read from the graph the two abscissa values corresponding to the two ordinate values selected: 0 and 327.

Divide 2.8451 by the difference between the two abscissa values: $2.8451/(0 - 327) = -0.008701$.

Take the absolute value: $|-0.008701| = 0.008701$.

Multiply this by 2.3026: $(0.008701)(2.3026) = 0.02$.

Therefore: $r = 0.02$, or 2%.

3. Sources of error

A number of problems occur in estimating r from laboratory experiments, not to mention in natural populations. One basic difficulty is that r itself is often variable, depending on the genetic history and age structure of the population and such environmental factors as temperature and humidity. Also, a lag often occurs in the beginning of the experiment due to life history changes and to the individuals adjusting to new conditions (Cole, 1954;

Smith, 1952). One also assumes that the mathematical model used (such as exponential or logistic) accurately describes true population behavior. Generally, however, these models are only simplifications. More sophisticated models have been proposed to consider natural phenomena more fully, but even these complex equations simplify many population processes.

Limitations of the logistic model as a description of population growth are seen by considering the theoretical assumptions of the model (see also Pielou, 1974, 1977; Poole, 1974; Vandermeer, 1981):

1. All individuals in the population have identical ecological properties; that is, they have equal opportunity to give birth, eat, or be eaten. This ignores the age structure of the population.
2. Changes in the population's birth and death rates occur instantaneously, without lag, with changes in population density.
3. All members of the population are equally affected by crowding; this would be true only if the population dispersion (section 4C) were uniform.
4. There is a constant upper limit to population size (that is, a constant carrying capacity), implying that the environment is constant, and that the population never "overshoots" K.
5. The population has a stable age distribution.
6. Birth and death rates are not affected by abiotic factors in the environment.

4. Suggested exercises

1. Culture a species of *Paramecium* in flasks for three weeks. Three days after inoculation with a known number of individuals, count the number per milliliter each day (see section 3E.4.5). Use equation 14 and/or the graphing of equation 15 or 19 to estimate r.
2. Use equation 14 and/or the graphing of equation 15 or 19 to estimate r for a laboratory population of *Tribolium* or *Drosophila*.
3. Estimate r, as in the above exercise 1 or 2, using populations maintained at different environmental conditions (e.g., temperature or moisture).

5. Selected references

Andrewartha, H. G., and L. C. Birch. 1954. The distribution and abundance of animals. University of Chicago Press, Chicago.
Begon, M., L. J. Harper, and C. R. Townsend. 1986. Ecology: Individuals, populations, and communities. Sinauer Associates, Sunderland, Mass.
Begon, M., and M. Mortimer. 1986. Population ecology: A unified study of animals and plants. 2d ed. Sinauer Associates, Sunderland, Mass.
Birch, L. C. 1948. The intrinsic rate of increase of an insect population. Anim. Ecol. 17:15–26.
Cole, L. C. 1954. The population consequences of life history phenomena. Quart. Rev. Biol. 29:103–137.
Gause, G. F. 1934. The struggle for existence. Waverly Press, Baltimore.
Harper, J. L. 1981. Population biology of plants. Academic Press, New York.
Hutchings, M. J. 1986. Plant population biology, 377–435. In P. D. Moore and S. B. Chapman (eds.), Methods in plant ecology. 2d ed. Blackwell Scientific Publications, Oxford, England.
Hutchinson, G. E. 1978. Introduction to population ecology. Yale University Press, New Haven, Conn.
Leslie, P. H., and T. Park. 1949. The intrinsic rate of natural increase of *Tribolium castaneum* Herbst. Ecology 30: 469–477.
May, R. M. 1981. Models for single populations, 5–29. In R. M. May (ed.), Theoretical ecology: Principles and applications. 2d ed. W. B. Saunders Co., Philadelphia.
Pearl, R. 1928. The rate of living. Alfred A. Knopf, New York.
Pearl, R., and J. L. Reed. 1920. On the rate of growth of the population of the United States since 1790 and its mathematical representation. Proc. Nat. Acad. Sci. 6:275–288.
Pianka, E. 1983. Evolutionary ecology. 3d ed. Harper & Row Publishers, New York.
Pielou, E. C. 1974. Population and community ecology. Gordon and Breach, New York.
Pielou, E. C. 1977. Mathematical ecology. John C. Wiley & Sons, New York.
Poole, R. W. 1974. An introduction to quantitative ecology. McGraw-Hill Book Co., New York.
Silverton, J. W. 1987. Introduction to plant population biology. 2d ed. John Wiley & Sons, New York.
Slobodkin, L. B. 1980. Growth and regulation of animal populations. Holt, Rinehart and Winston, New York.
Smith, F. E. 1952. Experimental methods in population dynamics: A critique. Ecology 33:441–450.
Tamarin, R. H. (ed.). 1978. Population regulation. Academic Press, New York.
Vandermeer, J. 1981. Elementary mathematical ecology. John Wiley & Sons, New York.
Wangersky, P. J. 1978. Lotka-Volterra population models. Annu. Rev. Ecol. Systemat. 9:189–218.
Whittaker, R. H. 1975. Communities and ecosystems. Macmillan Publishing Co., New York.

4c

population dispersion

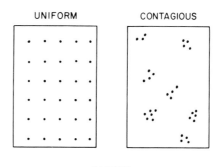

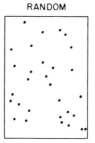

Figure 4C.1. The three basic types of spatial distribution of individuals within a habitat.

1. Introduction

Density alone gives an incomplete picture of how a population is distributed within a habitat, for two populations can have the same density but quite different spatial patterns of arrangement. The arrangement of members of a population within a habitat is referred to as *dispersion,* or the *distribution* or *pattern* of a population.[1]

There are three basic patterns of dispersion (figure 4C.1): *uniform* (or *regular*), *random,* and *contagious* (also called *clumped, clustered,* or *patchy*). Contagious and uniform dispersion patterns are sometimes called *overdispersed* and *underdispersed,* respectively. The distribution of organisms in nature is seldom as uniform as in an orchard or a cornfield but is generally contagious. A truly random dispersion (in which the position of an individual is completely independent of the position of any other individual in the population) may be approached in some species. If we were sampling with a three-dimensional plot (sampling plankton suspended in water, for instance), the two-dimensional presentations of figure 4C.1 could be expanded to consider dispersion in a habitat volume rather than in a habitat area.

Two major kinds of methods are used to describe spatial distribution and to assess departure from randomness. One involves the sampling of plots and comparing the data to the mathematical distribution described in section 4C.2. The other procedure does not use plot sampling but measures distances between plants or distances of plants from random points. Both procedures will be explained in section 4C.3.

2. The Poisson distribution

If a grid is superimposed on the studied habitat area (figure 4C.2), we could count the number of squares, or cells, in the grid having no individuals, the number having one individual, the number with two individuals, and so on. The grid in figure 4C.2 is, of course, a set of continuous plots. However, a grid is not needed for the considerations below; you may use data from any plot sampling (section 3A).

Let us define $f(X)$, the frequency of X, as the number of grid squares containing X individuals. Let us also define $P(X)$ as the proportion of the squares with X individuals, which is the same as saying that $P(X)$ is the probability of finding X individuals in a square. The probability of finding zero individuals in a square would be represented by $P(0)$, the probability of finding one individual in a square would be $P(1)$, and so on. Mathematical theory tells us that if the dispersion of individuals is random, then the probabilities $P(0)$, $P(1)$, $P(2)$, etc., will conform to what is known as the "Poisson distribution."

According to the Poisson distribution,

$$P(X) = \frac{e^{-\mu}\mu^X}{X!}. \qquad (1)$$

In this equation, μ (the lower case Greek letter mu) stands for the mean number of individuals per cell, and e is a constant, the base of natural logarithms (2.71828). The use of $X!$ to represent "X factorial" is standard notation: e.g., for $X = 4$, $X! = (4)(3)(2) = 24$. (Both 0! and 1! are defined as 1.) We do not usually know μ, which is the mean of an entire statistical population (the mean number of individuals per square for all squares that could possibly be superimposed on the entire biological population of interest). Therefore, we estimate μ by $\bar{X}$, the mean number

---

1. Dispersion should not be confused with *dispersal,* which is the movement of organisms or their propagules from one place to another. Unfortunately, some authors have used these two terms interchangeably for one or the other concept. As a result, Pielou (1974) recommends the word "diffusion" to denote the movement of biological entities (i.e., dispersal), but many biologists would prefer to reserve that term for passive rather than active movement.

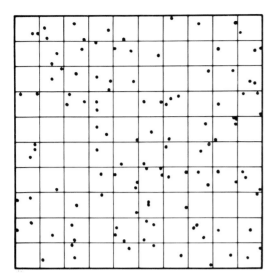

Figure 4C.2. A grid superimposed on a habitat. The 10×10 grid consists of 100 plots, or cells, and there are within the grid a total of 120 individuals from a population.

of individuals per square for the number of squares for which we have data. In figure 4C.2, 120 individuals are found in 100 squares, so:

$$\bar{X} = 120 \text{ individuals}/100 \text{ squares}$$
$$= 1.2 \text{ individuals/square},$$

which is considered a good estimate of μ, the mean number of individuals per square that we would have calculated had we used a grid large enough to cover the entire biological population in question.

The form of the Poisson distribution then will be determined by the mean population density. By examining equation 1 you can see that if the population of individuals is distributed at random, the proportion of squares without individuals in them will be:

$$P(0) = (e^{-\mu}\mu^0)/0! = e^{-\mu}, \quad (2)$$

and, also by examination of equation 1:

$$P(1) = (e^{-\mu}\mu^1)/1! = e^{-\mu}\mu, \quad (3)$$
$$P(2) = (e^{-\mu}\mu^2)/2! = (e^{-\mu}\mu^2)/2, \quad (4)$$
$$P(3) = (e^{-\mu}\mu^3)/3! = (e^{-\mu}\mu^3)/6, \quad (5)$$
$$P(4) = (e^{-\mu}\mu^4)/4! = (e^{-\mu}\mu^4)/24, \quad (6)$$

and so on. Table 4C.4 supplies the value of $P(X)$ for a variety of mean densities, μ. Figure 4C.3 shows the Poisson distribution for $\mu = 1, 2, 3,$ and 4.

In a random distribution of individuals, the values of $P(X)$ will be those predicted by the Poisson distribution (equations 2 through 6); there is an equal and independent chance of a member of the population occurring at any point in the habitat. If the occurrence of some individuals is enhanced by that of others, a contagious dispersion results. Examples of this would be social attraction in animals or germination of seeds close to parent plants. If the occurrence of an individual hinders that of another

Table 4C.1. *The observed data from figure 4C.2 and the probabilities expected from the Poisson distribution with a mean (μ) of 1.2 individuals per plot.*

number in plot X	observed frequency $f(X)$	observed probability $p(X)$	Poisson probability* $P(X)$
0	27	0.27	0.301
1	39	0.39	0.361
2	22	0.22	0.217
3	11	0.11	0.087
4	1	0.01	0.026
5	0	0.0	0.006

* From table 4C.4, for $\mu = 1.2$.

nearby, then the population dispersion tends toward a uniform distribution. Examples of this would be desert plants competing for scarce moisture, or nesting birds defending territories.

If n is the total number of cells in the grid, and $f(X)$ the number, or frequency, of cells containing X individuals, then the observed proportion of cells having X individuals is:

$$p(X) = f(X)/n. \quad (7)$$

Table 4C.1 shows $f(X)$ and $p(X)$ for the grid in figure 4C.2.

3. Determination of spatial distribution

By employing the Poisson distribution as a mathematical model for random dispersion of a biological population, we can infer whether an observed set of data deviates from a random distribution of individuals. For such testing, the mean number of individuals per sample plot should not exceed 10, should preferably be less than 5, and ideally should be in the neighborhood of 1 or 2. Plot size may be adjusted to obtain the desired magnitude of the mean. Table 4C.1 presents the data for figure 4C.2. One tabulates $f(X)$, the number of plots containing X individuals. Then, the values of $p(X)$ (the observed porportion of plots having X individuals) are calculated from equation 7. Plot data may be collected by the methods of section 3A, or analogous portions of sections 3D and 3E.

3.1. Graphical Examination   Once the estimate of mean density of the population is known, table 4C.4 (or equation 1) may be used to compute each $P(X)$, the predicted probability of X individuals occurring per plot if the distribution is random. For example, our observed data have a mean of 1.2 individuals per plot. By consulting table 4C.4 for $\mu = 1.2$, we can plot the predicted random $P(X)$ values as shown in figure 4C.4. Observed data may be plotted on the same graph, as shown, so as to compare them with the Poisson distribution. The vertical axis may be either the proportion of plots containing X, as shown, or the number of plots containing X (i.e., we may plot either $P(X)$ or $f(X)$).

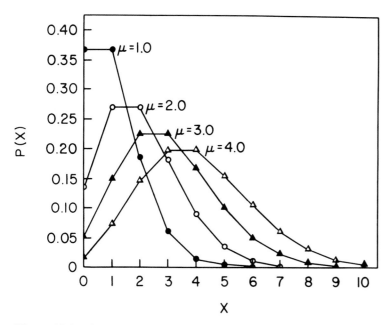

Figure 4C.3. The Poisson distribution for various mean population densities (μ).

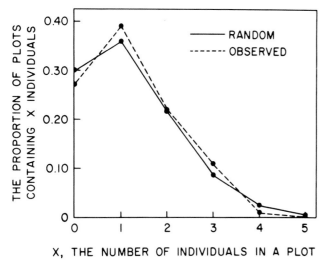

Figure 4C.4. The observed probabilities of occurrence of X and the random probabilities predicted from the Poisson distribution from table 4C.1 and figure 4C.2.

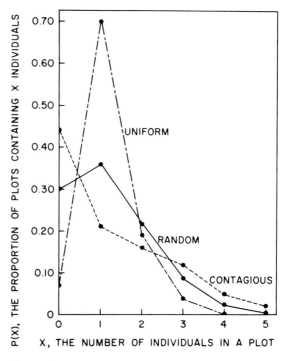

Figure 4C.5. Probabilities of occurrence of X for dispersions that are random, contagious, and tending toward uniform. ($\mu = 1.2$.)

Departures from randomness of dispersion will appear graphically as in figure 4C.5. In a contagious distribution with a mean population density of 1.2 individuals per plot, the observed proportion of empty cells would be higher than the $P(0)$ predicted by the Poisson distribution, and the observed proportions of aggregations ($p(2), p(3)$, etc.) would likewise be higher than the predicted random probabilities ($P(2), P(3)$, etc.). In a distribution with the same mean density but which tended toward uniform dispersion, the observed proportions would be lower than the random probabilities on the left side and right side of the mean, but higher than the random probabilities near the mean of the curve. Thus there would be a lower than random probability of finding either empty plots or aggregations. This should be clear in examining figure 4C.5. In a contagious distribution there is a greater than random probability of finding many individuals in a few plots. And in a relatively uniform distribution there is a higher than random probability of finding only a few individuals in most plots. Thus our example data (see figure 4C.4) appear to follow a random spatial distribution, with a slight tendency toward uniform dispersion.

3.2. Variance-to-Mean Ratio

In a population having a Poisson distribution, the population mean, μ, is equal to the population variance, σ^2. Therefore a randomly dispersed population would have a ratio of its variance to its mean of:

$$\sigma^2/\mu = 1.0. \qquad (8)$$

A ratio much less than 1.0 indicates a uniform distribution, and a ratio much greater than 1.0 indicates contagion.

While we cannot generally compute μ and σ^2 directly (see sections 1B.2.1 and 1B.2.2), we can estimate them by calculating $\bar{X}$ (the sample mean) and s^2 (the sample variance), respectively, for the n plots studied. For our 100 plots, these statistics are found to be:

$$\bar{X} = 1.20$$

and,

$$s^2 = 0.990$$

(see sections 1B.2.1 and 1B.2.2 for computational methods). Therefore, σ^2/μ is estimated by $s^2/\bar{X} = 0.990/1.20 = 0.825$, indicating a dispersion nearly random but with a tendency toward uniform.

The variance-to-mean ratio (sometimes called the *coefficient of dispersion* or the *relative variance*) has a potential disadvantage as a measure of aggregation: It is affected by population size and plot size. If one desires a measure that does not have this property (i.e., one that is unaffected by the random removal of some members of population), then see Morisita's index in section 3.4.

The significance of departure from randomness (i.e., departure from a variance-to-mean ratio of 1.0) may be assessed statistically by computing:

$$t = \frac{|s^2/\bar{X} - 1.0|}{\sqrt{2/(n-1)}} \qquad (9)$$

(Clapham, 1936), and comparing t to critical values on table 1B.1 for $n - 1$ degrees of freedom. For our data this would be:

$$t = \frac{|0.990/1.20 - 1.0|}{\sqrt{2/(100-1)}}$$

$$= \frac{0.175}{0.142}$$

$$= 1.23.$$

Because the critical value of t (for 99 degrees of freedom, at the 5% significance level) is 1.98, and because 1.23 is less than 1.98, we conclude that the dispersion is random.

A better method for determining significant departure from randomness uses the chi-square statistic:

$$\chi^2 = SS/\bar{X}, \qquad (10)$$

where SS, the "sum of squares," may be computed as in section 1B.2.2, or by:

$$SS = (n-1)(s^2). \qquad (11)$$

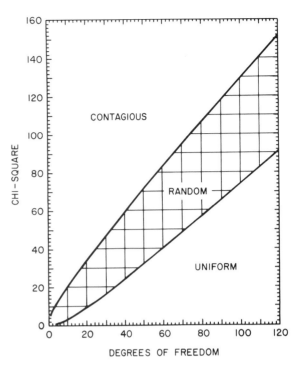

Figure 4C.6. Chi-square values for comparing the variance to the mean (see section 4C.3.2) at the 5% level of statistical significance.

For the present data, $SS = 98.0$, so:

$$\chi^2 = 98.0/1.20 = 81.67.$$

The statistical significance of this may be obtained by using chi-square tables more extensive than table 1B.3, or more easily by consulting figure 4C.6. We define the "degrees of freedom" (see section 1B.4) as:

$$DF = n - 1, \qquad (12)$$

or 99 in the present example. By examining figure 4C.6 for 99 degrees of freedom, we find a χ^2 of 81.67 to be within that portion of the graph indicating a random distribution. Although this derives from an old procedure (Clapham, 1936), it has been shown to be an excellent method (Heltsche and Ritchey, 1984).

3.3. Goodness of Fit

Another method to conclude objectively whether a distribution is random is to employ chi-square "goodness of fit" testing (section 1B.4). The observed frequencies, $f(X)$, from figure 4C.2 are in table 4C.2, while the expected frequencies $F(X)$, under the null hypothesis of randomness, are those predicted from the Poisson equation for $P(X)$, namely:

$$F(X) = [P(X)][n]. \qquad (13)$$

The values of $f(X)$ and $F(X)$ for our present example are in table 4C.2. The value of $F(>4)$ is obtained by subtracting the sum of all the other $F(X)$ values (namely 99.2) from 100. In chi-square goodness of fit testing for the Poisson distribution, we combine values of $f(X)$ and values

Table 4C.2. *Observed frequencies, and frequencies expected from the Poisson distribution, for the data in figure 4C.2 and table 4C.1. These frequencies are required for chi-square goodness of fit analysis.*

number in plot X	observed frequency $f(X)$	expected frequency $F(X)$
0	27	30.1
1	39	36.1
2	22	21.7
3	11	8.7
4	1 } 1	2.6 } 3.4
>4	0	0.8

of $F(X)$ so that no $F(X)$ is less than 1.0 (Zar, 1984:409); the results of this combining are shown by the brackets in table 4C.2. Then, the appropriate chi-square statistic is computed in the usual fashion (section 1B.4):

$$\chi^2 = \frac{(27-30.1)^2}{30.1} + \frac{(39-36.1)^2}{36.1}$$
$$+ \frac{(22-21.7)^2}{21.7} + \frac{(11-8.7)^2}{8.7} + \frac{(1-3.4)^2}{3.4}$$
$$= 0.32 + 0.23 + 0.00 + 0.61 + 1.69$$
$$= 2.85$$

For fitting Poisson distributions, the appropriate degrees of freedom are the number of frequency categories after combining (in our case 5) minus 2; therefore DF = 5 − 2 = 3. Consulting table 1B.3, we find the critical value of χ^2 for three degrees of freedom and a significance level of 5% is 7.815. As 2.85 is less than 7.815, we conclude that the observed frequency distribution is not significantly different from random. Goodness-of-fit testing is dependent upon quadrat size, a disadvantage that is avoided by using the Morisita measure in section 3.4.

Note that this method does not necessarily yield the same conclusions as the variance-to-mean method of section 4C.3.2. Therefore, many ecologists prefer to routinely apply both procedures to their data.

3.4. Morisita's Index of Dispersion   The dispersion measure proposed by Morisita (1959) has excellent qualities. We may calculate the Morisita index as:

$$I_d = n \frac{\Sigma X^2 - N}{N(N-1)}, \qquad (14)$$

where n is the number of plots, N is the total number of individuals counted on all n plots, and ΣX^2 is the squares of the numbers of individuals per plot, summed over all plots.[2] If the dispersion is random, then $I_d = 1.0$; if per-

2. I_d is computationally related to Simpson's index of dominance (section 5B.2.2) Thus a highly aggregated dispersion pattern is one in which a small number of plots "dominate" the distribution of data. I_d also is related very closely to Lloyd's (1967) "index of mean crowding":

$$\overset{*}{m} = \frac{1}{N}\Sigma X_i(X_i-1) = \frac{\Sigma X_i^2}{\Sigma X_i} - 1. \qquad (15)$$

fectly uniform, $I_d = 0$; and if maximally aggregated (i.e., all individuals in one plot), $I_d = n$.

For the data in table 4C.1, only the first two columns are needed:

$$n = \Sigma f(X) = 27 + 39 + 22 + 11 + 1 = 100;$$
$$N = \Sigma[f(X)][X] = (27)(0) + (39)(1)$$
$$+ (22)(2) + (11)(3) + (1)(4)$$
$$= 120;$$
$$\Sigma X^2 = \Sigma[f(X)][X^2] = (27)(0) + (39)(1)$$
$$+ (22)(4) + (11)(9) + (1)(16)$$
$$= 242.$$

From this, we calculate:

$$I_d = 100 \frac{242-120}{120(119)} = 0.85.$$

Thus we conclude that the dispersion in the observed data is nearly random, having a slight tendency toward uniformity.

If desired, the departure of an observed dispersion pattern from randomness may be assessed statistically, by computing:

$$\chi^2 = n\Sigma X^2/N - N, \qquad (16)$$

which, for our example, would be:

$$\chi^2 = (100)(242)/120 - 120$$
$$= 201.67 - 120 = 81.67.$$

This computed value of chi-square may then be compared to the appropriate critical value in table 1B.3, with $n − 1$ degrees of freedom. From the example given beneath that table, we find a critical value of 123.22; as our computed value 81.67 is smaller than this, we conclude that the dispersion pattern is not significantly different from random.

3.5. Ratio of Observed to Expected Density   This procedure has generally been used with plants but may also be applied to animal populations, especially sessile forms, or burrows, nests, and other animal signs. We define frequency (f) as the proportion of the total number of sampling plots in which the individuals of a species occur (see the introduction to unit 3). In the Poisson distribution, the proportion of plots containing no individuals is $e^{-\mu}$ (equation 2), so the proportion of plots having one or more individuals is $1 - e^{-\mu}$. For example, if 0.25, or 25%, of the plots contain no individuals, then $1 - 0.25 = 0.75$, or 75%, of the plots contain at least one individual. We estimate the value of μ by a density measurement, D, as described in unit 3. Thus D is the mean number of individuals per sampling plot, and we can write:

$$f = 1 - e^{-D} \qquad (17)$$

as the relationship that exists between frequency (f) and density (D), *if* the plants are distributed randomly within the habitat.

If we observe a particular frequency (f) for a plant population, then we can predict the density in a random plant distribution by solving for D in equation 16. Let us denote this predicted density of a randomly distributed population as D' and the density actually observed as D. Then, if $D/D' = 1.00$, we can say that the plant distribution is random. But if D/D' is much greater than 1.00 than we have evidence of contagion. And if D/D' is much less than 1.00 then a distribution more uniform than random is indicated.

The observed frequency (f) and observed density (D) measurements can be obtained as described in sections 3A or 3C. The predicted, or theoretical, density (D') can be obtained by solving equation 17 for D:

$$D' = -\ln(1 - f)$$
$$\text{or } D' = -2.3206 \log (1 - f) \qquad (17)$$

(Table D.1 in Appendix D gives common logarithms). For convenience, one may employ table 4C.3, which was produced using equation 17. In general, a D/D' ratio greater than 2.0 is evidence of a distinctly contagious distribution.

For the data in table 4C.1, we computed a density, D, of 1.2. The frequency, f, is the proportion of plots in which the species occurs, namely:

$$f = \frac{39 + 22 + 11 + 1}{100} = \frac{73}{100} = 0.73.$$

From table 4C.3, we see that for $f = 0.73$:

$$D' = 1.31;$$

so:

$$D/D' = 1.2/1.31 = 0.92.$$

The value of 0.92 is very near 1.00, indicating a nearly random distribution, with a slight tendency toward uniformity.

Plant ecologists have referred to the ratio of density to frequency as an *index of sociability:*

$$\text{index of sociability} = D/f. \qquad (19)$$

Table 4C.3. *Predicting density from frequency, using equation 18.*

f	D'	f	D'	f	D'
0.01	0.01	0.36	0.45	0.71	1.24
0.02	0.02	0.37	0.46	0.72	1.27
0.03	0.03	0.38	0.48	0.73	1.31
0.04	0.04	0.39	0.49	0.74	1.35
0.05	0.05	0.40	0.51	0.75	1.39
0.06	0.06	0.41	0.53	0.76	1.43
0.07	0.07	0.42	0.54	0.77	1.47
0.08	0.08	0.43	0.56	0.78	1.51
0.09	0.09	0.44	0.58	0.79	1.56
0.10	0.11	0.45	0.60	0.80	1.61
0.11	0.12	0.46	0.62	0.81	1.66
0.12	0.13	0.47	0.63	0.82	1.71
0.13	0.14	0.48	0.65	0.83	1.77
0.14	0.15	0.49	0.67	0.84	1.83
0.15	0.16	0.50	0.69	0.85	1.90
0.16	0.17	0.51	0.71	0.86	1.97
0.17	0.19	0.52	0.73	0.87	2.04
0.18	0.20	0.53	0.76	0.88	2.12
0.19	0.21	0.54	0.78	0.89	2.21
0.20	0.22	0.55	0.80	0.90	2.30
0.21	0.24	0.56	0.82	0.91	2.41
0.22	0.25	0.57	0.84	0.92	2.53
0.23	0.26	0.58	0.87	0.93	2.66
0.24	0.27	0.59	0.89	0.94	2.81
0.25	0.29	0.60	0.92	0.945	2.90
0.26	0.30	0.61	0.94	0.950	3.00
0.27	0.31	0.62	0.97	0.955	3.10
0.28	0.33	0.63	0.99	0.960	3.22
0.29	0.34	0.64	1.02	0.965	3.35
0.30	0.36	0.65	1.05	0.970	3.51
0.31	0.37	0.66	1.08	0.975	3.69
0.32	0.39	0.67	1.11	0.980	3.91
0.33	0.40	0.68	1.14	0.985	4.20
0.34	0.42	0.69	1.17	0.990	4.61
0.35	0.43	0.70	1.20	0.995	5.30

3.6. Plotless Methods   Setting up plots, either in a grid or in a random fashion, may be tedious (see section 3A for plot sampling). Thus procedures have been developed to describe and analyze the spatial distribution of individuals without a need for plots or Poisson mathematics. Two such methods will be described here.

In the Holgate (1965) method, a number of randomly selected points (see section 1A.2) are marked in the habitat. Then, from each point one measures the distance to the nearest plant (let us call this point-to-plant distance d) and the distance to the second nearest plant (d'). Then, we square d, square d', and calculate the ratio d^2/d'^2 for each point. Then all of the ratios are summed, and the sum is divided by n, the total number of points. The result

may be considered an "index of aggregation," called by some a "coefficient of aggregation":

$$A_1 = \frac{\Sigma(d^2/d'^2)}{n} - 0.5. \qquad (20)$$

If the distribution of individuals is random, A_1 will be 0. If the dispersion is contagious, then $A_1 > 0$; if it is uniform, then $A_1 < 0$.

Furthermore, the significance of the departure of A from 0.5 may be determined statistically by computing:

$$t = \frac{|A_1|}{\sqrt{n/12}}; \qquad (21)$$

This value of t may then be compared to the value of t for infinity degrees of freedom on table 1B.1. That is, if one performs the test at the 5% level of statistical significance, and a t from equation 21 is greater than or equal to 1.96 (from table 1B.1), then the conclusion is that the dispersion was not random.

The Hopkins (1954) index of aggregation is determined as follows: Mark n random points in the habitat, as for the Holgate method above. Then, as above, measure the distance of each point to the nearest plant (call this point-to-plant measurement d). Then select a total of n plants at random and measure the distance from each of them to the nearest plant (d'). The index of aggregation is:

$$A_2 = \Sigma d^2 / \Sigma d'^2 - 1. \quad (22)$$

A randomly distributed population will result in $A_2 = 0$. $A_2 > 0$ would signify aggregation, and $A_2 < 0$ would indicate uniform dispersion.

Determining A_2 usually is somewhat more difficult than determining A_1, as the selection of random plants may be tedious. (One may *not* consider the plant nearest a random point to be a randomly selected plant. A good but time-consuming procedure would be to label each plant in the habitat with a different number and select random numbers, as from table 1A.1.)

The statistical significance of A_2 is determined by calculating:

$$t = 2|(A_2 + 1)/(A_2 + 2) - 0.5| \sqrt{2n + 1}, \quad (23)$$

and the computed value is compared to the appropriate critical value of t for ∞ degrees of freedom (table 1B.1). An n of at least 50 is necessary for this significance testing to be valid.

The index of aggregation of Johnson and Zimmer (1985) is based on point-to-point distances (d) that are determined just as for the Holgate or Hopkins indices described above:

$$A_3 = \frac{(n + 1)\Sigma(d_i^2)^2}{(\Sigma d_i^2)^2} - 2 \quad (24)$$

and, as with A_1 and A_2, $A_3 = 0$ indicates a random distribution, $A_3 > 0$ indicates contagion, and $A_3 <$ indicates a uniform distribution.

Statistical significance of A_3 may be ascertained by

$$t = \frac{A_3}{\sqrt{\dfrac{4(n - 1)}{(n + 2)(n + 3)}}}, \quad (25)$$

comparing t to the critical value for ∞ degrees of freedom in Table 1B.1.

Other measures are available that employ only plant-to-plant distances (Sinclair, 1985).

4. Suggested exercises

1. Using the plot method of plant sampling (section 3A), and the methods of sections 3.1–3.5, above, determine if the total number of all species of trees in an area is distributed at random. Then examine each major species individually. Which species show random, uniform, or aggregated distributions? (Use any of the methods in section 3.)
2. Compare the plant distribution patterns of two different areas, such as a grassland and a forest, or a grassland and a desert. Describe and discuss the patterns discerned.
3. Study the patterns of distribution of the aquatic invertebrates at the bottom of a stream, using collection methods given in section 3E.
4. Using drop boards or plot methods (see section 3D), determine the pattern of distribution for soil or litter invertebrates.
5. In the habitat(s) chosen for the above exercise 1, 2, or 3, compare a Poisson method to a plotless method in assessing dispersion pattern.

5. Selected references

Barbour, M. G., J. H. Burk, and W. D. Pitts. 1980. Terrestrial plant ecology. Benjamin Cummings Publishing Co., Menlo Park, Calif.

Clapham, A. R. 1936. Overdispersion in grassland communities and the use of statistical methods in plant ecology. J. Ecol. 24:232–251.

Cole, L. C. 1946. A study of the cryptozoa of an Illinois woodland. Ecol. Monogr. 16:70–74.

Grieg-Smith, P. 1983. Quantitative plant ecology. 3d ed. University of California Press, Berkeley, Calif.

Heltsche, J. F., and T. A. Ritchey. 1984. Spatial pattern detection using quadrat samples. Biometrics 40:877–885.

Holgate, P. 1965. Some new tests of randomness. J. Ecol. 53:261–266.

Hopkins, B. 1954. A new method for determining the type of distribution of plant individuals. Ann. Bot. London 18:213–227.

Johnson, R. B., and W. J. Zimmer. 1985. A more powerful test for dispersion using distance measurements. Ecology 66:1084–1085.

Kershaw, K. A., and J. H. Looney. 1985. Quantitative and dynamic plant ecology. 3d ed. Edward Arnold, London.

Lloyd, M. 1967. Mean crowding. J. Animal Ecol. 36:1–30.

McGinnis, W. G. 1934. The relationship between frequency indices and abundance as applied to plant populations in a semiarid region. Ecology 15:263–282.

Morisita, M. 1959. Measuring the dispersion of individuals and analysis of the distributional patterns. Mem. Fac. Sci. Kyushu Univ., Ser. E (Biol.) 2:215–235.

Pielou, E. C. 1974. Population and community ecology. Gordon and Breach, New York.

Pielou, E. C. 1977. Mathematical ecology. John Wiley & Sons, New York.

Poole, R. W. 1974. An introduction to quantitative ecology. McGraw-Hill Book Co., New York.

Rohlf, F. J. and R. R. Sokal. 1981. Statistical tables. 2d ed. W. H. Freeman and Co., San Francisco.

Sinclair, D. F. 1985. On tests of spatial randomness using mean nearest neighbor distance. Ecology 66:1084–1085.

Southwood, T. R. E. 1978. Ecological methods. Chapman and Hall, London.

Vandermeer, J. 1981. Elementary mathematical ecology. John Wiley & Sons, New York.

Zar, J. H. 1974. Biostatistical analysis. Prentice-Hall, Englewood Cliffs, N.J.

Zar, J. H. 1984. Biostatistical Analysis. 2d ed. Prentice-Hall, Englewood Cliffs, N.J.

Table 4C.4. *Probabilities, predicted by the Poisson distribution, of finding* X *individuals per plot, if the mean density per plot is* μ.

X	$\mu=$ 0.1	0.2	0.3	0.4	0.5	0.6	0.7	0.8	0.9	1.0
0	0.905	0.819	0.741	0.670	0.607	0.549	0.497	0.449	0.407	0.368
1	0.090	0.164	0.222	0.268	0.303	0.329	0.348	0.359	0.366	0.368
2	0.005	0.016	0.033	0.054	0.076	0.099	0.122	0.144	0.165	0.184
3	0.000	0.001	0.003	0.007	0.013	0.020	0.028	0.038	0.049	0.061
4	0.000	0.000	0.000	0.001	0.002	0.003	0.005	0.008	0.011	0.015
5	0.000	0.000	0.000	0.000	0.000	0.000	0.001	0.001	0.002	0.003
6	0.000	0.000	0.000	0.000	0.000	0.000	0.000	0.000	0.000	0.001

X	$\mu=$ 1.1	1.2	1.3	1.4	1.5	1.6	1.7	1.8	1.9	2.0
0	0.333	0.301	0.273	0.247	0.223	0.202	0.183	0.165	0.150	0.135
1	0.366	0.361	0.354	0.345	0.335	0.323	0.311	0.298	0.284	0.271
2	0.201	0.217	0.230	0.242	0.251	0.258	0.264	0.268	0.270	0.271
3	0.074	0.087	0.100	0.113	0.126	0.138	0.150	0.161	0.171	0.180
4	0.020	0.026	0.032	0.039	0.047	0.055	0.064	0.072	0.081	0.090
5	0.004	0.006	0.008	0.011	0.014	0.018	0.022	0.026	0.031	0.036
6	0.001	0.001	0.002	0.003	0.004	0.005	0.006	0.008	0.010	0.012
7	0.000	0.000	0.000	0.001	0.001	0.001	0.001	0.002	0.003	0.003
8	0.000	0.000	0.000	0.000	0.000	0.000	0.000	0.000	0.001	0.001

X	$\mu=$ 2.1	2.2	2.3	2.4	2.5	2.6	2.7	2.8	2.9	3.0
0	0.122	0.111	0.100	0.091	0.082	0.074	0.067	0.061	0.055	0.050
1	0.257	0.244	0.231	0.218	0.205	0.193	0.181	0.170	0.160	0.149
2	0.270	0.268	0.265	0.261	0.257	0.251	0.245	0.238	0.231	0.224
3	0.189	0.197	0.203	0.209	0.214	0.218	0.220	0.222	0.224	0.224
4	0.099	0.108	0.117	0.125	0.134	0.141	0.149	0.156	0.162	0.168
5	0.042	0.048	0.054	0.060	0.067	0.074	0.080	0.087	0.094	0.101
6	0.015	0.017	0.021	0.024	0.028	0.032	0.036	0.041	0.045	0.050
7	0.004	0.005	0.007	0.008	0.010	0.012	0.014	0.016	0.019	0.022
8	0.001	0.002	0.002	0.002	0.003	0.004	0.005	0.006	0.007	0.008
9	0.000	0.000	0.000	0.001	0.001	0.001	0.001	0.002	0.002	0.003
10	0.000	0.000	0.000	0.000	0.000	0.000	0.000	0.000	0.001	0.001

X	$\mu=$ 3.1	3.2	3.3	3.4	3.5	3.6	3.7	3.8	3.9	4.0
0	0.045	0.041	0.037	0.033	0.030	0.027	0.025	0.022	0.020	0.018
1	0.140	0.130	0.122	0.113	0.106	0.098	0.091	0.085	0.079	0.073
2	0.216	0.209	0.201	0.193	0.185	0.177	0.169	0.162	0.154	0.147
3	0.224	0.223	0.221	0.219	0.216	0.212	0.209	0.205	0.200	0.195
4	0.173	0.178	0.182	0.186	0.189	0.191	0.193	0.194	0.195	0.195
5	0.107	0.114	0.120	0.126	0.132	0.138	0.143	0.148	0.152	0.156
6	0.056	0.061	0.066	0.072	0.077	0.083	0.088	0.094	0.099	0.104
7	0.025	0.028	0.031	0.035	0.039	0.042	0.047	0.051	0.055	0.060
8	0.010	0.011	0.013	0.015	0.017	0.019	0.022	0.024	0.027	0.030
9	0.003	0.004	0.005	0.006	0.007	0.008	0.009	0.010	0.012	0.013
10	0.001	0.001	0.002	0.002	0.002	0.003	0.003	0.004	0.005	0.005
11	0.000	0.000	0.000	0.001	0.001	0.001	0.001	0.001	0.002	0.002
12	0.000	0.000	0.000	0.000	0.000	0.000	0.000	0.000	0.001	0.001

The values in this table were computed from equation 1. More extensive tables of Poisson probabilities are found in Rohlf and Sokal (1981:164–165) and Zar (1974:543–548).

Table 4C.4—*(Continued)*

X	μ = 4.1	4.2	4.3	4.4	4.5	4.6	4.7	4.8	4.9	5.0
0	0.017	0.015	0.014	0.012	0.011	0.010	0.009	0.008	0.007	0.007
1	0.068	0.063	0.058	0.054	0.050	0.046	0.043	0.040	0.036	0.034
2	0.139	0.132	0.125	0.119	0.112	0.106	0.100	0.095	0.089	0.084
3	0.190	0.185	0.180	0.174	0.169	0.163	0.157	0.152	0.146	0.140
4	0.195	0.194	0.193	0.192	0.190	0.188	0.185	0.182	0.179	0.175
5	0.160	0.163	0.166	0.169	0.171	0.173	0.174	0.175	0.175	0.175
6	0.109	0.114	0.119	0.124	0.128	0.132	0.136	0.140	0.143	0.146
7	0.064	0.069	0.073	0.078	0.082	0.087	0.091	0.096	0.100	0.104
8	0.033	0.036	0.039	0.043	0.046	0.050	0.054	0.058	0.061	0.065
9	0.015	0.017	0.019	0.021	0.023	0.026	0.028	0.031	0.033	0.036
10	0.006	0.007	0.008	0.009	0.010	0.012	0.013	0.015	0.016	0.018
11	0.002	0.003	0.003	0.004	0.004	0.005	0.006	0.006	0.007	0.008
12	0.001	0.001	0.001	0.001	0.002	0.002	0.002	0.003	0.003	0.003
13	0.000	0.000	0.000	0.000	0.001	0.001	0.001	0.001	0.001	0.001

X	μ = 5.1	5.2	5.3	5.4	5.5	5.6	5.7	5.8	5.9	6.0
0	0.006	0.006	0.005	0.005	0.004	0.004	0.003	0.003	0.003	0.002
1	0.031	0.029	0.026	0.024	0.022	0.021	0.019	0.018	0.016	0.015
2	0.079	0.075	0.070	0.066	0.062	0.058	0.054	0.051	0.048	0.045
3	0.135	0.129	0.124	0.119	0.113	0.108	0.103	0.098	0.094	0.089
4	0.172	0.168	0.164	0.160	0.156	0.152	0.147	0.143	0.138	0.134
5	0.175	0.175	0.174	0.173	0.171	0.170	0.168	0.166	0.163	0.161
6	0.149	0.151	0.154	0.156	0.157	0.158	0.159	0.160	0.160	0.161
7	0.109	0.113	0.116	0.120	0.123	0.127	0.130	0.133	0.135	0.138
8	0.069	0.073	0.077	0.081	0.085	0.089	0.092	0.096	0.100	0.103
9	0.039	0.042	0.045	0.049	0.052	0.055	0.059	0.062	0.065	0.069
10	0.020	0.022	0.024	0.026	0.029	0.031	0.033	0.036	0.039	0.041
11	0.009	0.010	0.012	0.013	0.014	0.016	0.017	0.019	0.021	0.023
12	0.004	0.005	0.005	0.006	0.007	0.007	0.008	0.009	0.010	0.011
13	0.002	0.002	0.002	0.002	0.003	0.003	0.004	0.004	0.005	0.005
14	0.001	0.001	0.001	0.001	0.001	0.001	0.001	0.002	0.002	0.002
15	0.000	0.000	0.000	0.000	0.000	0.000	0.001	0.001	0.001	0.001

X	μ = 6.1	6.2	6.3	6.4	6.5	6.6	6.7	6.8	6.9	7.0
0	0.002	0.002	0.002	0.002	0.002	0.001	0.001	0.001	0.001	0.001
1	0.014	0.013	0.012	0.011	0.010	0.009	0.008	0.008	0.007	0.006
2	0.042	0.039	0.036	0.034	0.032	0.030	0.028	0.026	0.024	0.022
3	0.085	0.081	0.077	0.073	0.069	0.065	0.062	0.058	0.055	0.052
4	0.129	0.125	0.121	0.116	0.112	0.108	0.103	0.099	0.095	0.091
5	0.158	0.155	0.152	0.149	0.145	0.142	0.138	0.135	0.131	0.128
6	0.160	0.160	0.159	0.159	0.157	0.156	0.155	0.153	0.151	0.149
7	0.140	0.142	0.144	0.145	0.146	0.147	0.148	0.149	0.149	0.149
8	0.107	0.110	0.113	0.116	0.119	0.121	0.124	0.126	0.128	0.130
9	0.072	0.076	0.079	0.082	0.086	0.089	0.092	0.095	0.098	0.101
10	0.044	0.047	0.050	0.053	0.056	0.059	0.062	0.065	0.068	0.071
11	0.024	0.026	0.029	0.031	0.033	0.035	0.038	0.040	0.043	0.045
12	0.012	0.014	0.015	0.016	0.018	0.019	0.021	0.023	0.025	0.026
13	0.006	0.007	0.007	0.008	0.009	0.010	0.011	0.012	0.013	0.014
14	0.003	0.003	0.003	0.004	0.004	0.005	0.005	0.006	0.006	0.007
15	0.001	0.001	0.001	0.002	0.002	0.002	0.002	0.003	0.003	0.003
16	0.000	0.000	0.001	0.001	0.001	0.001	0.001	0.001	0.001	0.001
17	0.000	0.000	0.000	0.000	0.000	0.000	0.000	0.000	0.001	0.001

4d

competition

1. Introduction

Competition is the striving by two or more individuals for a common resource in short supply, with a disadvantage accruing to at least one of the competitors. Whether a resource (food, space, water, nutrients) is scarce depends not only on the number of species desiring it, but on how many individuals are in each species and how much of the resource is needed by each individual.

Intraspecific competition (competition between individuals of the same population) is one of the factors that in many species contributes to the decrease in dN/dt as logistic population growth progresses (section 4B.2.2). *Interspecific competition* (between two or more species) for *all* of the required resources tends to result in the survival of only one of the species. This important concept is the *competitive exclusion principle* (Hardin, 1960) stated also as "complete competitors cannot coexist."[1]

Much available literature concerns the theory of competition. The mathematical models, many of them complex, are difficult to apply in a simple laboratory exercise. In this section, we shall examine some of the basic properties of intra- and interspecific competition as may be revealed by simple laboratory experiments. Such laboratory studies are simplifications of events in natural habitats, yet they may help demonstrate the basic features of competition as an ecological process.

Competition is an ecological concept having significant ramifications with respect to resource utilization, niche theory, and evolution.

2. Procedure

Two approaches to the study of competition are given here. In one, population growth curves are observed as in section 4B.2.3, but for two different species. In the second, competition is analyzed by determining the effects of different controlled densities on the growth and survival of individuals of two species. In either approach, the experiment is designed so that some members of the two species are allowed to grow separately while others of the same two species are raised together. The population dynamics approach is recommended as a class study for species having short generation times, such as yeast, *Paramecium*, *Drosophila*, or *Tribolium*. (Procedures for culturing *Paramecium* are given in section 6D.2.) The second approach is useful for species having long generation times, such as seed plants. The studies may be performed on any suitable plant, animal, or microbial organisms, those that are easily grown in the laboratory. It is recommended that the two species be from related taxa.

2.1. Effects of Competition on Population Growth   The procedure of section 4B.2.3 may be used to study the effects of competition on population dynamics. Set up at least five replicates of each of three different experimental cultures: one with species A alone, one with species B alone, and one with both species together. Combine all replicate data for each of the three experiments and calculate the intrinsic rate of growth, r, and the carrying capacity, K, for each species grown alone. Compare this to the r and K observed for each species when grown together. Make a graph of population density as a function of time, on which the data for the two species grown alone are plotted together. Prepare a second graph of density versus time, in which the two species are in competition (as the solid lines in figures 4D.1 and 4D.2). In these plotted curves, look for evidence of population coexistence (e.g., figure 4D.1) or extinction (e.g., figure 4D.2).

If two species have requirements dissimilar enough so that the competitive impact of each on the other is low, then we say that both are weak competitors. They most likely can coexist in a habitat. Figure 4D.1 shows the growth of two such populations in each other's presence. If one of the two species is a strong competitor relative to the other (e.g., species A has a severe inhibiting effect on the population growth of species B), then typically it will eventually exclude the other (as shown in figure 4D.2). If two species are mutually strong competitors in a habitat (each has a strong competitive effect on the other), then one will competitively exclude the other. In the latter situation, the "winner" is a function of the very specific environmental conditions and of the values of K and N_o for each species.

1. An often-perpetuated misnaming of this is *Gause's principle,* after G. F. Gause, who never stated the general principle of competitive exclusion and who gave credit for the concept to the priority of V. Volterra and A. J. Lotka, in 1926 and 1932, respectively (Gilbert et al., 1952; Hardin, 1960). Although the principle is not his and he never claimed it to be, Gause (1934a, 1934b) may rightfully be given credit for some important early empirical testing of it.

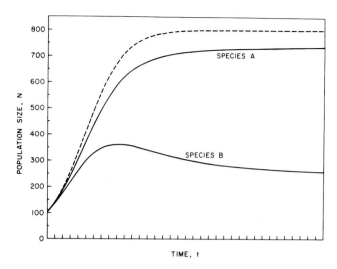

Figure 4D.1. Growth of two weakly competing populations with sustained coexistence. The dashed line is the growth of each population alone, where each population has the same r and K. The solid lines are for the two populations in competition.

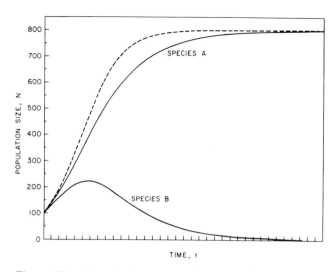

Figure 4D.2. Growth of two competing populations, one (A) a distinctly stronger competitor than the other (B), with competitive exclusion of the latter by the former. The dashed line is the growth of each population alone, where each population has the same r and K. The solid lines are for the two populations in competition.

What is observed in figures 4D.1 and 4D.2 is the expression of the effect of interspecific competition on logistic population growth. If each of two populations is allowed to grow alone (i.e., in the absence of competition from another population), equation 12 in section 4B.2 may be written as

$$\frac{dN_1}{dt} = r_1 N_1 \left(\frac{K_1 - N_1}{K_1} \right) \qquad (1)$$

for population 1 and

$$\frac{dN_2}{dt} = r_2 N_2 \left(\frac{K_2 - N_2}{K_2} \right) \qquad (2)$$

for population 2.

If the two populations occur in the same place at the same time, and are in competition (i.e., are striving for some of the same resources), then the growth of each is modified as follows (these being known as Gause, Gause-Volterra, or Lotka-Volterra, equations):

$$\frac{dN_1}{dt} = r_1 N_1 \left(\frac{K_1 - N_1 - \alpha_{12} N_2}{K_1} \right) \qquad (3)$$

$$\frac{dN_2}{dt} = r_2 N_2 \left(\frac{K_2 - N_2 - \alpha_{21} N_1}{K_2} \right), \qquad (4)$$

where the *competition coefficient*, α_{ij}, is the effect of population j on population i.

If $\alpha_{12} > K_1/K_2$ and $\alpha_{21} < K_2/K_1$, then species 1 will be eliminated and species 2 will persist. If $\alpha_{12} < K_1/K_2$ and $\alpha_{21} > K_2/K_1$, then species 2 will be eliminated and species 1 will persist (as in figure 4D.2). If $\alpha_{12} < K_1/K_2$, then both species will persist, reaching equilibrium population sizes (as in figure 4D.1). If $\alpha_{12} > K_1/K_2$ and $\alpha_{21} > K_2/K_1$, then one species will be eliminated and one will persist, the identity of each depending on the relative population sizes at the start.

In laboratory experimentation one species will frequently "win" in some experimental trials and "lose" in others, although the environmental conditions are kept as consistent as possible from one trial to another. Therefore, a number of replications are required to determine who the winner is most likely to be. Tabulate the number of replicates in which each species wins, loses, or coexists.

A determination of whether one species is a significantly better competitor under a given set of circumstances may be made by employing chi-square goodness-of-fit testing as described in section 1B.4. For this test, n, the number of replicates in which either species eliminated the other, should be at least 10. The null hypothesis is that there is no difference in competitive advantage between the two species. The expected frequencies are $F_1 = F_2 = n/2$; f_1 is the number of trials in which species A won, and f_2 is the number of trials in which species B eliminated species A.

For further investigations, this approach may be used in experiments that control such variables as the initial population densities of the competitors, or the temperature, humidity, light, nutrients, or salt concentrations in the environment. Competitive exclusion, or other aspects of population dynamics, might be found to depend on such factors. Temperature and humidity are good variables to control in studies with insects, whereas mineral nutrients and salts are conveniently varied with algae, higher plants, and aquatic invertebrates. Consult original literature when necessary to determine the best conditions to use for the species chosen.

2.2. Effects of Competition on Individual Vigor Green plants are most easily used in this second approach in studying competition, although populations of young animals also may be examined. In the case of plants, use fast-growing annuals or grasses in which a high percentage of germination is certain. Plants that produce an allelopathogen or have some other clear competitive mechanism will stand out in a demonstration. Set up plots or trays in which seeds are planted in different densities. Invertebrates may also be used; place equal numbers of very young individuals in each of several containers. Label each plant or animal container and make at least five replications of each of the two populations growing alone and of the two populations growing together.

Allow plant populations to grow for a month or two, depending on the amount of time needed to exhibit the effects of competition. Then, count the number of individuals in each container and estimate the percentage of the original number (of seeds or young) that survive. Measure the shoot lengths of individual plants, or the wet- and dry-weight biomass (section 6A). If animals are used, measure the total length or the biomass of each individual.

Observe the effect of inter- and intraspecific competition on the yield (in $g/m^2/day$; see section 6A) of each population in each experimental condition. In the containers with interspecific competition, assign one-half the area of each container to each of the two species in calculating $g/m^2/day$. This enables you to compare survivals and yields with those of the single-species containers. Graph numbers surviving (as well as lengths and yields) at various times for each of the three experimental conditions. Observe how survival, length, and yield are affected by intra- and interspecific competition.

Perform similar experiments using manipulations of a variety of environmental factors (e.g., temperature, humidity, light, nutrients). You can thus see what effects competition has on similar or dissimilar species in various environments.

3. Suggested exercises

1. Perform the experiment described in section 4D.2.1 and determine the outcome of the competition. Repeat the experiment using two species that are either more or less similar ecologically. Repeat the experiment under different environmental conditions (such as temperature, humidity).
2. Perform the experiment described in section 4D.2.2, and determine the effect of the competition. Repeat the experiment using two ecologically similar or dissimilar species. Repeat the experiment under different environmental conditions.

4. Selected references

Andrewartha, H. G., and L. C. Birch. 1954. The distribution and abundance of animals. University of Chicago Press, Chicago.

Barbour, M. G., J. H. Burk, and W. D. Pitts. 1980. Terrestrial plant ecology. Benjamin/Cummings Publishing Co., Menlo Park, Calif.

Clements, F. E., J. E. Weaver, and R. C. Hanson. 1929. Plant competition. Carnegie Institute Publ. 398, Washington, D.C.

Evans, R. 1960. Differential responses of three species of annual grassland type to plant competition and mineral nutrition. Ecology 41:305–310.

Gause, G. F. 1934a. Experimental analysis of Vito Volterra's mathematical theory of the struggle for existence. Science 79:16–17.

Gause, G. F. 1934b. The struggle for existence. Williams and Wilkins Co., Baltimore.

Gause, G. F., and A. A. Witt. 1935. Behavior of mixed populations and the problem of natural selection. Amer. Natur. 69:576–609.

Gilbert, O., T. B. Reynoldson, and J. Hobart. 1952. Gause's hypothesis: An examination. J. Animal Ecol. 21:310–312.

Hardin, G. 1960. The competitive exclusion principle. Science 131:1292–1297.

Harper, J. L. 1980. Population biology of plants. Academic Press, New York.

Hutchinson, G. E. 1978. Introduction to population ecology. Yale University Press, New Haven, Conn.

MacArthur, R. H. and J. H. Connell. 1966. The biology of populations. John Wiley & Sons, New York.

MacArthur, R. H. and R. Levins. 1964. Competition, habitat selection and character displacement in a patchy environment. Proc. Nat. Acad. Sci. 51:1207–1210.

May, R. M. 1981. Models for two interacting populations, 78–104, 197–227. In R. M. May (ed.), Theoretical ecology: Principles and applications. Sinauer Associates, Sunderland, Mass.

Miller, R. S. 1967. Pattern and process in competition. Adv. Ecol. Res. 4:1–74.

Overland, L. 1966. The role of allelopathic substances in the "smother crop" barley. Amer. J. Bot. 53:423–432.

Park, T. 1962. Beetles, competition, and populations. Science 138:1369–1375.

Pianka, E. R. 1981. Competition and niche theory, 167–196. In R. M. May (ed.), Theoretical ecology: Principles and applications. Sinauer Associates, Sunderland, Mass.

Pielou, E. C. 1974. Population and community ecology. Gordon and Breach, New York.

Pielou, E. C. 1977. Mathematical ecology. John Wiley & Sons, New York.

Poole, R. W. 1974. An introduction to quantitative ecology. McGraw-Hill Book Co., New York.

Sakai, K. 1955. Competition in plants and its relation to selection. Cold Springs Harbor Symp. Quant. Biol. 20:137–157.

Schoener, T. W. 1974. Resource partitioning in ecological communities. Science 185:27–39.

Silverton, J. W. 1987. Introduction to plant population ecology. 2d ed. John Wiley & Sons, New York.

Slobodkin, L. B. 1980. Growth and regulation of animal populations. Holt, Rinehart and Winston, New York.

Southwood, T. R. E. 1978. Ecological methods. Chapman & Hall, London.

Tamarin, R. H. (ed.). 1981. Population regulation. Benchmark papers in ecology no. 7. Academic Press, New York.

Vandermeer, J. 1981. Elementary mathematical ecology. John Wiley & Sons, New York.

Wangersky, P. J. 1978. Lotka-Volterra population models. Annu. Rev. Ecol. Systemat. 9:189–218.

Whittaker, R. H., and S. A. Levin. 1975. Niche: Theory and application. Benchmark papers in ecology no. 3. Academic Press, New York.

Wilson, E. O., and W. H. Bossert. 1971. A primer of population biology. Sinauer Associates, Stamford, Conn.

4e

Predation

1. Introduction

Predation is the process by which an animal consumes all or part of another organism. The feeding behavior of individual predators and the dynamics of predator and prey populations are the two primary aspects of predator-prey interactions. If a prey population increases in size, the predator population can respond by each predator consuming more prey (*functional response*), and/or by the predator population increasing in size through reproduction and/or immigration (*numerical response*). Thus, the functional response of a predator population describes the feeding behavior of individual predators in response to changes in prey density, and the numerical response describes the change in size of the predator population. The extent to which predator populations can regulate the density of prey populations has long interested ecologists. Mathematical models can be used to investigate parameters that contribute to stability or instability of predator and prey populations.

Optimal foraging theory uses a different approach to the study of feeding behavior of individual predators. It predicts how a predator would select from a variety of prey if it were to feed in an optimal manner. The theory is based on mathematical models that assume an animal forages for prey in a way that maximizes its energy return or minimizes its time spent capturing prey.

2. Functional response

A graph of the functional response of a predator shows the relationship between the number of prey eaten by an individual predator in a fixed period of time and prey density (figure 4E.1). Because there is a maximum rate of prey consumption by a predator, the curve should level off at some higher prey density. There generally have been three kinds of functional responses described: type I, where

there is an increasing linear response to a plateau; type II, where the response rises continually, but at a decreasing rate; and type III, where the response is sigmoid. To regulate the dynamics of a prey population, a predator population must cause *density-dependent* mortality. It must consume a greater proportion of the prey at higher prey densities that at lower densities. Density-dependent mortality by itself will not regulate prey density, but it is necessary for regulation.

In a type I functional response, a constant proportion of the prey are killed at the lower prey densities, and a constant number are killed when the curve plateaus (*density-independent mortality*). A decreasing proportion of the prey are killed as prey density increases in a type II functional response (*negatively density-dependent mortality*). The sigmoid shape of a type III functional response curve produces *positively density-dependent* prey mortality and is the only one of the three types of response that can lead to regulation of a prey population.

The type II functional response is the type most frequently observed for animals (Begon et al., 1986). An explanation for the shape of the type II curve was proposed by Holling (1959) based on the time a predator spent searching for and handling prey. First, let us say that N_p is the number of prey eaten by a predator spending a given length of time (T_s) searching for prey. N_p will depend on the density of Prey (D), the attack rate of the predator (A), and the length of the search time (T_s). The relationship can be written as

$$N_p = AT_sD, \qquad (1)$$

but the predator also has to spend some time handling (capturing and consuming) each prey item, so handling time per item (T_h) should be incorporated into the equation. If T is total predation time and T_hN_p is the total handling time for all the prey captured, their relationship with T_s can be written as

$$T_s = T - T_hN_p. \qquad (2)$$

Substituting this into equation 1 and rearranging terms we arrive at Holling's equation for a type II response:

$$N_p = AD(T - T_hN_p) \qquad (3)$$

or, rearranged:

$$N_p = \frac{AD}{1 + AT_hD}. \qquad (4)$$

This equation produces a functional response curve describing decreasing predation rates at higher prey densities. Holling suggested that the curve plateaus because handling time becomes the limiting factor to prey consumption at the higher prey densities. The number of prey eaten by the predator is limited by the rate at which it can handle (capture and consume) prey when they are very

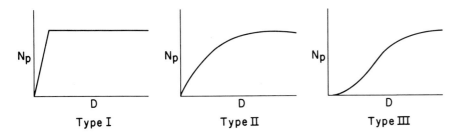

Figure 4E.1. Three types of predator-prey functional response curves. (D = density of prey; N_p = number of prey consumed by an individual predator.)

abundant. This equation assumes the denisty of prey remains constant (i.e., prey are replaced as removed), and that the number eaten is estimated for a fixed T^1.

For functional response data approximating a type II curve, it is possible to estimate coefficients for equation 3 by rearranging the equation to fit a straight line (linear) relationship to the data. Although this technique is the easiest to apply, it can produce biased and inaccurate estimates of parameters. Therefore, its use should be restricted to class exercise where the demonstration of the phenomenon of a functional response is the purpose.[2]

To use the linear method we must rearrange equation 3. If we divide both sides of the equation by D:

$$N_p/D = AT - AT_hN_p, \qquad (6)$$

we have an equation in the form of a straight line (equation 20 in Section 1B.6.1), where N_p/D is the dependent variable, N_p is the independent variable, AT is the Y-intercept, and $-AT_h$ is the slope (i.e., a negative slope) of the line. The values of AT and AT_h can be obtained by drawing or calculating a "best fit" straight line (Section 1B.6.1) through a plot of N_p/D vs. N_p, and determining where the line intercepts the Y axis and X axis, respectively. As T is fixed by the investigator, the equations below can be used to solve for A and T_h. For the straight line, a is the Y-intercept and b is the slope:

$$a = AT \qquad (7)$$

so

$$A = a/T \qquad (8)$$

1. Royama (1971) and Rogers (1972) proposed a modification of equation 3 for situations in which prey density cannot be held constant:

$$N_p = D\{1 - e^{[-A(T - T_hN_p)]}\} \qquad (5)$$

2. Williams and Juliano (1985) found a direct nonlinear regression method gave the most consistent and least biased estimates of parameters compared to linearization techniques. For situations in which nonlinear regression methods are unavailable, however, they also proposed a nonparametric technique. (See also Juliano and Williams, 1987.) Trexler et al. (1988) suggest logit analysis (logistic regression) is more reliable than nonlinear curve-fitting in determining the type of functional response a data set best fits.

and

$$b = -AT_h \qquad (9)$$

so

$$T_h = -b/A. \qquad (10)$$

3. Optimal foraging

Optimization models have been applied to many aspects of predator behavior. The basis of their use in behavioral ecology is that given an animal has certain morphological and physiological characteristics (constraints), we wish to ask what behavior(s) maximize(s) or minimize(s) the individual's fitness under that set of constraints. Optimization refers to the maximum or minimum associated with a particular set of constraints, not the optimum under all possible sets of conditions (Krebs and McCleary 1984). One of the earliest applications to predation theory was a situation in which a predator has a choice of feeding on alternative prey types.

Following Charnov (1976) and Krebs and McCleary (1984), we can develop an optimal choice model that considers the ratio of energy obtained from capturing prey and the cost of the time required to "handle" (capture and consume) and search for a prey item. Assume a predator can feed on two prey types and spends T_s time searching for prey and encounters them at rates of λ_1 and λ_2 prey per second, respectively. The energy gained from each prey type (E_1 and E_2, respectively) divided by the respective handling times (T_{h_1} and T_{h_2}) gives the profitabilities of each prey item for the predator (E_1/T_{h_1} and E_2/T_{h_2}). If the predator feeds unselectively on both prey items for a period of time, T_s, then its energy return for that period would be E, where

$$E = T_s(\lambda_1E_1 + \lambda_2E_2) \qquad (11)$$

and its total time to do this would be

$$T = T_s + T_s(\lambda_1T_{h_1} + \lambda_2T_{h_1}). \qquad (12)$$

Notice that the first term on the right side of equation 12 is total search time and the second is the total handling time for the two prey. By taking the ratio of equations 11 and 12 we get the overall energy intake (E/T) of the predator (with T_s canceling out); this a form of a two-species version of Holling's (1959) equation 4:

$$\frac{E}{T} = \frac{\lambda_1 E_1 + \lambda_2 E_2}{1 + \lambda_1 T_{h_1} + \lambda_2 T_{h_2}}. \quad (13)$$

If we now could rank the two prey items in terms of their profitability and found prey type 1 was more profitable than prey type 2 ($E_1/T_{h_1} > E_2/T_{h_2}$), could we then predict the diet the predator should have to maximize E/T? Yes, the predator should specialize on prey type 1 as long as it is more profitable than preying on a combination of prey type 1 and prey type 2. In terms of equations it can be written as

$$\frac{\lambda_1 E_1}{1 + \lambda_1 T_{h_1}} > \frac{\lambda_1 E_1 + \lambda_2 E_2}{1 + \lambda_1 T_{h_1} + \lambda_2 T_{h_2}} \quad (14)$$

where the left hand side is E/T from selecting only prey 1 and the right side is E/T from selecting both prey. Rearranging inequality 14 (where, in manipulation, the $>$ sign in an inequality can be treated in the same fashion as an $=$ sign in an equation), we derive the following relationship:

$$\frac{1}{\lambda_1} < \frac{E_1 T_{h_2}}{E_2} - T_{h_1}, \quad (15)$$

which is expressed in terms of the search time for the next prey item ($1/\lambda_1$). Thus, we have used these simple equations to make predictions on how a predator should feed if it were to feed in an optimal manner. The assumptions involved in this equation and predictions that can be developed from it are listed by Krebs and McCleary (1984).

4. Suggested exercises

1. Estimate the functional response curve for a predator exposed to different densities of prey for a fixed time period. Possible exercises are: (a) a blindfolded student finding and retrieving pieces of sandpaper (tacked to corkboard); (b) dragonfly larvae feeding on midge larvae; (c) damselfly larvae feeding on zooplankton (e.g., *Daphnia*); fish feeding on mealworms, midge larvae, or tubificid worms.
2. For the optimal foraging model try using some of the same predators and prey suggested in exercise 1: (a) Repeat the exercise with prey of significantly different sizes (of the same or different species), so there will be a difference in energy content and handling time. For a particular system, determine the mean dry weight (an an index of energy content), mean handling time, and mean search time for each of the

two prey sizes. Substitute these data into the model to predict whether the predator should feed on both prey sizes or just one size. (b) Conduct a prey selection experiment in which both prey are presented to the predator. Do the results agree with the theory? This kind of experiment also could be done with squirrels in a local park. Two different kinds of food items could be used (e.g., walnuts and peanuts).

5. Selected references

Begon, M., J. L. Harper, and C. R. Townsend. 1986. Ecology: Individuals, populations, and communities. Sinauer Associates, Sunderland, Mass.

Begon, M., and M. Mortimer. 1986. Population ecology: A unified study of animals and plants. 2d ed. Sinauer Associates, Sunderland, Mass.

Charnov, E. L. 1976. Optimal foraging: Attack strategy of a mantid. Amer. Natur. 110:141–151.

Gause, G. F. 1934. The struggle for existence. Williams and Wilkins Co., Baltimore. (Also, 1964, Hafner Publishing Co., New York.)

Hassell, M. P. 1978. The dynamics of arthropod predator-prey systems. Princeton University Press, Princeton, N.J.

Holling, C. S. 1959. Some characteristics of simple types of predation and parasitism. Can. Entomol. 91:385–389.

Juliano, S. A., and F. M. Williams. 1987. A comparison of methods for estimating the functional response parameter of the random predator equation. J. Anim. Ecol. 56:641–653.

Krebs, J. R., and N. B. Davies (eds.). 1984. Behavioral ecology: An evolutionary approach. Sinauer Associates, Sunderland, Mass.

Krebs, J. R., and R. H. McCleary. 1984. Optimization and behavioral ecology, 91–121. In J. R. Krebs and N. B. Davies (eds.), Behavioral ecology: An evolutionary synthesis. Sinauer Associates, Sunderland, Mass.

MacArthur, R. H., and E. R. Pianka. 1966. On optimal use of patchy environment Amer. Natur. 100:603–609.

Rogers, D. T. 1972. Random search and insect population models. J. Anim. Ecol. 41:369–383.

Rose, M. R. 1987. Quantitative ecological theory: An introduction to basic models. Johns Hopkins University Press, Baltimore, Md.

Royama, T. 1971. A comparative study of models for predation and parasitism. Res. Pop. Ecol. (Suppl. 1):1–91.

Stephens, D. W., and J. R. Krebs. 1986. Foraging theory. Princeton University Press, Princeton, N.J.

Trexler, J. C., C. D. McCulloch, and J. Travis. 1988. How can the functional response best be determined? Oecologia 76:206–214.

Vandermeer, J. 1981. Elementary mathematical ecology. John Wiley & Sons, New York.

Williams, F. M., and S. A. Juliano. 1985. Further difficulties in the analysis of functional response experiments and a resolution. Can. Entomol. 117:631–640.

introduction

A *community* is an assemblage of interacting populations that constitute a relatively self-sufficient ecological unit. Often we consider only a portion of a community, that is, an assemblage of interacting populations that lack self-sufficiency. This is called a *subcommunity*. For example, we may examine the ground subcommunity or a rotting log subcommunity, within a deciduous forest community.

Although a community does not always have easily delineated spatial boundaries, it is a basic ecological concept. Communities show organization and homeostasis, and they have unique structural and functional attributes not possessed by their individual component populations. We shall examine aspects of community structure in sections 5A, 5B, and 5C, and consider an important dynamic community process in section 5D. Considerations of community biomass and productivity are covered in unit 6.

unit 5

analysis of communities

5a

community structure

1. Introduction

A community is too large and complex an ecological unit to be studied in detail during one or a few field trips. However, you can study some basic features of community structure and function in a relatively short time. You can also analyze in greater detail one or two specific aspects of community organization, such as species diversity, zonation, or stratification. Or, you can subdivide a community and analyze only certain subcommunities within it, such as the plant components (section 3A, 3B, or 3C) or those animals or plants of certain strata within the community. Examples of commonly studied subcommunities include rotting logs, leaf litter, pond nekton, and stream benthos.

In terrestrial habitats, vegetation greatly influences physical and chemical factors in the habitat and thus the resident biological populations. Microclimate, light penetration, and soil conditions are largely determined by the dominant plants, which also afford protection and feeding and nesting sites for animals. We are here concerned not with a species description of the plant community, but with a summary of the vegetation features that affect the habitat. Aspects of plant community analysis are treated in sections 3A, 3B, 3C, and 5A. In this section we shall examine those features common to many types of communities and subcommunities. Although emphasis is placed on terrestrial situations, most of the concepts and many of the approaches discussed apply also to aquatic communities.

2. Vegetation analysis

Three different methods have been used to describe the plant portion of the community. First are detailed floristic lists, but these exclude many considerations useful to habitat analysis and generally require a well-trained taxonomist. A second approach involves a broad classification of community types using the dominant species names such as *mixed hemlock and sugar maple forest,* or *big bluestem prairie.* However, this approach characterizes only one aspect of the community and provides very little useful detail other than in a macrohabitat description (see section 2A.7). The third approach, *physiognomy,* consists of description and measurement of the form and appearance of the vegetation and is the one used in this section.

The idea that the form, structure, and spatial arrangement of vegetation affects the ecology of organisms is an important ecological concept. Therefore, it is not surprising that ecologists have turned to this type of habitat analysis. Physiognomic aspects of vegetation play a greater role in affecting the environment than does the species composition in the habitat. Physiognomic description of vegetation is a botanical procedure easily used by a nonspecialist; it results in a description of the basic organization, general appearance, and specific forms of the vegetation.

At least six important features of vegetation affect the community structure: dominant species, life form, stratification, foliage density, coverage, and plant dispersion. When combined with measurements of physical variables, physiognomic description has the advantages of being detailed yet nontechnical, accurate yet not quantitatively overwhelming, and organized yet flexible. The system used here is based on those used by Emlen (1956) and Kuchler (1949). For more details on various physiognomic systems consult Phillips (1959), Dansereau (1957), and Whittaker (1970).

Dominant species are the most influential in the habitat. They control the structure and species composition of the community by affecting physical and chemical factors such as temperature, wind, and humidity, as well as the availability of light, water, and nutrients. Record such species and note how they might affect others in the community. As you will see in section 5B, the degree of dominance has been quantified by some ecologists by using diversity indices.

The *horizontal pattern,* if any, should be recorded as a feature of the community structure. Within a community, organisms might be distributed along an environmental gradient, resulting in a local gradation in species distribution called *zonation.* For example, a hill may exhibit zones of vegetation comprised of different species from the top of the hill to the bottom. If the gradient is very large, such as on a mountain side, the zonation may be so prominent that distinct communities form. Outline any zonation patterns apparent in your study area and attempt to correlate them with observable environmental gradients. Often the plants found in such zones are important indicators of specific soil and moisture conditions.

At the junction of two communities you may find an *ecotone,* or transition area, sometimes very distinct. An ecotone typically contains some species characteristic of each adjoining community but may also have some species not found in either of them.

Stratification of vegetation was discussed in section 2A.7 as part of a macrohabitat analysis. Section 2A.8 noted its contribution to the description of habitat diversity. This layering of vegetation should also be considered in an analysis of community structure. Before engaging in the details of the community analysis, acquaint yourself with the overall features of the community type being studied, as described in section 2A.7 and table 2A.4.

3. Plant form

Terrestrial plant life forms, foliage forms, and seasonal conditions commonly are described by terms such as in table 5A.1. For example, a white oak-shagbark hickory forest might contain plants of the following descriptions: green broad-leaved deciduous trees, budding broad-leaved thorny shrubs, green broad-leaved vines, and green elongated-leaved herbs. For more detail, the relative abundances of these categories can be quantified by the considerations of sections 3A through 3C. A subjective quantification of dominant, abundant, common, uncommon, or rare is adequate for a general study. If taxonomic detail is required, then a brief list of the common plants can be included (see section 3A.6 for guidance).

4. Foliage density and screening efficiency

Foliage density is the density of leaves within a given volume of the habitat. This vegetative feature has a large influence on light intensity, temperature, soil moisture, and habitat space for animals. Unfortunately, there is no simple direct measure of foliage density, as either numbers, volume, or weight of leaves per volume of habitat. Usually the best we can do is measure the mean thickness or height of the foliage of each stratum (see section 2A.7).

Screening efficiency is the relative amount of shading or concealment of the ground by the vegetation. It may be estimated as a percentage of the background obscured by a layer of foliage of a given thickness. The visible background may be a percentage of bare soil visible in a field, or the percentage of the sky visible from the forest floor. A simple method for determining screening efficiency uses a 0.5-m² clear plastic square (approximately 70 × 70 cm) marked off in a 10 × 10 grid. One holds the grid directly overhead and counts either the number of grid squares that do or do not contain visible sky. After taking 20 random readings, one can calculate the proportion of squares concealed from the sky. This proportion (a value from 0 to 1), or its corresponding percentage (0 to 100%), is an expression of screening efficiency.

Measurement of light intensity, inversely related to screening efficiency, must be standardized as it is subject to other factors as well. When using a light meter one should measure the light intensity in an open area and compare it to an area under the vegetation at the same time of day and under the same cloud conditions. Record the screening efficiency as the percent of light transmitted in the habitat divided by the light intensity in the open. See section 2B.4.1 for further discussion of light measurement.

Table 5A.1. *Descriptions of plant form and condition.*

Life form	Foliage form	Seasonal condition
fungus	broad-leaved	green
lichen	needle-leaved	yellow/brown
moss	palmlike	defoliated
liverwort	fernlike	budding
fern	grasslike	flowering
herb	thorny or spiny	fruiting
sod grass	sclerophyllous	
bunch grass		
broad-leaved		
vine, or liana		
succulent		
cactoid		
woody		
vine, or liana		
succulent		
cactoid		
bush		
shrub		
tree		
deciduous		
evergreen		
others		
epiphyte		

5. Coverage

Another measure of the quantity and distribution of foliage is *coverage,* the amount of an area covered by a perpendicularly projected outline of vegetation. The terms *sparse, medium,* and *dense* may be used in a general analysis, as: *dense,* a species or plant life form whose foliage outline covers more than 75% of the habitat area; *medium-dense,* 50 to 75%; *medium,* 25 to 50%; *medium-sparse,* 5 to 25%; and *sparse,* less than 5%. As coverage is an outline measurement and does not reflect the height or density of foliage, it does not measure light penetrability well and, therefore, is not the same as screening efficiency (see section 4). For a more detailed analysis, quantitative measurement of coverage may be performed as described in sections 3A, 3B, and 3C.

6. Dispersion

Spatial distribution of plants may be one of these three types: *even* or *uniform* (as in rows); *random; clumped* or *aggregated.* Further, the plants may be said to be widely spaced (sparse), or closely spaced (dense). For a quantitative assessment of dispersion, consult section 4C. A distinct zonation of vegetation may occur within a habitat as a result of topography, moisture, or succession (see section 2A.7). Record and describe the presence of such zonation.

7. Graphical examination

Graphical presentations of certain aspects of community structure are often helpful in community analysis. Obtain species abundance data by any of the appropriate methods in unit 3. Generally, one uses data only from a particular subcommunity, taxonomic group, or trophic level.

7.1. Relative Abundance Curve   This type of graph, also called a *dominance-density curve,* or a *species importance curve,* may be prepared using density, coverage, biomass, frequency, productivity, or importance value. One ranks the species in a sequence from 1 to *s,* where *s* is the total number of species being considered. The most abundant species (or the one with the greatest coverage, biomass, etc., depending on the measured variable) is assigned rank 1, the second most abundant is given rank 2, and so on, with the least abundant receiving rank *s.* Then, the abundance (or coverage, biomass, etc.) is plotted on a logarithmic scale against the corresponding rank, as in figure 5A.1. Logarithms permit convenient placement of a large range of values on the graph and result in a curve such as C in figure 5A.1 being a straight line.

A community with a high degree of diversity will tend to have more species and a more even abundance in each species than will a community of low diversity. (Section 5B discusses the concept of species diversity more fully.) In figure 5A.1, curve A is a species assemblage with extremely high diversity, and curve D is one with very low diversity. But neither of these situations is likely to be found in nature. Curves B and C are found, however (e.g., trees in a deciduous forest community); they are intermediate between curves A and D. Curve C is a situation where the most abundant species has twice as many individuals as the next most abundant species, which has twice the numbers of the next most abundant, and so on. Communities with low species diversity and/or a high degree of dominance tend to have very steep curves on such a graph. Those with high species diversity and/or low dominance assume a more horizontal aspect. As discussed in section 2D.5.1, species diversity has been considered an important community characteristic in detecting polluted waters. Relative abundance curves are also useful in comparing empirical data to mathematical models (e.g., Magurran, 1988:11–32; Whittaker, 1965, 1977), but this will not be discussed here.

7.2. Species-Area Curve   The species-area, or species-sample, curve was introduced in section 1A.3. This type of graph is useful in evaluating the *species richness* (number of species) in a habitat. It describes the relationship between habitat area and species richness and therefore is of interest in biogeographical studies, especially

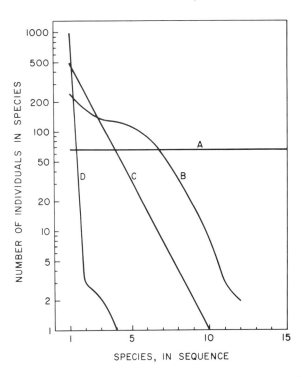

Figure 5A.1. A relative abundance curve, also called a dominance-diversity curve, or a species importance curve. Curve A exemplifies the highest diversity and lowest dominance; curve D is the lowest of the four in diversity and highest in dominance.

those of islands. Both axes are commonly made logarithmic to arrive at a straight-line relationship between number of species and area.

7.3. Lognormal Curve.   The number of species of a given abundance is often found related in a predictable way to the logarithm of the abundance (Preston, 1948, 1962). A common plot of this relationship is shown in figure 5A.2. Here the abscissa is divided into geometric abundance intervals (1 to 2, 2 to 4, 4 to 8, 8 to 16, and so on), and the ordinate shows the number of species with these abundances.[1] If a species has an abundance on an abundance interval borderline (an abundance of 1, 2, 4, 8, etc.), then assign half a species to the interval on each side of the borderline. For example, if five species had an abundance of 8, then we would add 2.5 to the 4–8 interval and 2.5 to the 8–16 interval; if six species had an abundance of 1, then we would add 3 to the 1–2 interval.

The resultant curve (figure 5A.2) is often a normal curve (see figure 1B.2) with part of its left side cut off (that is, "truncated"). (See Pielou, 1975, for mathematical aspects of this curve.) The curve will be displaced to the right one abundance interval (one "octave") for each doubling of the sample size. Thus if the number of data are large enough the curve will not appear truncated.

---

1. Each abundance interval is called an *octave.* Its width is a multiple of 2, so that the scale of the abscissa is actually the logarithm of the abundance to the base 2.

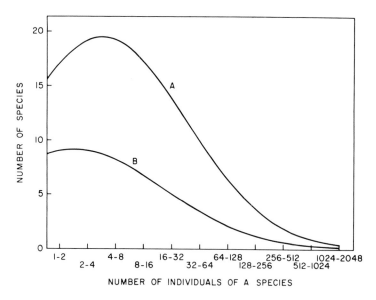

Figure 5A.2. A lognormal curve, plotting the number of species having certain abundances against logarithmic abundance intervals.

Lognormal curves for different communities or subcommunities may be placed on the same graph, as shown in figure 5A.2. Curve B represents a less diverse species assemblage than does curve A (perhaps due to pollution or other disturbance). There are fewer species in B, as seen by the height of the curves; there are fewer members of each species, as seen by curve B being farther to the left than A; and there are more dominants, or more unevenness in the distribution of individuals among species, as seen from the greater spread (greater standard deviation) of curve A.

8. Trophic structure

The trophic structure of a community refers to the pathways of obtaining energy (by photosynthesis or by feeding). One way of examining trophic structure is to construct a pyramid of biomass, as described in section 6A.6. Usually one studies the trophic relationships in only a portion of a community, because the entire community is too large and complex. For example, we may sample the invertebrates from a part of a community (section 3D or 3E) and, with the aid of literature sources, categorize the feeding habits of each species. Then, we may assign each species to a trophic level (primary consumer, secondary consumer, or tertiary consumer, as defined in section 6A.6).

A second approach to studying the trophic structure in a community is to construct a *food web*. This diagrammatically shows which populations feed on which. The various species can be listed, those of higher trophic levels appearing above those in lower levels. Then, arrows may be drawn from each population to those populations that feed on it. Thicker arrows may be used to designate major energy transfer pathways, thinner arrows for lesser ones.

The stability of a community is related to the number and complexity of pathways of energy and nutrient flow. The greater the complexity of the food web, the more stable a community tends to be. Thus a higher diversity of species is usually associated with greater community stability (see section 5B).

9. Temporal difference

Obviously no one can study the seasonal or daily patterns of distribution in a community in a single short field trip. But you may conduct a series of field examinations during different portions of the day, and divide the activities of both animals and plants into those that are diurnal, nocturnal, and crepuscular (daytime, nighttime, and dawn and/or dusk, respectively).

Seasonal variation also occurs in a community. We can examine floral and faunal data from previous times of year and describe seasonal changes in the biotic characteristics of the community (e.g., species composition and diversity, screening efficiency, coverage), as well as the abiotic factors influenced by them (e.g., wind, light, humidity). In section 5D, we will discuss ecological succession—gradual change from one community to another over a long period of time.

10. Suggested exercises

1. Compare the plant life forms in two similar habitats, such as a field and prairie, or an oak forest and maple forest.
2. Select a series of forests (or other community type) of different sizes and compare species-area curves for these communities. What is the effect of habitat size on the number of species present?
3. Compare the relative abundance curves for two or more communities.

4. Describe a community by its dominants, life forms, stratification, zonation, and dispersion pattern.
5. Construct lognormal curves for two different communities and interpret the results in terms of richness, relative abundance, and sample size.
6. Construct a food web for a pond, stream, soil, or litter community or subcommunity. Relate the structure of this web to the results of the graphical procedures of section 5A.3.

11. Selected references

Barbour, M. G., and W. D. Billings (eds.). 1987. North American terrestrial vegetation. Cambridge University Press, New York.

Braun-Blanquet, J. 1932. Plant sociology. Oxford University Press, Oxford, England.

Clapham, W. B., Jr. 1983. Natural ecosystems. 2d ed. Macmillan Co., New York.

Cody, M. L., and J. M. Diamond (eds.). 1975. Ecology and evolution of communities. Belknap Press, Cambridge, Mass.

Conner, E. F., and E. D. McCoy. 1979. The statistics and biology of the species-area relationship. Amer. Natur. 113:791–833.

Dansereau. P. 1957. Biogeography, an ecological perspective. Ronald Press Co., New York.

Daubenmire, R. 1974. Plant and environment: A textbook of plant autecology. 3d ed. Harper & Row, New York.

Emlen, J. T., Jr. 1956. A method for describing and comparing avian habitats. Ibis 98:565–576.

Frontier, S. 1985. Diversity and structure in aquatic ecosystems. Oceanogr. Mar. Biol. Annu. Rev. 23:253–312.

Gray, J. S. 1988. Species abundance patterns. In Organization of communities: Past and present. Blackwell Scientific Publications, Oxford, England.

Kershaw, K. A., and J. H. H. Looney. 1985. Quantitative and dynamic plant ecology. 3d ed. Edward Arnold, London.

Kuchler, A. W. 1949. A physiognomic classification of vegetation. Ann. Assoc. Amer. Geogr. 39:201–209.

Ludwig, J. A., and J. F. Reynolds. 1988. Statistical ecology: A primer on methods and computing. John Wiley & Sons, New York.

MacArthur, R. H. 1972. Geographical ecology. Harper & Row, New York.

MacArthur, R. H., and J. W. MacArthur. 1961. On bird species diversity. Ecology 42:594–598.

McIntosh, R. P. 1978. Phytosociology. Benchmark papers in ecology. Volume 6. Dowden, Hutchinson & Ross, Stroudsburg, Penn.

Magurran, A. E. 1988. Ecological diversity and its measurement. Princeton University Press, Princeton, N.J.

May, R. M. 1975. Patterns of species abundance and diversity, 81–120. In M. L. Cody and J. M. Diamond (eds.), Ecology and evolution of communities. Harvard University Press, Cambridge, Mass.

Miller, A. H. 1942. Habitat selection among higher vertebrates and its relation to interspecific variation. Amer. Natur. 76:25–35.

Mueller-Dombois, D., and H. Ellenberg. 1974. Aims and methods of vegetation ecology. John Wiley & Sons, New York.

Oosting, H. J. 1956. The study of plant communities. W. H. Freeman and Co., San Francisco.

Phillips, F. A. 1959. Methods of vegetation study. Holt, Rinehart and Winston, New York.

Pielou, E. C. 1975. Ecological diversity. John Wiley & Sons, New York.

Preston, F. W. 1948. The commonness, and rarity, of species. Ecology 29:254–283.

Preston, F. W. 1962. The canonical distribution of commonness and rarity. Ecology 43:185–215, 410–432.

Raunkiaer, C. 1934. Life forms of plants and statistical plant geography. Clarendon Press, Oxford, England.

Shelford, V. E. 1963. The ecology of North America. University of Illinois Press, Urbana, Ill.

Sugihara, G. 1980. Minimal community structures: An explanation of species abundance patterns. Amer. Natur. 116:770–787.

Van der Maarel. 1975. The Braun-Blanquet approach in perspective. Vegetatio 30:213–219.

Walter, H. 1985. Vegetation of the earth and ecological systems of the biosphere. 3d ed. Springer-Verlag, New York.

Whittaker, R. H. 1965. Dominance and diversity in land plant communities. Science 147:250–260.

Whittaker, R. H. 1970. Communities and ecosystems. Macmillan Publishing Co., New York.

Whittaker, R. H. 1977. Evolution of species diversity in land communities, 1–67. In M. K. Hecht, W. C. Steere, and B. Wallace (eds.), Evolutionary biology, Volume 10. Plenum Press, New York.

5b

species diversity

1. Introduction

Species diversity (sometimes called *species heterogeneity*), a characteristic unique to the community level of biological organization, is an expression of community structure. A community is said to have a high species diversity if many equally or nearly equally abundant species are present. On the other hand, if a community is composed of a very few species, or if only a few species are abundant, then species diversity is low. For example, if a community had 100 individuals distributed among 10 species, then the maximum possible diversity would occur if there were 10 individuals in each of the 10 species (example A in table 5B.1). The minimum possible diversity would occur if there were 91 individuals belonging to 1 of the species and only 1 individual in each of the other 9 species (example C in table 5B.1). In the latter case, the typical species in the community is relatively rare, so that Patil and Taillie (1982) refer to species diversity as average rarity of species within a community, and relate diversity measures to the probability of interspecific encounters.

High species diversity indicates a highly complex community, for a greater variety of species allows for a larger array of species interactions. Thus, population interactions involving energy transfer (food webs), predation, competition, and niche apportionment are theoretically more complex and varied in a community of high species diversity. This is still the subject of considerable discussion; some ecologists have supported the concept of species diversity as a measure of community stability (the ability of community structure to be unaffected by disturbance of its components) while others have concluded that there is no simple relationship between diversity and stability. Some ecologists have also used diversity as an index of the maturity of a community on the premise that communities become more complex and more stable as they mature. However, this assumption is probably applicable only in certain ecological communities.

Table 5B.1. *Various diversity indices computed for hypothetical situations of* N *individuals distributed among* s *species, with* n_i *individuals in the* ith *species. Data sets A, B, and C have identical values of* N *and* s. *Set D has the same* s *and species distribution as A, but with a larger* N. *Set E has the same* N *and evenness as set A, but a smaller* s.

	hypothetical set of data				
	A	B	C	D	E
n_1	10	29	91	100	20
n_2	10	19	1	100	20
n_3	10	14	1	100	20
n_4	10	11	1	100	20
n_5	10	9	1	100	20
n_6	10	7	1	100	
n_7	10	5	1	100	
n_8	10	3	1	100	
n_9	10	2	1	100	
n_{10}	10	1	1	100	
s, number of species	10	10	10	10	5
N, number of individuals	100	100	100	1000	100
D_a, Margalef diversity	4.50	4.50	4.50	3.00	2.00
D_b, Menhinick diversity	1.00	1.00	1.00	0.32	0.50
l, Simpson dominance	0.09	0.16	0.83	0.10	0.19
D_s, Simpson diversity	0.91	0.84	0.17	0.90	0.81
d_s, inverse of l	11.00	6.23	1.21	10.09	5.21
H, Brillouin diversity	0.92	0.79	0.18	0.99	0.66
H', Shannon diversity	1.00	0.86	0.22	1.00	0.70
S', equally abundant species	10.0	7.2	1.7	10.00	5.0
D_{max}, maximum D_s	0.91	0.91	0.91	0.90	0.81
E_s, evenness, using D_s	1.00	0.92	0.19	1.00	1.00
d_{max}, maximum d_s	11.00	11.00	11.00	10.09	5.21
e_s, evenness, using d_s	1.00	0.57	0.11	1.00	1.00
H_{max}, maximum H	0.92	0.92	0.92	0.99	0.66
J, evenness, using H	1.00	0.86	0.20	1.00	1.00
$1 - J$, dominance, using H	0.0	0.14	0.80	0.0	0.0
H_{max}', maximum H'	1.00	1.00	1.00	1.00	0.70
J', evenness, using H'	1.00	0.86	0.22	1.00	1.00
$1 - J'$, dominance, using H'	0.0	0.14	0.78	0.0	0.0

On the following pages we shall discuss *species* diversity, assuming that all individuals in a biological collection can be identified to species. If such identification is not possible or practical (for example, in a class exercise), then other taxonomic groups may be used. (For example, we may speak of genus or family diversity.) Indeed, specific identification is not needed for most comparative studies; the individuals collected may simply be identified as *taxon 1, taxon 2,* and so on, as long as such nomenclature is consistent from collection to collection. If you want to compare diversity indices of different communities or subcommunities, try to obtain the same sized sample from each. This is because all measures of diversity depend to some extent on the number of species collected, which depends in turn on the sample size.

Diversity is usually considered for only certain subcommunities at a time rather than for an entire ecological community. Great differences among organism sizes make diversity measures difficult to interpret in large-scale studies. Thus, we speak of the species diversity of birds, insects, or algae, or the species diversity in the soil, or on tree trunks. Section 2A.8 discusses quantifying the diversity of habitats, and section 5B.4 refers to diversity as a measure of niche width.

Section 5A.3 presents the *relative abundance curve* and the *lognormal curve,* which express the distribution of individuals among species. These plots may be used to show species diversity graphically. Typically, however, it is best to express quantitative measures of diversity, as discussed below.

A large number of measures of diversity have been proposed and many are in contemporary use. Of those mentioned below, we recommend that the student concentrate on the Simpson index (D_s, in section 5B.2.2) and the information-theoretic indices (H and H', in section 5B.2.3).

2. Measures of species diversity

2.1. Numbers of Species and Individuals
The simplest measure of species diversity is the number of species (s), or the species *richness.* Several indices of diversity have been proposed that incorporate both s and N, the total number of individuals in all the species; for example, Margalef's index:

$$D_a = (s - 1)/\log N \qquad (1)$$

(Margalef, 1957; essentially the same as the index of Gleason, 1922); and Menhinick's index:

$$D_b = s/\sqrt{N} \qquad (2)$$

(Menhinick, 1964). But measures such as s, D_a, and D_b are inadequate because they do not allow us to differentiate between the diversities of different communities having the same s and N. (For instance, collections A, B, and C in table 5B.1 are declared equally diverse by such indices.) And species richness is directly related to sample size. A good measure of diversity should take into account both the number of species and the evenness of occurrence of individuals in the various species.

2.2. Simpson's Index
Simpson[1] (1949) considered not only the number of species (s) and the total number of individuals (N), but also the proportion of the total that

Table 5B.2. *A hypothetical set of species abundance data, used in the text to illustrate the calculation of various diversity indices.*

species (i)	abundance (n_i)	relative abundance (p_i)
1	50	50/85 = 0.588
2	25	25/85 = 0.294
3	10	10/85 = 0.118
s = 3	N = 85	

occurs in each species. He showed that if two individuals are taken at random from a community, the probability that the two will belong to the same species is:

$$l = \frac{\Sigma n_i(n_i - 1)}{N(N - 1)}. \qquad (3)$$

The quantity l is, therefore, a measure of *dominance.*[2] A collection of species with high diversity will have low dominance, and,

$$D_s = 1 - l, \qquad (4)$$

namely:

$$D_s = 1 - \frac{\Sigma n_i(n_i - 1)}{N(N - 1)} \qquad (5)$$

is a good measure of diversity.[3] For the data of table 5B.2,

$$D_s = 1 - \frac{50(49) + 25(24) + 10(9)}{85(84)}$$
$$= 1 - 3140/7140$$
$$= 1 - 0.44$$
$$= 0.56.$$

Some ecologists have inverted Simpson's dominance index to arrive at a measure of diversity:

$$d_s = \frac{1}{l} = \frac{N(N - 1)}{\Sigma n_i(n_i - 1)}. \qquad (6)$$

This diversity index is an expression of the number of times one would have to take pairs of individuals at random from the entire aggregation to find a pair from the same species. It is also an expression of how many equally abundant species would have a diversity equal to that in the observed collection. The index d_s is preferable to D_s in comparing collections in which the values of D_s are very close to 1.0 and nearly the same. (For example, two collections yielding values of D_s of 0.96 and 0.98 would give

1. Simpson's approach was to apply to ecology a diversity measure ($\Sigma n_i^2/N$) introduced in an econometric context in 1912 by Gini (Bhargava and Uppuluri, 1975; Rao, 1982), so that one sees occasionally reference to the "Gini-Simpson index."

2. The quantity $\Sigma n_i(n_i - 1)$ in the numerator of l may be computed as $\Sigma n_i^2 - N$, which may prove simpler on some calculating machines. Morisita's measure of dispersion (section 4C.3.4) is computationally related to l. That is, a large l implies an aggregation of individuals in only a few species, whereas a small value of l denotes a more uniform distribution of individuals among species.

3. Hurlburt (1971) computes D_s as $\Sigma(n_i/N)[(N - n_i)/(N - 1)]$; but equation 5 is simpler. McIntosh (1967) has proposed $\sqrt{\Sigma n_i^2}$ as a dominance measure from which diversity indices may be derived.

us d_s values—with more discrimination—of 25.00 and 50.00, respectively.) Levins (1968) proposed the inverse of Simpson's diversity index as a measure of niche breadth, but others have since presented other measures that are preferable for that purpose (see, e.g., Feinsinger et al., 1981).

Hurlburt (1971) severely criticizes most diversity indices (including those below), but praises the characteristics of the above indices as being biologically meaningful, with D_s referring to the *probability of interspecific encounter* (which he calls PIE). This is the probability of an individual in the community encountering a member of another species. He relates this concept to specific kinds of encounters, such as competition and predation.

The above considerations of l, D_s, and d_s assume that the data at hand are a random sample from a community or subcommunity. There are occasions when this is not the case, as when we have data from an entire community or subcommunity (e.g., a laboratory culture of animals or scavengers at an animal carcass), rather than a sample; or when we do have a sample but it is known to be a nonrandom representation of a community or subcommunity. In such a case, the appropriate Simpson measure of dominance is

$$\lambda = \frac{\Sigma n_i^2}{N} \qquad (7)$$

and the diversity indices analogous to D_s and d_s are

$$\Delta_s = 1 - \lambda = 1 - \frac{\Sigma n_i^2}{N} \qquad (8)$$

and

$$\delta_s = \frac{1}{\lambda} = \frac{N}{\Sigma n_i^2}, \qquad (9)$$

respectively.

2.3. Information-Theoretic Indices Measures of species diversity based on information theory (introduced to ecologists by MacArthur, 1955, and Margalef, 1958) are related to the concept of "uncertainty." In a species aggregation of low diversity (e.g., sample C in table 5B.1), we can be relatively certain of the identity of a species chosen at random. (In this example, it will probably be a member of species 1.) In a highly diverse community, however (e.g., sample A in table 5B.1), it is difficult to predict the identity of a randomly picked individual. Thus high diversity is associated with high uncertainty and low diversity with low uncertainty.

Information-theoretic measures also allow us to consider and calculate measures of "hierarchical diversity." Consider example A in table 5B.1. If the ten species were each of a different genus, this would intuitively imply a greater diversity than if they were all of the same genus. To learn more about measuring hierarchical diversity,

taking into account the distribution of species within genera, genera within families, and so on, consult Pielou (1975, 1977).

Two kinds of ecological collections must be considered. The first is where our species abundance data comprise a sample taken at random from a community or subcommunity. In the second kind of collection (e.g., with a rotting log subcommunity or some laboratory situations), we know the total number of individuals in a collection of species without resorting to samples. Or, we may have obtained a nonrandom sample (thus nonrepresentative of its community or subcommunity); in these cases, the sample must also be considered a complete enumeration. For example, trap (sections 3D.5 and 3D.7), artificial substrate (section 3E.2.5), or seine (section 3E.3.4) sampling typically favors the collection of certain species. Thus these collections do not exhibit species compositions and abundances that accurately reflect those of the sampled community.

If our data are a random sample of species abundances from a larger community or subcommunity of interest, then we may appropriately use the Shannon[4] diversity index:

$$H' = - \Sigma p_i \log p_i, \qquad (10)$$

where,

$$p_i = n_i/N; \qquad (11)$$

that is, p_i is the proportion of the total number of individuals occurring in species i. For this calculation, one may use any logarithmic base; bases 10 and e are the commonest, although communications engineers (from whom the index has been borrowed) use base 2. The selection of a particular logarithmic base is immaterial as long as we are consistent; and H' computed in one base may be converted to the H' for another base by consulting table 5B.3. In Appendix D, table D.1 gives logarithms, and table D.2 gives logarithms of proportions. For the data in table 5B.2:

$$
\begin{aligned}
H' &= -[0.588 \log 0.588 + 0.294 \log 0.294 \\
&\quad + 0.118 \log 0.118] \\
&= -[0.588(-0.231) + 0.294(-0.532) \\
&\quad + 0.118(-0.928)] \\
&= -[-0.136 - 0.156 - 0.110] \\
&= 0.40.
\end{aligned}
$$

A little algebraic manipulation arrives as the equivalent equation:

$$H' = (N \log N - \Sigma n_i \log n_i)/N. \qquad (12)$$

---

4. This often is less properly called the *Shannon-Weaver* or *Shannon-Wiener* index, for C. E. Shannon's equation received some inspiration from N. Wiener and some clarification from W. W. Weaver (Perkins, 1982).

Table 5B.3. *Factors to convert between logarithmic bases 2, e, and 10. (For example, a value of 0.86 computed using base 10 is equivalent to a value of (0.86)(2.3026) = 1.98 using base e.)*

to convert to	to convert from		
	2	e	10
2	1.0000	1.4427	3.3219
e	0.6931	1.0000	2.3026
10	0.3010	0.4343	1.0000

This equation allows us to compute H' without first converting abundances (n_i) to proportions (p_i), both saving time and avoiding rounding errors. Table 5B.4 is very conveniently used with this equation. For the above example data,

$$H' = [85 \log 85 - (50 \log 50 + 25 \log 25 + 10 \log 10)]/85$$
$$= [164.001 - (84.949 + 34.949 + 10.000)]/85$$
$$= 34.103/85$$
$$= 0.40.$$

As noted above, the Shannon diversity index, H', is appropriate when you have a random sample of species abundances from a larger aggregation, say a random sample of an entire community. Such a sample (unless extremely large) will probably not contain representatives of each species in the entire community. So, typically our observed value of s is biased, an underestimate of the number of species in the entire community. However, the lack of data on rare species has little effect on the value of H' (although it has serious effect on H_{max}' and J', discussed in section 5B.2.4).

H' may also be calculated for data other than abundances: for example, to express habitat heterogeneity (section 2B.8), or the diversity of biomass (section 6A), or coverage (sections 3A, 3B, and 3C). Also note that equation 10 may be used with relative measures (e.g., relative abundance or relative biomass).

Another way of depicting species diversity is to express the number of equally abundant species that would produce the value of H' of the observed sample. This measure may be represented as

$$S' = B^{H'}, \tag{13}$$

where B is the logarithmic base used in computing H' (e.g., 10, e, or 2). For example, for data set B in Table 5B.1,

$$S' = 10^{0.86} = 7.2.$$

Now let us consider a set of species abundance data considered a nonrandom sample. In such a case, or if in fact our collected data are the entire community or sub-

community, then do not use H' (Pielou, 1966, 1967, 1975); instead use the Brillouin (1962) index:

$$H = \left(\log \frac{N!}{\Pi n_i!}\right)/N, \tag{14}$$

where Π (capital Greek pi) means to take the product, just as Σ means to take the sum; thus we can write equation 14 as:

$$H = \left(\log \frac{N!}{n_1! n_2! \ldots n_s!}\right)/N. \tag{15}$$

The computation of factorials—such as $6! = (6)(5)(4)(3)(2) = 720$—is tedious, and the numbers typically become unwieldy. Therefore, H is much more conveniently calculated using logarithms:

$$H = (\log N! - \Sigma \log n_i!)/N, \tag{16}$$

with the aid of table 5B.5. For our table 5B.2 example,

$$H = [\log 85! - (\log 50! + \log 25! + \log 10!)]/85$$
$$= [128.450 - (64.483 + 25.191 + 6.560)]/85$$
$$= 32.216/85$$
$$= 0.38.$$

As with H', the logarithmic base used is immaterial as long as we are consistent. H values in one base may be converted to those in another by using table 5B.3.

The units of H and H' are unimportant (and probably meaningless) to the ecologist. These indices are used only in a relative fashion, that is, to determine which species assemblages are more or less diverse than others.

2.4. Evenness The diversity indices in sections 2.2 and 2.3 above take into account both the species *richness* (the number of species) and the *evenness* of the individuals' distribution among the species. Separate measures of these two components of diversity are often desirable. Richness can be expressed simply as the number of species. Evenness may be expressed by considering how close a set of observed species abundances are to those from an aggregation of species having maximum possible diversity for a given N and s.

The maximum possible diversity for a collection of N individuals in a total of s species exists when the N individuals are distributed as evenly as possible among the s species, that is, when each $n_i = N/s$. The maximum possible values of D_s, d_s, H, and H' are as follows:

$$D_{max} = \left(\frac{s-1}{s}\right)\left(\frac{N}{N-1}\right) \tag{17}$$

$$d_{max} = s\left(\frac{N-1}{N-s}\right) \tag{18}$$

$$H_{max} = [\log N! - (s-r)\log c! - r\log(c+1)!]/N \tag{19}$$

(see below), and:

$$H_{max}' = \log s. \tag{20}$$

Table 5B.4. *Values of* n_i *log* n_i *(or N log N) for use in equation 14.**

n_i	0	1	2	3	4	5	6	7	8	9	n_i
0		0.000	0.602	1.431	2.408	3.495	4.669	5.916	7.225	8.588	0
10	10.000	11.455	12.950	14.481	16.046	17.641	19.266	20.918	22.595	24.296	10
20	26.021	27.767	29.533	31.320	33.125	34.949	36.789	38.647	40.520	42.410	20
30	44.314	46.232	48.165	50.111	52.070	54.042	56.027	58.023	60.032	62.052	30
40	64.082	66.124	68.176	70.239	72.312	74.395	76.487	78.589	80.700	82.820	40
50	84.949	87.086	89.232	91.387	93.549	95.720	97.899	100.085	102.279	104.480	50
60	106.689	108.905	111.128	113.358	115.596	117.839	120.090	122.347	124.611	126.881	60
70	129.157	131.439	133.728	136.023	138.323	140.630	142.942	145.260	147.583	149.913	70
80	152.247	154.587	156.933	159.283	161.639	164.001	166.367	168.738	171.114	173.496	80
90	175.882	178.273	180.668	183.069	185.474	187.884	190.298	192.717	195.140	197.568	90
100	200.000	202.436	204.877	207.322	209.771	212.225	214.682	217.144	219.610	222.079	100
110	224.553	227.031	229.512	231.998	234.487	236.980	239.477	241.978	244.482	246.990	110
120	249.502	252.017	254.536	257.058	259.584	262.114	264.647	267.183	269.723	272.266	120
130	274.813	277.363	279.916	282.472	285.032	287.595	290.161	292.731	295.303	297.879	130
140	300.458	303.040	305.625	308.213	310.804	313.398	315.996	318.596	321.199	323.805	140
150	326.414	329.026	331.640	334.258	336.878	339.501	342.127	344.756	347.388	350.022	150
160	352.659	355.299	357.941	360.587	363.234	365.885	368.538	371.194	373.852	376.513	160
170	379.176	381.842	384.511	387.182	389.856	392.532	395.210	397.891	400.575	403.261	170
180	405.949	408.640	411.333	414.029	416.726	419.427	422.129	424.834	427.542	430.251	180
190	432.963	435.677	438.394	441.113	443.834	446.557	449.282	452.010	454.740	457.472	190
200	460.206	462.942	465.681	468.422	471.165	473.910	476.657	479.406	482.157	484.911	200
210	487.666	490.424	493.183	495.945	498.709	501.474	504.242	507.012	509.784	512.557	210
220	515.333	518.111	520.890	523.672	526.456	529.241	532.029	534.818	537.609	540.402	220
230	543.197	545.994	548.793	551.594	554.397	557.201	560.007	562.815	565.625	568.437	230
240	571.251	574.066	576.883	579.702	582.523	585.346	588.170	590.996	593.824	596.654	240
250	599.485	602.318	605.153	607.989	610.828	613.668	616.509	619.353	622.198	625.045	250
260	627.893	630.743	633.595	636.448	639.303	642.160	645.019	647.879	650.740	653.603	260
270	656.468	659.335	662.203	665.072	667.944	670.816	673.691	676.567	679.444	682.324	270
280	685.204	688.086	690.970	693.856	696.742	699.631	702.521	705.412	708.305	711.199	280
290	714.095	716.993	719.892	722.792	725.694	728.597	731.502	734.409	737.316	740.226	290
300	743.136	746.049	748.962	751.877	754.794	757.711	760.631	763.551	766.474	769.397	300
310	772.322	775.248	778.176	781.105	784.036	786.968	789.901	792.836	795.772	798.709	310
320	801.648	804.588	807.530	810.472	813.417	816.362	819.309	822.257	825.207	828.157	320
330	831.110	834.063	837.018	839.974	842.931	845.890	848.850	851.811	854.774	857.738	330
340	860.703	863.669	866.637	869.606	872.576	875.548	878.520	881.494	884.470	887.446	340
350	890.424	893.403	896.383	899.364	902.347	905.331	908.316	911.303	914.290	917.279	350
360	920.269	923.260	926.253	929.246	932.241	935.237	938.234	941.232	944.232	947.233	360
370	950.235	953.238	956.242	959.247	962.254	965.262	968.271	971.281	974.292	977.304	370
380	980.318	983.332	986.348	989.365	992.383	995.402	998.423	1001.444	1004.467	1007.490	380
390	1010.515	1013.541	1016.568	1019.596	1022.626	1025.656	1028.687	1031.720	1034.753	1037.788	390
400	1040.824	1043.861	1046.899	1049.938	1052.978	1056.019	1059.062	1062.105	1065.149	1068.195	400
410	1071.241	1074.289	1077.338	1080.387	1083.438	1086.490	1089.543	1092.597	1095.652	1098.708	410
420	1101.765	1104.823	1107.882	1110.942	1114.003	1117.065	1120.128	1123.193	1126.258	1129.324	420
430	1132.391	1135.460	1138.529	1141.599	1144.671	1147.743	1150.816	1153.890	1156.966	1160.042	430
440	1163.119	1166.197	1169.277	1172.357	1175.438	1178.520	1181.603	1184.687	1187.773	1190.859	440
450	1193.946	1197.034	1200.123	1203.212	1206.303	1209.395	1212.488	1215.582	1218.676	1221.772	450
460	1224.869	1227.966	1231.065	1234.164	1237.264	1240.366	1243.468	1246.571	1249.675	1252.780	460
470	1255.886	1258.993	1262.101	1265.209	1268.319	1271.429	1274.541	1277.653	1280.767	1283.881	470
480	1286.996	1290.112	1293.229	1296.346	1299.465	1302.585	1305.705	1308.827	1311.949	1315.072	480
490	1318.196	1321.321	1324.447	1327.574	1330.701	1333.830	1336.959	1340.089	1343.220	1346.352	490

*If values for n_i or N larger than 499 are needed, consult Lloyd et al. (1968) or Zar (1974:401–404), or use Appendix D, table D.1.

Table 5B.5. *Values of log* n$_i$! *(or log* N!) *for use in equation 14.**

n_i	0	1	2	3	4	5	6	7	8	9	n_i
0	0.000	0.000	0.301	0.778	1.380	2.079	2.857	3.702	4.606	5.560	0
10	6.560	7.601	8.680	9.794	10.940	12.116	13.321	14.551	15.806	17.085	10
20	18.386	19.708	21.051	22.412	23.793	25.191	26.606	28.037	29.484	30.947	20
30	32.424	33.915	35.420	36.939	38.470	40.014	41.571	43.139	44.719	46.310	30
40	47.912	49.524	51.148	52.781	54.425	56.078	57.741	59.413	61.094	62.784	40
50	64.483	66.191	67.907	69.631	71.363	73.104	74.852	76.608	78.371	80.142	50
60	81.920	83.706	85.498	87.297	89.103	90.916	92.736	94.562	96.394	98.233	60
70	100.078	101.930	103.787	105.650	107.520	109.395	111.275	113.162	115.054	116.952	70
80	118.855	120.763	122.677	124.596	126.520	128.450	130.384	132.324	134.268	136.218	80
90	138.172	140.131	142.095	144.063	146.036	148.014	149.996	151.983	153.974	155.970	90
100	157.970	159.974	161.983	163.996	166.013	168.034	170.059	172.089	174.122	176.160	100
110	178.201	180.246	182.295	184.349	186.405	188.466	190.531	192.599	194.671	196.746	110
120	198.825	200.908	202.995	205.084	207.178	209.275	211.375	213.479	215.586	217.697	120
130	219.811	221.928	224.049	226.172	228.299	230.430	232.563	234.700	236.840	238.983	130
140	241.129	243.278	245.431	247.586	249.744	251.906	254.070	256.237	258.408	260.581	140
150	262.757	264.936	267.118	269.302	271.490	273.680	275.873	278.069	280.268	282.469	150
160	284.673	286.880	289.090	291.302	293.517	295.734	297.954	300.177	302.402	304.630	160
170	306.861	309.094	311.329	313.567	315.808	318.051	320.296	322.544	324.795	327.048	170
180	329.303	331.561	333.821	336.083	338.348	340.615	342.885	345.157	347.431	349.707	180
190	351.986	354.267	356.550	358.836	361.124	363.414	365.706	368.000	370.297	372.596	190
200	374.897	377.200	379.505	381.813	384.123	386.434	388.748	391.064	393.382	395.702	200
210	398.025	400.349	402.675	405.004	407.334	409.666	412.001	414.337	416.676	419.016	210
220	421.359	423.703	426.049	428.398	430.748	433.100	435.454	437.810	440.168	442.528	220
230	444.890	447.253	449.619	451.986	454.355	456.727	459.099	461.474	463.851	466.229	230
240	468.609	470.991	473.375	475.761	478.148	480.537	482.928	485.321	487.715	490.112	240
250	492.510	494.909	497.311	499.714	502.119	504.525	506.933	509.343	511.755	514.168	250
260	516.583	519.000	521.418	523.838	526.260	528.683	531.108	533.534	535.962	538.392	260
270	540.824	543.257	545.691	548.127	550.565	553.004	555.445	557.888	560.332	562.777	270
280	565.225	567.673	570.124	572.575	575.029	577.483	579.940	582.398	584.857	587.318	280
290	589.780	592.244	594.710	597.177	599.645	602.115	604.586	607.059	609.533	612.009	290
300	614.486	616.964	619.444	621.926	624.409	626.893	629.379	631.866	634.354	636.844	300
310	639.336	641.828	644.323	646.818	649.315	651.813	654.313	656.814	659.317	661.820	310
320	664.326	666.832	669.340	671.849	674.360	676.872	679.385	681.899	684.415	686.932	320
330	689.451	691.971	694.492	697.014	699.538	702.063	704.589	707.117	709.646	712.176	330
340	714.708	717.240	719.774	722.310	724.846	727.384	729.923	732.464	735.005	737.548	340
350	740.092	742.637	745.184	747.732	750.281	752.831	755.382	757.935	760.489	763.044	350
360	765.600	768.158	770.716	773.276	775.837	778.400	780.963	783.528	786.094	788.661	360
370	791.229	793.798	796.369	798.941	801.513	804.087	806.663	809.239	811.817	814.395	370
380	816.975	819.556	822.138	824.721	827.305	829.891	832.478	835.065	837.654	840.244	380
390	842.835	845.427	848.021	850.615	853.210	855.807	858.405	861.003	863.603	866.204	390
400	868.806	871.410	874.014	876.619	879.225	881.833	884.441	887.051	889.662	892.273	400
410	894.886	897.500	900.115	902.731	905.348	907.966	910.585	913.205	915.826	918.449	410
420	921.072	923.696	926.321	928.948	931.575	934.204	936.833	939.463	942.095	944.727	420
430	947.361	949.995	952.631	955.267	957.905	960.543	963.183	965.823	968.465	971.107	430
440	973.751	976.395	979.040	981.687	984.334	986.983	989.632	992.282	994.933	997.586	440
450	1000.239	1002.893	1005.548	1008.204	1010.861	1013.519	1016.178	1018.838	1021.499	1024.161	450
460	1026.824	1029.487	1032.152	1034.818	1037.484	1040.152	1042.820	1045.489	1048.160	1050.831	460
470	1053.503	1056.176	1058.850	1061.525	1064.200	1066.877	1069.555	1072.233	1074.913	1077.593	470
480	1080.274	1082.956	1085.639	1088.323	1091.008	1093.694	1096.381	1099.068	1101.757	1104.446	480
490	1107.136	1109.827	1112.519	1115.212	1117.906	1120.600	1123.296	1125.992	1128.689	1131.387	490

*If values for n_i (or N) larger than 499 are needed, consult Lloyd et al. (1968), Pearson and Hartley (1966: Table 51).
Or, one may use Appendix D, table D.1 to compute log n_i!
(or log N!) by "Stirling's approximation": log n_i! = (n_i + 0.5) log n_i − 0.4343 n_i + 0.3991.

In equation 19, c is the integer portion of N/s, and r is the remainder. For the table 5B.2 data, for example,

$$N/s = 85/3$$
$$= 28.33;$$

that is, the quotient of $85/3$ is 28 with a remainder of 1. Therefore, $c = 28$ and $r = 1$; so:

$$H_{max} = [\log 85! - (3 - 1) \log 28! - (1) \log 29!]/85$$
$$= [128.450 - 2(29.484) - 1(30.947)]/85$$
$$= 38.535/85$$
$$= 0.45.$$

The *evenness* of the distribution of N individuals among the s species in a set of data is then expressed as the nearness of the diversity index for the observed data to the index of maximum diversity. We use the following expressions for evenness:

$$E_s = D_s/D_{max} \qquad (21)$$

$$e_s = d_s/d_{max} \qquad (22)$$

$$J = H/H_{max} \qquad (23)$$

$$J' = H'/H_{max}'. \qquad (24)$$

Evenness measures are sometimes called measures of *relative diversity*. See table 5B.1 for some computed examples.

Remember that when you examine only a sample from a community or subcommunity you typically will underestimate the number of species in the entire community or subcommunity (unless the sample is extremely large). Therefore, expressions of maximum diversity will be underestimates, and the computed evenness measures will be overestimates of the evenness in the actual collection sampled. In addition, we should be aware that these measures of evenness are not independent of the number of species (i.e., species richness) (DeBendictis, 1973; Alatalo, 1981), so comparisons of evenness measures are most trustworthy when s is the same, or at least similar, in those collections being compared.

The quantity $1 - J$ or $1 - J'$ (or $1 - E_s$ or $1 - e_s$) may be used as an expression of dominance; it will have a low value (zero being the minimum) when dominance is low and a high value (with a maximum of 1.0) when dominance is high (see table 5B.1). The inverse of J or J' would also be a measure of dominance, but $1 - J$ or $1 - J'$ possesses the appeal of having values between 0 and 1. As J and J' are ratios of two quantities with the same units, they are unitless and unaffected by the choice of logarithmic base (as long as both numerator and denominator have the same base). That is, J (or J') will be the same regardless of the base in which H and H_{max} (or H' and H_{max}') were calculated.

Sheldon (1969) proposed $B^{H'}/s$ as an evenness index, but Heip (1974) concluded $(B^{H'} - 1)/(s - 1)$ to be much better; B is the logarithmic base used in computing H'.

The quantities $d_s/B^{H'}$ and $(d_s - 1)/(B^{H'} - 1)$ have also been proposed as evenness measures (Alatalo, 1981).

"Redundancy" indices (e.g., see Zar, 1968; Hurlbert, 1971) are another approach to the expression of evenness.

3. Correlation among diversity indices

It has been observed that two major groups of diversity indices are those that are most affected by the occurrence of rare species (i.e., that are based heavily on species richness) and those that are most sensitive to the relative abundances of the species (i.e., that depend heavily on the dominance versus evenness of the species abundances) (Magurran, 1988:73–75; Peet, 1974). In the first group are s, D_a, D_b, H, and H'; and the second group includes l, D_s, d_s, E_s, e_s, J, and J'. Measures in the first group tend to discriminate better between communities that have different diversities, but they also tend to be more seriously affected by sample size (Magurran, 1988:80). Although the literature on using species diversity indices has increased impressively, there is no agreement as to which index is "best."

4. Measuring niche width

By width, or breadth, of a niche an ecologist means the diversity of resources used by a species. Typically, the context is the resource utilization of a population in a particular habitat. Aspects of a niche may be measured by diversity indices. For example, the diversity of food items utilized measures a component of an animal's niche width. This can be quantified by determining the proportion (p_i) of each food item in the animal's diet and determining the Shannon diversity index (H'). Similarly, the diversity of the habitat's structure used by the species is a measure of its niche width; so it might be determined what proportion of a population of plants is in each of several habitat types, or what proportion of time a population of animals is in each of several habitat types; and H' could then be calculated.

5. Comparing diversity measurements

Oftentimes we have two samples of data and a diversity index is calculated for each of them. It may then be desired to compare objectively the two indices and ask whether the two sampled taxonomic aggregations are equally diverse. (See section 1B.3.1 for basic concepts of two-sample testing.)

5.1. Comparing Simpson Indices   If $(D_s)_1$ is the Simpson diversity index for one sample, and $(D_s)_2$ is the index for a second sample, then we can test the null hypothesis that these two samples come from aggregations having the same diversity (Keefe and Bergerson, 1977), but first we must calculate s^2, the variance of D_s, for each sample, which is, approximately[5]:

$$s^2 = 4[\Sigma p_i^3 - (\Sigma p_i^2)^2]/N, \tag{26}$$

where p_i is as in equation 11. Then, we compute the following statistic:

$$t = \frac{(D_s)_1 - (D_s)_2}{\sqrt{s_1^2 + s_2^2}} \tag{27}$$

and compare it to the critical value of Student's t (table 1B.1) for infinity degrees of freedom (DF $= \infty$). If we are dealing with data comprising an entire community or subcommunity, then $(D_s)_1$ and $(D_s)_2$ may be compared by inspection, instead of statistically.

5.2. Comparing Shannon Indices   To compare Shannon diversity indices from two collections of data (H_1' and H_2'), we need the variance of H' for each collection[6]:

$$s^2 = \frac{\Sigma f_i \log^2 f_i - (\Sigma f_i \log f_i)^2/n}{n^2} \tag{28}$$

(Basharin, 1959), where $\log^2 f_i$ is mathematical notation for $(\log f_i)^2$. Then we compute

$$t = \frac{H_1' - H_2'}{\sqrt{s_1^2 + s_2^2}} \tag{29}$$

and compare it to the critical value of Student's t (table 1B.1) for degrees of freedom as follows (rounded down if not a whole number):

$$DF = \frac{(s_{H_1'}^2 + s_{H_2'}^2)^2}{\dfrac{(s_{H_1'}^2)^2}{n_1} + \dfrac{(s_{H_2'}^2)^2}{n_2}} \tag{30}$$

(Hutcheson, 1970). See Zar (1984:146–148) for examples of this procedure.

If, instead of one H' for each of the two communities to be compared, we had an H' for each of several replicate random samples from each community, then we could employ the t test as in section 1B.3.1 (i.e., using equations 10 through 12 in that section). This is because replicated estimates of Shannon diversity from the same community tend to be normally distributed (see section 1B.3.3). Indeed, if we had replicate H' values from three or more communities, hypothesis testing could be done via the procedures of analysis of variance and multiple comparisons alluded to in section 1B.3.2. Alternatively, we could analyze replicate values of H' by nonparametric testing (see section 1B.3.4).

If we were using Brillouin's index (H) for data that comprise an entire community, a statistical procedure such as that above would not be necessary to compare two indices; the comparison would simply be made by inspection. If Brillouin's H is used because you have a nonrandom sample, then several such samples may be collected from each of two communities and the community diversities may be compared using the statistical procedures of section 1B.3.1, 1B.3.2, or 1B.3.4.

---

6. Suggested exercises

1. Examine the effect of sample size on the Menhinick, Simpson, and information-theoretic diversity indices by plotting the value of each index as a function of cumulative sample size. Which is most and least affected by sample size?
2. Determine which of two communities exhibits more evenness (or dominance) using an index from section 5B.2.4 above. (a) Compare your conclusions to the relative abundance curves (section 5A.3.1) for the two communities. (b) Test these differences statistically, using section 5B.3.

---

7. Selected references

Alatalo, R. V. 1981. Problems in the measurement of evenness in ecology. Oikos 37:199–204.

Basharin, G. P. 1959. On a statistical estimate for the entropy of a sequence of independent variables. Theory Prob. Appl. 4:333–336.

Bhargava, T. N. and V. R. R. Uppuluri. 1975. On an axiomatic derivation of Gini diversity with applications. Metron 33:41–53.

Brillouin, L. 1956, 1962. Science and information theory. Academic Press, New York.

Brookhaven National Laboratory. 1969. Diversity and stability in ecological systems. Brookhaven Symposia in Biology 22.

Connell, J. H. 1978. Diversity in tropical rainforests and coral reefs. Science 199:1302–1310.

DeBenedictis, P. A. 1973. On the correlation between certain diversity indices. Amer. Natur. 107:295–302.

Feinsinger, P., E. E. Spears, and R. W. Poole. 1981. A simple measure of niche breadth. Ecology 62:27–32.

Fisher, R. A., A. S. Corbet, and C. B. Williams. 1943. The relation between the number of species and the number of individuals in a random sample of an animal population. J. Animal Ecol. 12:42–58.

Frontier, S. 1985. Diversity and structure in aquatic ecosystems. Oceanogr. Mar. Biol. Annu. Rev. 23:253–312.

Gleason, H. A. 1922. On the relation between species and area. Ecology 3:158–162.

Goodman, D. 1975. The theory of diversity-stability relationships in ecology. Quart. Rev. Biol. 50:237–266.

Hair, J. D. 1980. Measurement of ecological diversity, 269–275. In S. D. Schemnitz (ed.), Wildlife Management techniques manual. 4th ed. Wildlife Society, Washington, D.C.

Heip, C. 1974. A new index measuring evenness. J. Mar. Biol. Assoc. U.K. 54:555–557.

Heip, C., and P. Engels. 1974. Comparing species diversity and evenness indices. J. Mar. Biol. Assoc. U.K. 54:559–564.

Hurlburt, S. H. 1971. The nonconcept of species diversity: A critique and alternative parameters. Ecology 52:577–586.

---

5. This approximation is good for large N; for small N one can use

$$s^2 = \frac{4N(N-1)(N-2)\Sigma p_i^3 + 2N(N-1)\Sigma p_i^2 - 2N(N-1)(2N-3)(\Sigma p_i^2)^2}{[N(N-1)]^2} \tag{25}$$

(Simpson, 1949).

6. This approximation is good for large N; for small N, subtract $(S-1)/N$ from equation 28.

Hutcheson, K. 1970. A test for comparing diversities based on the Shannon formula. J. Theoret. Biol. 29:151–154.

Hutchinson, G. E. 1959. Homage to Santa Rosalina, or why are there so many kinds of animals? Amer. Natur. 93:145–159.

Keefe, T. J., and E. P. Bergersen. 1977. A simple diversity index based on the theory of runs. Water Res. 11:689–691.

Laxton, R. R. 1978. The measures of diversity. J. Theoret. Biol. 70:51–67.

Levins, R. 1968. Evolution in changing environments: Some theoretical considerations. Princeton University Press, Princeton, N.J.

Lloyd, M., J. H. Zar, and J. R. Karr. 1968. On the calculation of information theoretical measures of diversity. Amer. Midland Natur. 79:257–272.

MacArthur, R. H. 1955. Fluctuations of animal populations and a measure of community stability. Ecology 36:533–536.

MacArthur, R. H. 1965. Patterns of species diversity. Biol. Rev. 40:510–533.

McIntosh, R. P. 1967. An index of diversity and the relation of certain concepts to diversity. Ecology 48:392–404.

Magurran, A. E. 1988. Ecological diversity and its measurement. Princeton University Press, Princeton, N. J.

Margalef, R. 1957. La teoría de la información en ecología. Mem. R. Acad. Cien. Artes 32:373–449.

Margalef, R. 1958. Information theory in ecology. Gen. Systems 3:36–71.

Margalef, R. 1963. On certain unifying principles in ecology. Amer. Natur. 97:357–374.

Margalef, R. 1975. Perspectives in ecological theory. University of Chicago Press, Chicago.

May, R. M. 1975. Patterns of species abundance and diversity, 81–120. In M. L. Cody and J. M. Diamond (eds.), Ecology and evolution of communities. Harvard University Press, Cambridge, Mass.

May, R. M. 1984. An overview: Real and apparent patterns in community structure, 3–18. In D. R. Strong, D. Simberloff, L. G. Abele, and A. B. Thistle (eds.), Ecological communities: Conceptual issues and the evidence. Princeton University Press, Princeton, N.J.

Menhinick, E. F. 1964. A comparison of some species-individuals diversity indices applied to samples of field insects. Ecology 45:859–861.

Patil, G. P. and C. Taillie. 1982. Diversity as a concept and its measurement. J. Amer. Statist. Assoc. 77:548–561.

Pearson, E. S. and H. O. Hartley. 1966. Biometrika tables for statisticians. Cambridge University Press, Cambridge.

Peet, R. K. 1974. The measurement of diversity. Annu. Rev. Ecol. Systemat. 5:285–307.

Peet, R. K. 1975. Relative diversity indices. Ecology 56:496–498.

Perkins, J. L. 1982. Shannon-Weaver or Shannon-Wiener? J. Water Pollut. Contr. Fed. 54:1049–1050.

Pianka, E. R. 1983. Evolutionary ecology. Harper & Row Publishers, New York.

Pielou, E. C. 1966. The measurement of diversity in different types of biological collections. J. Theoret. Biol. 13:131–144.

Pielou, E. C. 1967. The use of information theory in the study of the diversity of biological populations. Proc. Fifth Berkeley Symp. Math. Prob. 4:163–177.

Pielou, E. C. 1974. Population and community ecology. Gordon and Breach, New York.

Pielou, E. C. 1975. Ecological diversity. John C. Wiley & Sons, New York.

Pielou, E. C. 1977. Mathematical ecology. John C. Wiley & Sons, New York.

Pimm, S. L. 1984. The complexity and stability of ecosystems. Nature 307:321–326.

Poole, R. W. 1974. An introduction to quantitative ecology. McGraw-Hill Book Co., New York.

Preston, F. W. 1948. The commonness, and rarity, of species. Ecology 29:254–283.

Rao, C. R. 1982. Diversity and dissimilarity coefficients: A unified approach. Theoret. Pop. Biol. 21:24–43.

Shannon, C. E. 1948. A mathematical theory of communication. Bell System Tech. J. 27:379–423, 623–656.

Sheldon, A. L. 1969. Equitability indices: Dependence on the species count. Ecology 50:466–467.

Simpson, E. H. 1949. Measurement of diversity. Nature 163:688.

Usher, M. B., and M. H. Williamson. 1974. Ecological stability. Halsted Press, New York.

Van Voris, P., R. V. O'Neill, W. P. R. Emanuel, and H. H. Shugart, J.R. 1980. Functional complexity and ecosystem stability. Ecology 61; 1352–1360.

Whittaker, R. H. 1972. Evolution and measurement of species diversity. Taxon 21:213–251.

Whittaker, R. H. 1977. Evolution of species diversity in land communities, 1–67. In M. K. Hecht, W. C. Steere, and B. Wallace (eds.), Evolutionary biology. Volume 10. Plenum Press, New York.

Williams, C. B. 1964. Patterns in the balance of nature. Academic Press, New York.

Williamson, M. H. 1973. Species diversity in ecological communities, 325–334. In M. S. Bartlett and R. W. Hiorns (eds.). The mathematical theory of the dynamics of biological populations. Academic Press, New York.

Woodwell, G. M. and H. H. Smith. 1969. Diversity and stability in ecological systems. Brookhaven National Laboratory, Upton, N.Y.

Zar, J. H. 1968. Computer calculation of information-theoretic measures of diversity. Trans. Ill. State Acad. Sci. 61:217–219.

Zar, J. H. 1974. Biostatistical analysis. Prentice-Hall, Englewood Cliffs, N.J.

Zar, J. H. 1984. Biostatistical analysis. 2d ed. Prentice-Hall, Englewood Cliffs, N.J.

Zaret, T. M. 1982. The stability/diversity controversy: A test of hypotheses. Ecology 63:721–731.

Zaret, T. M. 1984. Ecology and epistemology. Bull. Soc. Amer. 65:4–7.

1. Introduction

After tabulating the species composition of each of two communities, you might wonder how similar or dissimilar they are. Or, you might wish to compare the species composition of the same community at two different times. Although the comparison of communities will be discussed, subcommunities (indeed, any collection of species) may be compared in the same fashion.

A large number of quantitative measures of community similarity have been proposed. Of the several in contemporary use, we recommend the percent similarity (section 2.2), Morisita's index (section 2.4), or Horn's index (section 2.5); or the Jaccard coefficient of community when the data consist of presence or absence of species. It is essential to realize that for a given amount of similarity between communities, these several similarity measures do not necessarily express the same quantitative measures of similarity. Thus, any one of these indices may be used to express relative similarities between communities, but different indices should not be compared with each other. Hurlbert (1978) and Wallace (1981) discuss the application of similarity indices to the measurement and quantification of niche overlap, an area of contemporary ecological interest that generates much discussion and disagreement among ecologists.

2. Indices of community similarity

2.1. Coefficient of community To quantify community similarity, ecologists for decades have used *coefficients of community*. The Jaccard coefficient, dating from the beginning of this century (Mueller-Dombois and Ellenberg, 1974:213), is

$$CC_J = \frac{c}{s_1 + s_2 - c} \qquad (1)$$

or, equivalently,

$$CC_J = \frac{c}{S},$$

where s_1 and s_2 are the number of species in communities 1 and 2, respectively, c is the number of species common to both communities, and S is the total number of species found in the two communities.

The Sørensen coefficient[1] (Sørensen, 1948; also known as the "quotient of similarity") is also used:

$$CC_S = \frac{2c}{s_1 + s_2}. \qquad (2)$$

---

1. This measure may have been proposed in 1913 by Czekanowski (Clifford and Stephenson, 1975:55 Wolda, 1981).

5c

community similarity

For example, if community 1 contained 20 species, community 2 contained 18 species, and both contained 12 species in common, then (by equation 2):

$$CC_S = \frac{2(12)}{20 + 18}$$
$$= 0.63, \text{ or } 63\%.$$

If, as in the above example, $s_1 = 20$, $s_2 = 18$, but $c = 16$, rather than 12, then:

$$CC_S = \frac{2(16)}{20 + 18}$$
$$= 0.84, \text{ or } 84\%,$$

indicating a greater community similarity in the latter case.

The value of CC_J or CC_S ranges from 0 (when no species are found in both communities) to 1.0 (when all species are found in both communities).

Unfortunately, coefficients of community are inadequate in many circumstances, for they do not take into account the relative abundances of the various species. Thus for example, communities 1 and 2 are judged identical in both examples A and B in table 5C.1. The coefficient of community is a useful measure only when our major interest can be satisfied by consideration of presence or absence of species. This index is used in some water pollution studies where we want to measure differences in community structure expressable as the presence or absence of species (as between areas with and without a pollutant).

2.2. Proportional Similarity Species abundances in each of two (or more) communities may be tabulated as percentages, as shown in table 5C.2. If x_i is the number (or density) observed for species i in community 1, and y_i is the number (or density) of individuals of that species in community 2, then $p_i = x_i/N_i \times 100$ is the percent of the individuals in community 1 that belong to species 1, and $q_i = y_i/N_2 \times 100$ is the percent in community 2 that

167

Table 5C.1. *Various indices of community similarity calculated for three hypothetical situations. "Groups" may refer to communities, subcommunities, or other species assemblages.*

	EXAMPLE A			EXAMPLE B			EXAMPLE C	
	Number of individuals in species i			Number of individuals in species i			Number of individuals in species i	
Species (i)	Group 1 (x_i)	Group 2 (y_i)	Species (i)	Group 1 (x_i)	Group 2 (y_i)	Species (i)	Group 1 (x_i)	Group 2 (y_i)
1	50	60	1	50	3	1	50	0
2	25	30	2	25	7	2	25	7
3	12	15	3	12	15	3	12	15
4	6	7	4	6	30	4	6	30
5	3	3	5	3	60	5	0	60
$S = 5$	$s_1 = 5$	$s_2 = 5$	$S = 5$	$s_1 = 5$	$s_2 = 5$	$S = 5$	$s_1 = 4$	$s_2 = 4$
	$N_1 = 96$	$N_2 = 115$		$N_1 = 96$	$N_2 = 115$		$N_1 = 93$	$N_2 = 112$
	$CC_J = 1.00$			$CC_J = 1.00$			$CC_J = 0.60$	
	$CC_s = 1.00$			$CC_s = 1.00$			$CC_s = 0.75$	
	$PS = 99\%$			$PS = 30\%$			$PS = 25\%$	
	$I_1 = 11.62$			$I_1 = 79.79$			$I_1 = 83.72$	
	$I_2 = 5.20$			$I_2 = 35.68$			$I_2 = 37.44$	
	$I_3 = 0.08$			$I_3 = 0.69$			$I_3 = 0.74$	
	$I_{BC} = 0.91$			$I_{BC} = 0.29$			$I_{BC} = 0.24$	
	$I_{CM} = 0.93$			$I_{CM} = 0.37$			$I_{CM} = 0.33$	
	$I_S = 1.00$			$I_S = 0.22$			$I_S = 0.13$	
	$I_M = 1.02$			$I_M = 0.22$			$I_M = 0.14$	
	$R_O = 1.00$			$R_O = 0.52$			$R_O = 0.37$	

Table 5C.2. *The percent composition of the communities in example C of table 5C.1.*

(i)	Community 1	Community 2
1	$50/93 = 54\%$	$0/112 = 0\%$ *
2	$25/93 = 27\%$	$7/112 = 6\%$ *
3	$12/93 = 13\%$ *	$15/112 = 13\%$
4	$6/93 = 6\%$ *	$30/112 = 27\%$
5	$0/93 = 0\%$ *	$60/112 = 54\%$

* The lower of the two percentages for each species.

belong to that species. Then for each species, look for the lowest percentage among the communities. The "percentage of similarity"[2] is defined as:

$$PS = \Sigma[\text{lowest percentage for the species}]. \quad (3)$$

or, equivalently, as

$$PS = 1 - \frac{\Sigma|p_i - q_i|}{2}. \quad (4)$$

If there are two communities being compared, then p_i is the proportion composition of species i in the first community and q_i is the proportion composition of that species in the second community, and

$$PS = \Sigma(p_i \text{ or } q_i, \text{ whichever is lower}), \quad (5)$$

2. Huhta (1979) and Wolda (1981) attribute this method to Renkonen in 1938.

which for the table 5C.2 data, is:

$$PS = 0 + 0.06 + 0.13 + 0.06 + 0$$
$$= 0.25 = 25\%.$$

or, by equation 4:

$$PS = 1 - \frac{0.54 + 0.21 + 0 + 0.21 + 0.54}{2}$$
$$= 1 - 0.75 = 0.25 = 25\%.$$

2.3. Differences in Species Abundance   Similarities between communities may be expressed by measuring the difference between the abundances of each species present, where x_i is the abundance (or density) of species i in community 1 and y_i is the abundance of that species in the other community.

The following measure (used by Odum, 1950, and commonly attributed to Bray and Curtis, 1957) operates in that fashion:

$$I_{BC} = 1 - \frac{\Sigma|x_i - y_i|}{\Sigma(x_i + y_i)}. \quad (6)$$

For example C in table 5C.1, this would be computed as

$$I_{BC} = 1 - \frac{|50 - 0| + |25 - 7| + |12 - 15|}{(50 + 0) + (25 + 7) + (12 + 15)}$$

$$\frac{+ |6 - 30| + |0 - 60|}{+ (6 + 30) + (0 + 60)}$$

$$= 1 - 155/205$$

$$= 1 - 0.76 = 0.24.$$

The so-called *Canberra metric* (introduced by Lance and Williams, 1966) is

$$I_{CM} = 1 - \frac{1}{S} \sum \frac{|x_i - y_i|}{(x_i + y_i)}. \qquad (7)$$

In using it, replace any zero value of x_i or y_i with a small number, such as 0.2 (Clifford and Stephenson, 1975:59, 92). Thus, for example C in Table 5C.1, we would calculate

$$I_{CM} = 1 - \frac{1}{5} \left[\frac{|50 - 0.2|}{(50 + 0.2)} + \frac{|25 - 7|}{(25 + 7)} \right.$$

$$\left. + \frac{|12 - 15|}{(12 - 15)} + \frac{|6 - 30|}{(6 - 30)} + \frac{|0.2 - 60|}{(0.2 + 60)} \right]$$

$$I_{CM} = [0.99 + 0.56 + 0.11 + 0.67 + 0.99]/5$$

$$= 1 - 3.32/5 = 1 - 0.66 = 0.34.$$

Both of these similarity measures take values from 0 (when the two communities are vastly different) to 1 (when the two are identical in species composition and abundance). Some authors (see Huhta, 1979) have suggested that better measures are obtained if a logarithmic transformation is used (e.g., use $\log(x_i + 1)$ in place of x_i and $\log(y_i + 1)$ in place of y_i). It should be noted that I_{BC} is more inflated than I_{CM} by the occurrence of very dominant species. I_{BC} and PS are identical when the two sample sizes (n_1 and n_2) are equal; and when $n_1 \neq n_2$, PS is preferable (Kohn and Riggs, 1982).

2.4. Dominance Indices   Morisita's index of community similarity (Horn, 1966; Morisita, 1959) is based on Simpson's index of dominance (l; see section 5B.2.2). The probability that two randomly selected individuals from a community will be of the same species:

$$l_1 = \frac{\sum x_i(x_i - 1)}{N_1(N_1 - 1)} \qquad (8)$$

is the Simpson dominance index for community 1, where x_i is the number of individuals in species i in community 1, and N_1 is the total number of individuals in community 1 ($N_1 = \sum x_i$); likewise:

$$l_2 = \frac{\sum y_i(y_i - 1)}{N_2(N_2 - 1)}, \qquad (9)$$

where, for community 2, l_2 is Simpson's dominance index, y_i is the abundance of species i, and $N_2 = \sum y_i$, the total number of individuals in community 2.

The Morisita index of community similarity (also called Morisita's index of overlap; Horn, 1966) is:

$$I_M = \frac{2\sum x_i y_i}{(l_1 + l_2)N_1 N_2}. \qquad (10)$$

It may range from 0 (no similarity) to approximately 1.0 (identical). It has the desirable characteristic of being little affected by the sizes (Morisita, 1959; Wolda, 1981) and diversities (Wolda, 1981) of the samples. For example C in table 5C.1, we can compute:

$$l_1 = [(50)(49) + (25)(24) + (12)(11)$$
$$+ (6)(5)]/(93)(92)$$
$$= 0.375$$
$$l_2 = [(7)(6) + (15)(14) + (30)(29)$$
$$+ (60)(59)]/(112)(111)$$
$$= 0.375$$
$$\sum x_i y_i = (50)(0) + (25)(7) + (12)(15)$$
$$+ (6)(30) + (0)(60)$$
$$= 535$$
$$I_M = (2)(535)/[(0.375 + 0.375)(93)(112)]$$
$$= 1070/7812.000$$
$$= 0.14.$$

This index of community similarity refers to the probability that individuals randomly drawn from each of the two communities will belong to the same species, relative to the probability of randomly selecting a pair of specimens of the same species from one of the communities.

Another index that is related to Simpson's index of dominance is that attributed to Stander (e.g., by Johnson and Miller, 1982; Sullivan, 1975):

$$I_S = \frac{\sum x_i y_i}{\sqrt{\sum x_i^2 y_i^2}}, \qquad (11)$$

which has found some favor among algal ecologists.

2.5. Information-Theoretic Index   Horn (1966) proposed an index of community similarity ("community overlap") based on information theory (see section 5B.2.3). First calculate the Shannon diversity index for each community, using either equation 10 or 12 in section 5B. Let us use the latter equation for example C in table 5C.1; signify the diversity indices for communities 1 and 2 as H_1' and H_2', respectively, with the previously defined symbols:

$$H_1' = (N_1 \log N_1 - \sum x_i \log x_i)/N_1 \qquad (12)$$
$$= [93 \log 93 - (50 \log 50 + 25 \log 25$$
$$+ 12 \log 12 + 6 \log 6)]/93$$
$$= 0.490$$

$$H_2' = (N_2 \log N_2 - \sum y_i \log y_i)/N_2 \qquad (13)$$
$$= [112 \log 112 - (7 \log 7 + 15 \log 15$$
$$+ 30 \log 30 + 60 \log 60)]/112$$
$$= 0.491.$$

Then, calculate a value of H' for the sums of the species abundances for each species; that is, an H' considering all data to be from the same collection:

$$H_3' = -\Sigma\left(\frac{x_i + y_i}{N_1 + N_2} \log \frac{x_i + y_i}{N_1 + N_2}\right), \qquad (14)$$

which can more readily be worked with as:

$$H_3' = [N \log N - \Sigma(x_i + y_i) \log (x_i + y_i)]/N, \qquad (15)$$

where $N = N_1 + N_2$, the total number of individuals in both communities. So, for example C in table 5C.1:

$$N = 93 + 112 = 205$$
$$\begin{aligned}H_3' = &\ 205 \log 205 - (50 \log 50 + 32 \log 32 \\ &+ 27 \log 27 + 36 \log 36 + 60 \log 60)/205 \\ = &\ 0.680.\end{aligned}$$

The value of H_3' will be lowest if there is much community similarity, or overlap; highest with the most community dissimilarity.

Then, we calculate what H' would have been if each species abundance (each x_i and y_i value) from the two communities would have been from a different species—the maximum value of H' obtainable from the given species abundances:

$$H_4' = -\Sigma\left(\frac{x_i}{N_1 + N_2} \log \frac{x_i}{N_1 + N_2}\right)$$
$$- \Sigma\left(\frac{y_i}{N_1 + N_2} \log \frac{y_i}{N_1 + N_2}\right), \qquad (16)$$

which is equivalent to:

$$H_4' = (N \log N - \Sigma x_i \log x_i - \Sigma y_i \log y_i)/N. \qquad (17)$$

For the present data:

$$\begin{aligned}H_4' = &\ 205 \log 205 - (50 \log 50 + 25 \log 25 \\ &+ 12 \log 12 + 6 \log 6 + 7 \log 7 \\ &+ 15 \log 15 + 30 \log 30 + 60 \log 60)/205 \\ = &\ 0.789.\end{aligned}$$

Lastly, we calculate:

$$\begin{aligned}H_5' &= (N_1 H_1' + N_2 H_2')/N \qquad (18) \\ &= [(93)(0.490) + (112)(0.491)]/205 \\ &= 0.491.\end{aligned}$$

This is a measure of the minimum H' obtainable (with maximum overlap of the x_i and y_i values).

Then, Horn's index of community overlap is:

$$R_0 = \frac{H_4' - H_3'}{H_4' - H_5'}, \qquad (19)$$

which, for our data, is:

$$\begin{aligned}R_0 &= \frac{0.789 - 0.680}{0.789 - 0.491} \\ &= 0.37.\end{aligned}$$

Horn's index, R_0, is 0 when the two communities have no species in common and is a maximum of 1.0 when the species compositions and relative abundances are identical in both communities. (Rejmánek (1981) provides a correction to R_0 applicable in the uncommon instance when either H_1' or H_2' is greater then H_3'.)

3. Suggested exercises

1. Determine the similarity (i.e., overlap) between the fauna at two stream or two pond sites, or at the same site at two different times of year. Use sampling methods of section 3E and any of the above similarity indices.
2. Determine the similarity (i.e., overlap) between two plant communities, using data collected by methods in section 3A, 3B, or 3C.
3. Determine the similarity (i.e., overlap) between two terrestrial invertebrate communities (or for the same community at two different times of year), using methods in section 3D.
4. Determine the similarity (i.e., overlap) between food items in the diets of two species of animals. (If the two species occur in the same community, then the overlap index for diet items is a measure of how the species share food resources.)

4. Selected references

Barbour, M. G., J. H. Burk, and W. D. Pitts. 1980. Terrestrial plant ecology. Benjamin Cummings Publishing Co., Menlo Park, Calif.

Bray, J. R., and J. T. Curtis. 1957. An ordination of the upland forest communities of southern Wisconsin. Ecol. Monogr. 27:325–334, 337–349.

Cheetam, A. H., and J. E. Hazel. 1969. Binary (presence-absence) similarity coefficients. J. Paleontol. 43:1130–1146.

Clifford, H. T., and W. Stephenson. 1975. An introduction to numerical classification. Academic Press, New York.

Digby, P. G. N., and R. A. Kempton. 1987. Multivariate analysis of ecological communities. Chapman and Hall, London.

Ghent, A. W. 1983. *Tau* as an index of similarity in community comparisons: An approach permitting the hypothesis of unequal species abundance. Can. J. Zool. 61:687–690.

Grieg-Smith, P. 1983. Quantitative plant ecology. 3d. ed. University of California Press, Berkeley, Calif.

Horn, H. S. 1966. Measurement of "overlap" in comparative ecological studies. Amer. Natur. 100:419–424.

Huhta, V. 1979. Evaluation of different similarity indices as measures of succession in arthropod communities of the forest floor after clear-cutting. Oecologia 4:11–13.

Hurlbert, S. H. 1978. The measurement of niche overlap and some relatives. Ecology 59:67–77. (See also: Abrams, P. A. 1980. Some comments on measuring niche overlap. Ecology 61:44–49. Hurlbert, S. H. 1982. Notes on the measurement of overlap. Ecology 63:252–253. Abrams, P. A. 1982. Reply to a comment by Hurlbert. Ecology 63:253–254.)

Janson, S., and J. Vegelius. 1981. Measures of ecological association. Oecologia 49:371–376.

Johnson, B. E., and D. F. Miller. 1982. The estimation and applicability of confidence intervals for Stander's similarity index (SIM2) in algal assemblage comparisons. Hydrobiologia 89:3–8.

Kaesler, R. L., and J. Cairns, Jr. 1972. Cluster analysis of data from limnological surveys of the upper Potomac River. Amer. Midland Natur. 88:56–67.

Kershaw, K. A., and J. H. H. Looney. 1985. Quantitative and dynamic plant ecology. 3d ed. Edward Arnold, London.

Kohn, A. J., and A. C. Riggs. 1982. Sample size dependence in measures of proportional similarity. Mar. Ecol. Progr. Ser. 9:147–151.

Lance, G. N., and W. T. Williams. 1966. Computer programs for hierarchical polythetic classification ("similarity analysis"). Comput. J. 9:60–64.

Ludwig, J. A., and J. F. Reynolds. 1988. Statistical ecology: A primer on methods and computing. John Wiley & Sons, New York.

McIntosh, R. P. (ed.). 1978. Phytosociology. Benchmark papers in ecology. Volume 6. Dowden, Hutchinson & Ross, Stroudsburg, Penn.

Morisita, M. 1959. Measuring of interspecific association and similarity between communities. Mem. Fac. Sci. Kyushu Univ., Ser. E (Biol.) 3:65–80.

Mueller-Dombois, D., and H. Ellenberg. 1974. Aims and methods of vegetation ecology. John Wiley & Sons, New York.

Odum, E. P. 1950. Bird populations of the Highlands (North Carolina) Plateau in relation to plant succession and avian invasion. Ecology 31:587–605.

Pielou, E. C. 1984. The interpretation of ecological data: A primer on classification and ordination. John Wiley & Sons, New York.

Rejmánek, M. 1981. Corrections to the indices of community similarity based on species diversity measures. Oekologia 48:290–291.

Sneath, P. H. A., and R. R. Sokal. 1973. Numerical taxonomy. W. H. Freeman Co., San Francisco.

Sørensen, T. 1948. A method of establishing groups of equal amplitude in plant sociology based on similarity of species content. K. Dansk. Vidensk. Selsk. Biol. Skrift 5(4):2–16, 34. (McIntosh, 1978:234–249).

Southwood, T. R. E. 1978. Ecological methods. John Wiley & Sons, New York.

Sullivan, M. J. 1975. Diatom communities from a Delaware salt marsh. J. Phycol. 11:384–390.

Wallace, R. K., Jr. 1981. An assessment of diet-overlap indices. Trans. Amer. Fish. Soc. 110:72–76.

Whittaker, R. H. 1967. Gradient analysis of vegetation. Biol. Rev. 49:207–264.

Wilson, M. V., and C. L. Mohler. 1983. Measuring compositional change among gradients. Vegetatio 54:129–141.

Wilson, M. V., and A. Shmida. 1984. Measuring beta diversity with presence-absence data. J. Ecol. 72:1055–1064.

Wolda, H. 1981. Similarity indices, sample size and diversity. Oekologia 50:296–302.

5d

ecological succession

1. Introduction

The progressive, orderly, and somewhat predictable series of replacements of one community by another is called *succession*. Not only does the species composition of communities change as succession proceeds, but life forms and habitats also change, there is an increased accumulation of biomass, increased productivity, and the development of a more complex community structure. Environmental changes, such as in soil chemistry and moderation of microclimates, also accompany succession, and there are changes in species composition and diversity, and a reduction in temperature extremes. In terrestrial succession there is typically an increase in soil organic matter, a reduction in fluctuations of temperature and humidity, the development of stratification and shade, and a reduction of light and wind penetration.

A *sere* is the sequence of communities that develops by the process of ecological succession. The individual communities in a given sere, prior to the final (or *climax*) stage, are called *seral stages*. A sere is named for the habitat occurring at the start of succession. Examples of seres beginning on dry land (*xeroseres*) are sand, rock, and clay. A pond sere is an example of succession having an aquatic origin (*hydrosere*). *Primary succession* (exemplified by the above-mentioned seres) begins on sterile substrate. *Secondary succession* begins on substrate that previously supported life (e.g., flooded, burned, grazed, logged, or farmed areas). Since the substrate in the latter case is already well developed, secondary succession usually proceeds much faster than primary succession.

During the first stages in a sere, especially in primary succession, the environment is relatively harsh, and members of very few specialized taxa can invade the area. The occurrence of pioneer plants, however, gradually adds organic matter and wind-blown soil particles to the meager substrate. Such occurrence also provides a modified regime of light, wind, and temperature, eventually allowing members of other taxa to gain admittance to the habitat.

Thus in each seral stage, various plant species modify the environment enough to permit the invasion of others. In general, those that eventually invade are better adapted to the changed environment, and they exclude and replace their predecessors.

The final stage of a sere is the *climax* community—typically that community characteristic of the biome in which the sere is located. That is, while the character and composition of the early ("pioneer") seral stages depend heavily on substrate and other microhabitat characteristics, the climax community is largely a function of the climate of the area. For example, all ecological succession in the temperate deciduous forest biome typically climaxes in a deciduous forest, regardless of whether the start of the sere took place on rock, sand, floodplain, or cornfield. This is called *convergence* of seres. Local soil and landform conditions may result in climax communities differing somewhat from those typical of the geographic area, and these are called *edaphic* and *topographic* climaxes, respectively. Continual disturbance of an area may maintain indefinitely a nonclimax community known as a *disclimax;* for example the grazing of domestic animals may maintain a grassland where a forest might otherwise naturally develop as the climax. In moist temperate climates the following sequence is found, beginning from the initial substrate condition: pioneer species, secondary grasses or forbs (nongrassy herbs), shrubs, pioneer trees, secondary trees, subclimax trees, and climas trees. However, some of these stages may be accelerated or skipped over; or some may be retarded and remain for extremely long periods of time.

Often the sequence of stages in a sere is not entirely predictable, as there may be several possible pathways from pioneer to climax communities. The occurrence of one pathway rather than another may depend on factors such as local topographic and substrate conditions, physiographic processes (e.g., erosion or deposition), availability of seed sources for colonization, or habitat disturbance (e.g., fire, flood, or clearing of vegetation). These factors will often result in a mosaic pattern of seral communities over the landscape, rather than a simple linear sequence of stages across the land.

Excellent descriptions of various seres are given by Bird (1930), Dansereau and Segadas-Vianna (1952), Horn (1975), Kormondy (1969), Olson (1958), Shelford (1963), and the classical work of Clements (1916), Cowles (1899), and Shelford (1913). Consult appropriate literature references for comparison with your observations.

2. Procedures

The study of succession is essentially the study of habitats and communities, thus employing the techniques of units 2, 3, and 5. The problem in such investigations is that a sere progresses from stage to stage over a period of years, decades, or even longer. The rate of progression depends on the original substrate, the climate, and other habitat factors.

Fortunately, it is frequently possible to locate several successional stages of a given sere in the same geographic area, because succession may have begun on different portions of the same substrate at different times. For example, sand succession may have begun at different times on different dunes, various agricultural fields may have been abandoned at different times, and various ponds in an area may have been artificially constructed at different times. As a result, the age of the sere is often discernible (e.g., the date of last farming, the date of pond construction, the date of last strip mining, or the date of the last fire).

2.1. Habitat Analysis   In each successional stage studied, perform a biotic and an abiotic analysis of the habitat using the appropriate portions of unit 2. Important measurable habitat variables are air temperature, relative humidity, light penetration, soil temperature, soil moisture, soil organic carbon, and soil nitrate (see sections 2B, 2C, and 2E). Note the thickness of the A1 soil horizon and the soil texture. Observe how all these factors change as succession progresses. For aquatic seral stages, the variables described in section 2D.3 should be measured. Section 2E describes chemical analysis of water, in which dissolved oxygen, nitrate, and phosphate determinations are the most important.

2.2. Qualitative Analysis   Much can be learned about succession through qualitative description of the vegetation in the seral and climax stages (section 5A). Record dominant species and their forms and changes in the physiognomies of the communities during succession. Pay particular attention to changes in stratification, coverage, and screening efficiency, and relate these to changes in habitat factors (e.g., wind, light, humidity, and temperature), and to the habitat requirements of animals and nondominant plants.

2.3. Quantitative Analysis   The process of succession may be studied quantitatively by considering changes in density, coverage, frequency, importance value, biomass, or productivity. Species diversity (section 5B) may be determined, and it often increases during the early stages of succession.

Use section 3A, 3B, or 3C to sample plants, and section 3D or 3E to sample animals in each community to be studied. Unit 6 describes the measurement of biomass and productivity.

After plant sampling, select a few common species in the communities of the sere. For each of these species, compute the relative density in each stage of the sere, as:

$$RD_j = D_j/\Sigma D, \qquad (1)$$

where RD_j is the relative density of a species in successional stage j, D_j is the density of the species in stage j, and ΣD is the sum of the densities of that species in all stages (table 5D.1). Then for each species, graph relative density (RD_j) against the successional stage (j), as in figure 5D.1. Observe during which stages the species increases in abundance, when it reaches its peak, and when it decreases in abundance. Do this for several species on the same graph. Similar plots may be prepared using species, or other taxa, within major animal groups (e.g., birds or insects).

Similar figures may result from relative coverage (RC_j) of each major species, computed from that species' coverage (C_j) in stage j and the sum of the coverage values (ΣC) for the species over all stages. With coverage data, only those species whose data were comparably collected (e.g., as aerial coverage, basal coverage, or basal area) should be plotted together on the same graph. A similar graphical presentation may be prepared using relative frequencies ($Rf_j = f_j/\Sigma f$) or relative importance values:

$$RIV_j = IV_j/\Sigma IV, \qquad (2)$$

where RIV_j is the relative importance value of a species in stage j, IV_j is the importance value for that species in stage j, and ΣIV is the sum of the importance values of the species in all stages.

Measures of similarity (section 5C) may be used to compare successional stages, and in this fashion it may be concluded at which stage there are greatest transitions from one community to another. For example, we could apply a measure of community similarity to compare the adjacent seral stages in table 5D.1. The results of such an examination are shown in table 5D.2, from which it can be seen that there is greatest similarity in species composition between seral stages 1 and 2, between stages 3 and 4, and between stages 5 and 6.

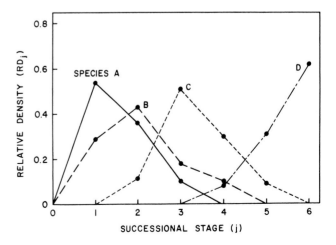

Figure 5D.1. The data of table 5D.1, plotting relative density versus successional stage for each of four species.

Table 5D.1. *Hypothetical succession data for preparing figure 5D.1. D_j is the density of a species in stage j. The relative density of the species in stage j is computed as $RD_j = D_j/\Sigma D$, where ΣD is the sum of the densities of that species over all stages.*

Species	Successional stage												
	1		2		3		4		5		6		
	D_1	RD_1	D_2	RD_2	D_3	RD_3	D_4	RD_4	D_5	RD_5	D_6	RD_6	ΣD
A	42	0.54	28	0.36	8	0.10	0	0.0	0	0.0	0	0.0	78
B	25	0.29	37	0.43	16	0.18	9	0.10	0	0.0	0	0.0	87
C	0	0.0	6	0.11	29	0.51	17	0.30	5	0.09	0	0.0	57
D	0	0.0	0	0.0	0	0.0	3	0.08	12	0.31	24	0.62	39

Table 5D.2. *Comparing successional stages, using community similarity measure I_M from section 5C.2.4 and the data of table 5D.1.*

Stage:	1 vs. 2	2 vs. 3	3 vs. 4	4 vs. 5	5 vs. 6
I_M:	0.93	0.64	0.99	0.50	0.91

3. Suggested exercises

1. Examine one or a few seres in your region. Record the dominant species in each stage and the initial physiographic feature on which the sere began. Trace the development of the sere by preparing a simple diagram showing the sequence of dominant life forms for each stage (pioneer plants, secondary grasses and forbs, etc.). Compare the sere to literature descriptions. Are any typical stages skipped over in the sere under study? Are the dominants of a particular stage different from those expected? If so, explain. Is there evidence of a stage being retarded or of one reverting back to a less advanced stage due to habitat alteration? How is the progress toward the climax being affected by topography or soil?

2. For each stage in a xerosere, determine as many of the following characteristics of the dominant plant life forms as time allows: species, density, coverage, biomass, and species diversity. Plot each quantitative characteristic as a function of successional stage (or as a function of age of the sere). Often diversity and biomass reach a peak prior to the climax community. Describe and interpret where the maximum values of the above variables occur during the successional process.

3. For each stage in a xerosere, record the progressive changes in as many of the following habitat variables as time permits: air temperature, relative humidity, light penetration (by measuring light or screening efficiency), soil temperature, soil moisture, soil organic carbon, soil nitrate, soil phosphate, and thickness of the A1 soil horizon. Plot each variable as a function of successional stage or time, and interpret.

4. For each age community in a hydrosere, determine as many of the following as time permits (preferably separately for benthos, phytoplankton, zooplankton, nekton, and rooted plants): number of species, species abundance, and species diversity. Plot each against successional time. Interpret these graphs. When and why does each variable peak? How does rate of invasion of new species change with time? From where do the new species come?

5. For each age community in a hydrosere, record the progressive changes in as many of the following habitat variables as time allows: water temperature, depth, turbidity, substrate type, conductivity, dissolved oxygen, nitrate, and phosphate. Plot each quantitative variable against time, and interpret trends and peaks.

4. Selected references

Barbour, M. G., and W. D. Billings (eds.). 1987. North American terrestrial vegetation. Cambridge University Press, New York.

Barbour, M. G., J. H. Burk, and W. D. Pitts. 1980. Terrestrial plant ecology. Benjamin/Cummings Publishing Co., Menlo Park, Calif.

Bird, R. D. 1930. Biotic communities in the aspen parkland of central Canada. Ecology 11:356–442.

Cain, S. A. 1947. Characteristics of natural areas and factors in their development. Ecol. Monogr. 17:185–200.

Clements, F. E. 1916. Plant succession. An analysis of the development of vegetation. Carnegie Institute Publ. 242, Washington, D.C.

Cowles, H. C. 1889. The ecological relationships of the vegetation of the sand dunes of Lake Michigan. Bot. Gaz. 27:45–117, 167–202, 281–308, 361–391.

Dansereau, P. 1974. Types of succession, 123–135. In R. Knapp, Vegetation dynamics. Dr W. Junk Publishers, Boston.

Dansereau, P., and F. Segadas-Vianna. 1952. Ecological study of the peat bogs of eastern North America. Can. J. Bot. 30:490–520.

Golley, F. B. (ed.). 1977. Ecological succession. Academic Press, New York.

Horn, H. S. 1974. The ecology of secondary succession. Annu. Rev. Ecol. Systemat. 5:25–37.

Horn, H. S. 1975. Forest succession. Sci. Amer. 232:90–98.

Huhta, V. 1979. Evaluation of different similarity indices as measures of succession in arthropod communities of the forest floor after clear-cutting. Oecologia 41:11–23.

Kershaw, K. A., and J. H. H. Looney. 1985. Quantitative and dynamic plant ecology. 3d ed. Edward Arnold, London.

Kormondy, E. J. 1969. Comparative study of sandspit ponds. Amer. Midland Natur. 82:2861.

Margalef, R. 1963. On certain unifying principles of ecology. Amer. Natur. 97:357–374.

Mueller-Dombois, D. and H. Ellenberg. 1974. Aims and methods of vegetation ecology. John Wiley & Sons, New York.

Olson, J. S. 1958. Rates of succession and soil changes on southern Lake Michigan sand dunes. Bot. Gaz. 119:125–170.

Oosting, H. J. 1960. The study of plant communities. W. H. Freeman and Co., San Francisco.

Orlóci, L. 1978. Multivariate analysis in vegetation research. Dr W. Junk Publishers, Boston.

Shelford, V. E. 1913. Animal communities in temperate America. University of Chicago Press, Chicago.

Shelford, V. E. 1963. Ecology of North America. University of Illinois Press, Urbana, Ill.

Walker, D. 1970. Direction and rate in some British postglacial hyroseres, 117–139. In D. Walker and R. West (eds.), The vegetation history of the British Isles. Cambridge University Press, Cambridge, England.

West, D. C., H. H. Shugart, and D. Botkins (eds.). 1981. Forest succession: Concepts and applications. Springer-Verlag, New York.

Whitaker, R. A. 1953. A consideration of climax theory: The climax as a population and pattern. Ecol. Monogr. 23:41–78.

Whittaker, R. H. 1967. Gradient analysis of vegetation. Biol. Rev. 42:207–264.

Whittaker, R. 1974. Climax concepts and recognition, 137–154. In R. Knapp (ed.), Vegetation dynamics. Dr W. Junk Publishers, Boston.

Whittaker, R. H. (ed.). 1978. Ordination of plant communities. Dr W. Junk Publishers, Boston.

introduction

In previous sections we have been concerned with quantitative determinations of various population and community measures based on estimates of numbers of organisms. However, a large variety of ecological studies need estimates of biomass, production, or productivity. *Biomass* is simply the weight of living organisms. *Standing crop* is the biomass present, usually expressed as the weight of a population in a given area at a given time. *Production* is the amount of biomass produced over a given time, and *productivity* is the *rate* of biomass production per unit time. Thus one might measure a standing crop of 150 kilograms of fish per hectare and find that productivity of fish biomass is 15 kg/ha/yr. The amount of this productivity harvested commercially is called *yield.*

Gross primary production is the amount of organic matter photosynthesized by the producers (autotrophic organisms) in the ecosystem. *Net primary production* is the gross primary production minus the amount of biomass lost by the producers by way of respiration. Biomass accumulated in organisms feeding on producers is called *secondary production.*

As will be seen in later discussions, measures of production may be expressed as biomass equivalents of energy or of oxygen or carbon dioxide involved in photosynthesis or respiration. But because biomass production is discussed here, we shall refer to gross primary productivity and net primary productivity as *gross biomass productivity* (*GBP*) and *net biomass productivity* (*NBP*), respectively. We shall also indicate the biomass lost through respiration as *RB*. Sometimes ecologists express *GBP* relative to standing crop biomass (*B*), by reporting the *GBP/B* ratio. Similarly, *NBP/B* is the ratio of net biomass productivity to standing crop biomass.

A significant relative measure of productivity is the *turnover time:* the time needed for the ecosystem to produce an amount of biomass equal to that of the standing

unit 6

analysis of production

crop. It is calculated as standing crop biomass (*B*) divided by the gross biomass productivity (*GBP*):

$$T = \frac{B}{GBP}. \qquad (1)$$

Since, in equation 1, *B* is expressed as units of weight per area (or volume) of habitat, and *GBP* as units of weight per unit area (or volume) per unit time, *T* is the number of time units (days, months, years) required to produce as much biomass as is present at the time of measuring the standing crop. Young or simple ecosystems tend to have shorter turnover times than do older complex systems.

Another informative productivity index is the ratio *NBP/B*. Net biomass productivity tends to decrease as an ecosystem matures. In a steady state system this ratio approaches 0; also the ratio *RB/GBP* approaches 1, as the amount of biomass leaving the ecosystem through respiration is balanced by the amount of biomass entering the ecosystem through primary production (photosynthesis). If *RB* < *GBP*, then biomass is accumulating in the system, a common occurrence in young or agricultural ecosystems, or ecosystems early in succession.

1. Introduction

Values of biomass are often more useful to know than are values of population density. The sizes of individuals of different species vary greatly. However, carbon fixation as well as energy and nutrient transfers are largely dependent on the biomass, rather than just on the number of individuals. Because direct measurement of biomass cannot be made on an entire community or population, samples must be taken from a portion of a community or population. Sampling methods for determining biomass are essentially the same as those for estimating density. One may use the sampling procedures of unit 3 for the specific situation under study.

Biomass expressed as weight per unit area is commonly estimated from either of the following equations:

$$B = \Sigma W/A \qquad (1)$$

or

$$B = (D)(\bar{W}) = (D)(\Sigma W/n), \qquad (2)$$

where B is the biomass (e.g., as g/m^2 or kg/ha), ΣW is the sum of the weights of the individual organisms in a sample, and A is the total area sampled; D is the density (e.g., number of individuals per square meter or per hectare), $\bar{W}$ is the mean weight per individual taken from a random sample, and n is the number of individuals in the sample.

For organisms that vary greatly in moisture content, express biomass as dry weight rather than fresh wet weight. Also, organisms may vary in the amount of inorganic skeletal material; if so, measure their ash-free dry weight, nitrogen, carbon, or caloric content. Odum (1983:92) gives caloric equivalents of biomass for various types of organisms.

2. Dry weight

When comparing the biomass of different species, the *fresh* or *wet weight* of the population is often meaningless. As water contents of various taxa may vary considerably, *dry weights* usually afford better comparison of biologically important material. Also, accurate wet weights are difficult to obtain on dead or preserved specimens. Biomass expressed as wet weight should be limited to live or freshly killed, taxonomically similar organisms. Bottom-feeding and soil organisms, such as catfish and earthworms, may contain significant amounts of sediment or soil in their alimentary canals, therefore one should remove the gut contents so that no sediment or soil is part of the measured weight. However, food material is often only a small proportion of the total weight of the organism and frequently can be ignored.

Dry weights of biological material or soil are obtained by oven-drying samples at 105°C until they no longer lose weight. Take care not to cook or char the samples. (Bound water and water of crystallization are generally not lost

at this temperature.) As an alternative to oven-drying, you may dry the samples in a freeze-drier, especially if you are worried about losing volatile substances such as lipids, or heat-altering the chemical and caloric content of the tissues. Temperatures under 105°C may also dry samples sufficiently while alleviating heat-associated adverse effects, but this will require increased drying time.

Twenty-four hours is usually sufficient drying time for 1- to 10-gram samples. Samples smaller than 1 g may often be dried in four to twelve hours; samples larger than 10 g may take days. The duration of the required drying period may have to be determined by the investigator for each particular type and mass of material. It is good practice to cut up large items into smaller pieces of only a few grams to ensure drying of internal parts. Leaves, other herbaceous material, or soil should be spread out to expose a large surface area during drying.

Estimate the average weight per individual by the following method. Weigh a clean dry container such as a crucible, beaker, or aluminum dish. Then add 1 to 10 g of sample, weigh it and the container, and oven dry at 105°C for twenty-four hours. Using tongs, remove the container from the oven and place it in a desiccator containing a drying agent such as anhydrous $CaCl_2$, and allow the sample and container to cool to room temperature. After cooling, weigh the sample with its container. Be sure the balance is free of dust and that weights are determined in a low humidity.

The dry weight of the sample (W_d) is the weight of the container with the oven-dried sample (W_o) minus the weight of the container when empty (W_c):

$$W_d = W_o - W_c. \qquad (3)$$

The mean dry weight per individual then is the sum of the dry weights divided by the number of individuals in the sample.

The weight of water in the sample is, of course, the difference between the fresh weight and dry weight. Therefore, the percentage of water in the sample is the weight of water divided by the fresh weight.

3. Ash-free dry weight

For species having similar body compositions, dry weights are very useful, but for organisms having various types of skeletons, shells, or silica and calcium content, dry weights may be of little comparative value. Standard procedures used to offset this problem determine *ash-free dry weight, organic carbon, protein nitrogen,* or *caloric content,* measures that also may be used for humus and soil. Ash-free dry weights are sufficient for most introductory purposes. Dry weights are also inaccurate where sediments are difficult to remove.

To obtain ash-free dry weights, first determine the dry weight of the sample using a clean, dry, and preweighed ceramic crucible as described above. Determine the weight of the crucible and the sample to the nearest 0.1 mg using an analytical balance. Large samples of dried material over 10 g should be ground up and thoroughly homogenized. Place a few subsamples of this ground sample in separate preweighed crucibles and weigh; samples of 0.5 to 5.0 g are recommended for efficient ashing. Then place the crucible in a muffle furnace and completely oxidize at 500°C until only inorganic ash is left. The duration of ashing depends on the size and type of material being ashed. Once the furnace has reached a temperature of 500°C, four to twelve hours may be required to oxidize the organic matter. If black charcoal deposits are still visible, then continue ashing. Care must be taken not to exceed 550°C since sodium and potassium would be volatilized and $CaCO_3$ converted to CaO, thus resulting in a biased ash-free dry weight.

Using tongs, remove the crucible containing the ashed samples only after the furnace has been allowed to cool for several hours. After removal from the furnace, the crucible and its sample are cooled to room temperature in a desiccator with a drying agent and then weighed.

The ash-free dry weight (W_f) is calculated as the difference between the weight of the crucible containing the oven-dried sample (W_o) and the weight of the crucible with the ash (W_b):

$$W_f = W_o - W_b. \qquad (4)$$

It is a measure of the dry weight of organic matter in the sample. The biomass is then estimated from equations 1 or 2, where ΣW represents the sum of the ash-free dry weights of the organisms. The difference between the weight of the crucible with the ash (W_b) and the weight of the empty crucible (W_c) is the amount of ash, i.e., inorganic matter, in the sample (W_a), often expressed as a percentage of the dry weight of the sample:

$$\text{percent ash} = (W_a/W_d)(100\%). \qquad (5)$$

4. Indirect methods

Situations often exist where weights cannot be determined or are troublesome to obtain. In these instances,

one estimates lengths, areas, or volumes of the organisms as an index of biomass. Often these indices can be standardized to a known set of biomass measurements. Trees, for example, are too large to weigh but tree heights and basal areas can be conveniently measured. Biomass of phytoplankton species is also impossible to determine because individual species cannot be separated and weighed. For many arthropods lengths are more efficiently measured than weights.

4.1. Weight-length Curve   Body length (L) is often closely correlated with body weight (W). If this is so for a species of interest, then measure the length of each individual in a sample in the field or laboratory. Then from a subsample of these individuals, weigh different-sized organisms from the entire size range of this species. Once the relationship between weight and length has been calculated, you may predict body weights from lengths. A line may be fit to the data visually if the deviation of data points from a straight line is very small; but it is better to use the method of least squares (sections 1B.6.1 and 1B.6.2) to compute a regression line.

With many species (e.g., fishes), a plot of the logarithm of weight against the logarithm of length often results in a straight line (figure 6A.1), namely:

$$\log W = a + b \log L. \qquad (6)$$

(Table D.1 in appendix D gives common logarithms, that is, logarithms to base 10; other logarithmic bases could

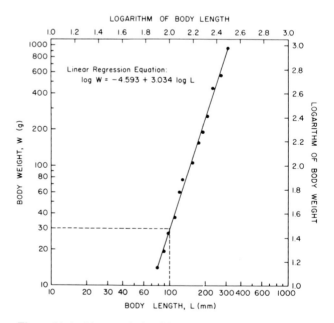

Figure 6A.1. Linear relationship between the logarithm of body weight *(W)* and the logarithm of body length *(L)* of members of a sample from a hypothetical fish population. The raw data may be plotted on graph paper with logarithmic scales (left and bottom axes), or the logarithms of the data may be calculated and then plotted on linear axes (top and right). Once the relationship has been determined, weight may be predicted from length measurements (see section 4.1).

also be employed.) For fishes, the regression coefficient, *b*, typically is very close to 3.0 (Carlander, 1969). For example, figure 6A.1 shows such a linear relationship, for which the regression line has been computed to be:

$$\log W = -4.593 + 3.034 \log L.$$

Thus if one had a 100-mm-long fish of this species, the regression equation would predict a body weight as follows:

$$\log W = -4.593 + (3.034)(\log 100)$$
$$= -4.593 + (3.034)(2.0000)$$
$$= 1.475;$$

by determining the antilogarithm of 1.475,

$$W = 29.9 \text{ g}.$$

Or, if such accuracy is not required, the value of *W* could be read directly off the graph as the value on the regression line at a length of 100 mm (see broken lines on figure 6A.1).

In many species (e.g., invertebrates), the logarithm of body weight is linearly related to body length, so that the regression line would be:

$$\log W = a + bL \qquad (7)$$

(see figure 6A.2). The data in figure 6A.2, for a hypothetical invertebrate species, result in the regression equation:

$$\log W = -1.118 + 0.0338 L.$$

Once we derive this equation, we can predict weights from given lengths of animals of this species. For example, the weight of a 30-mm individual from this population would be estimated as:

$$\log W = -1.118 + (0.0338)(30)$$
$$= -0.104;$$

and,

$$W = \text{antilog} (-0.104)$$
$$= 0.787 \text{ g},$$

a value which could also be read directly but less accurately off the graph (see broken lines in figure 6A.2).

Standard errors and confidence intervals may be computed for predictions from regression equations. The precision of such a prediction is directly related to the number of data used for calculation of the regression line and inversely related to the amount of scatter of data points around the line (see section 1B.6.4, equations 29 and 30).

4.2. Biovolume   In phytoplankton studies, algae are not easily separated from the microscopic animals for biomass determination. However, the larger zooplankton species may be separated using different sizes of mesh nets or sieves. Aside from estimating total biomass of the

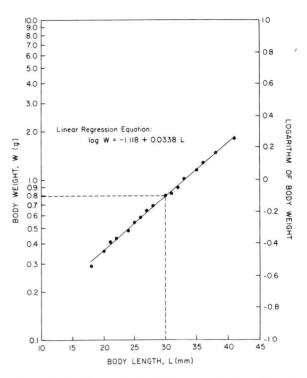

Figure 6A.2. Linear relationship between the logarithm of body weight (W) and the length (L) of members of a sample from a hypothetical population of invertebrates. The raw data may be plotted on semilogarithmic graph paper (the vertical axis is a logarithmic scale, as on the left), or the logarithms of the weights may be calculated and then plotted on a linear scale (as on the right). Once the relationship has been determined, weights may be predicted from length measurements (see section 4.1).

sample, analysis of individual biomass of species involves microscopic estimation of biovolume of individual microorganisms.

Using the methods for estimating density for plankton in section 3E, you may estimate the length and diameter of each individual organism observed in the microscope field. This involves calibrating a microscope using an ocular micrometer and a stage micrometer. Lengths (*L*) and diameters of each organism are then measured and volume approximated by assuming a geometrical form such as a sphere ($V = (4/3)\pi r^3$) or a cylinder ($V = \pi r^2 L$), where *r* is the radius of the organism (½ the diameter). For more accurate estimates, more complex geometrical figures may be assumed, but for introductory purposes these two will suffice.

In forests, use the basal area, height of the trees, and other variables to approximate biomass. Generally, procedures for an accurate estimate of forest biomass are too complex to treat adequately in an introductory discussion. For comparative purposes, however, the volume of an individual tree (V) can be approximated from estimates of tree height and basal area (Newbould, 1967):

$$V = \pi r^2 h/2, \qquad (8)$$

where r is the basal radius (at "breast height") and h is the height of an individual tree. V is, of course, only the volume of the trunk and does not include consideration of roots, branches, or foliage.

5. Herbaceous vegetation

Special problems occur in biomass analysis of vegetation. This measurement is difficult to perform on many terrestrial plants due to the problem of collecting and weighing the whole plant (as roots are often impossible to collect). Therefore, only a fraction of the total biomass can be estimated directly. But in certain communities the estimate of plant biomass is easier to determine than is density. Populations of grasses and vegetatively reproducing herbs cannot be defined adequately as the number of individuals per unit area, and in these cases the concept of an individual organism is vague. Determining the total weight of such plants, or at least their aerial parts, is usually easier than estimating the number of shoots, stems, or clumps.

Agricultural yield at harvest or primary productivity in wild fields can be found through plant biomass studies. Plant biomass can also be used to compare early stages of succession (section 5D), as by studying a field or a series of fields at different stages of succession, or fields having different soil or moisture conditions. Lay out plots of 1 m² (e.g., 0.71×1.41 m), as described in section 3A. Carefully dig up the soil to collect the roots, stems, and leaves of all grasses and forbs in each plot. (Collection of intact roots for certain deep-growing grasses may be impossible.) Separate and identify each species and gently dislodge the dirt from the roots. Label each sample and afterward, in the laboratory, carefully wash the roots free of all dirt. Oven dry the samples and record the dry weights. If you want ash-free dry weights, then determine them as described in section 3.

In an alternative procedure, crop all aerial parts close to the ground and place each sample in a plastic bag. Label and return it to the laboratory for determination of wet weights, dry weights, and/or ash-free dry weights as described above. Sample roots separately. Sampling the root biomass has many problems, such as taking a soil sample, extraction of roots from the soil, and identification of roots from different species. For species having shallow root systems, excavate the entire plant and wash the roots free of soil. But this is impractical for most woody plants, plants with deep roots, and sod-forming plants. In these cases take a soil core or a plot of soil of a given thickness. Then extract the roots by placing the sample in a tray of water and gently agitating it so the roots float free of the soil. Then estimate the biomass of the total root sample, expressed as weight of roots per cubic meter of soil.

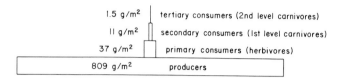

Figure 6A.3. A pyramid of biomass. This is essentially a horizontal histogram showing biomasses in the various trophic levels in a community. (Data from Odum, 1957.)

6. Standing crop pyramid

The species in a community may be grouped as producers, primary consumers, secondary consumers, and tertiary consumers. These are the different trophic levels in a community. (A *trophic level* is a level of energy transfer in a community, a particular link in a food chain.) *Producers* (predominantly photosynthetic organisms) obtain their energy requirements from abiotic sources (e.g., light for photosynthesis). By contrast, *consumers* must obtain their energy from organic sources. *Primary consumers* are those feeding directly on producers; *secondary consumers* feed upon primary consumers; and *tertiary consumers* feed upon secondary consumers. Primary consumers are also known as herbivores; other consumers may be termed carnivores, insectivores, piscivores, etc., depending on their source of food. A saprovore is an organism feeding on dead organic matter, and an omnivore is an organism whose feeding habits place it in more than one trophic level.

In a given community, determine the dry weight or ash-free dry weight for each species, and estimate the biomass. An *ecological pyramid of biomass* may be constructed by making a horizontal histogram of the total amount of biomass for each trophic level (figure 6A.3). However, since most consumers seldom take food from only one trophic level, make an approximation of the relative amount of food taken from the primary, secondary, and tertiary levels. If, for example, the biomass of an omnivore is 45 g/m², and algae make up 33% of its diet and primary consumers make up the remaining 67%, then 33% of its biomass (0.33×45 g/m² = 15 g/m²) is in the primary consumer trophic level and 67% (0.67×45 g/m² = 30 g/m²) is in the secondary consumer trophic level. If the organism is predominantly either a secondary consumer or a herbivore, then it arbitrarily can be treated 100% as a secondary or primary consumer, respectively.

7. Relative biomass

Community studies can be aided by calculation of the relative biomass (RB) for each species. It may be estimated as:

$$RB_i = B_i/\Sigma B, \qquad (9)$$

where B_i is the biomass for species i and ΣB is the sum of the biomasses of all species. Considerable information exists in a graph of the relative biomass as a function of the rank of decreasing biomass from the largest to the smallest value (see section 5A.7.1). Usually the influence of a species in a subcommunity is directly related to the species' biomass. Also, values of biomass can be used instead of density in computing some diversity indices (section 5B), resulting in measures of biomass diversity.

8. Suggested exercises

1. Sample a plant or animal community and graph the biomass of each major species as a function of the rank of the species. Discuss the concept of dominance as it pertains to your results.
2. Compare two similar community types with respect both to standing crop biomass of all major species and to relative species biomass. Evaluate likely environmental factors related to the differences seen.
3. Determine the wet weights, dry weights, and ash-free dry weights of all major species collected from a given community. Evaluate the sources of error in each of these types of measurements.
4. From a sample of organisms of a given species, determine the lengths or volumes, and the dry weights. Plot the logarithm of the dry weight as a function of the length or the wet weight as a function of the volume determined. Evaluate the reliability and efficiency of using such a measurement as an index of biomass.
5. Sample a population of a given species and determine the wet weights or dry weights of each individual. Using at least 100 individuals, construct a size-frequency histogram of the number of individuals in each size class. Size classes may be in intervals of 0.1 g, 0.5 g, or 1.0 g, depending on the size of the species. Evaluate this procedure for use in determining age classes of the population (section 4A). What size classes contain the greatest biomass?
6. Convert the values of dry weight or ash-free dry weight to caloric equivalents of biomass from tables found in the literature. Evaluate the advantages of using the caloric standing crop rather than the biomass standing crop.
7. Construct a pyramid of biomass for a given animal community sampled.
8. Estimate the net biomass productivity of a field or grassland community by sampling the standing crop biomass at weekly intervals. Use plot sampling (section 3A) to obtain biomass in g/m², and express net biomass productivity in g/m² per week or per month.

9. Selected references

Brown, J. K. 1976. Estimating shrub biomass from basal stem diameters. Can. J. For. Res. 6:153–158.
Carlander, K. C. 1969. Handbook of freshwater fishery biology. Volume One. Iowa State University Press, Ames, Iowa.
Carlander, K. C. 1977. Handbook of freshwater fishery biology. Volume Two. Iowa State University Press, Ames, Iowa.
Chapman, S. B. 1976. Production ecology, and nutrient budgets, 157–228. In S. B. Chapman (ed.), Methods in plant ecology. Blackwell Scientific Publications, Oxford, England.
Edmondson, W. T., and G. G. Winberg (eds.). 1971. A manual on methods for the assessment of secondary productivity in fresh waters. IBP handbook no. 17. Blackwell Scientific Publications, Oxford, England.
Newbould, P. J. 1967. Methods for estimating the primary production of forests. IBP handbook no. 2. Blackwell Scientific Publications, Oxford, England.
Odum, H. T. 1957. Trophic structure and productivity of Silver Springs, Florida. Ecol. Monogr. 27:55–112.
Odum, E. P. 1983. Basic ecology. Saunders College Publishing, Philadelphia.
Phillipson, J. (ed.). 1971. Methods of study in quantitative soil ecology. IBP handbook no. 18. Blackwell Scientific Publications, Oxford, England.
Rohlf, F. J., and R. R. Sokal. 1981. Statistical tables. 2d ed. W. H. Freeman, San Francisco.
Rosenberg, N. J., B. L. Blad, and S. B. Verma. 1983. Microclimate: The biological environment. 2d ed. John Wiley & Sons, New York.
Southwood, T. R. E. 1978. Ecological methods. Chapman and Hall, London.
Vollenweider, R. A. (ed.). 1969. A manual on methods for measuring primary production in aquatic environments. IBP handbook no. 12. Blackwell Scientific Publications, Oxford, England.
Winberg, G. G. (ed.). 1971. Methods for the estimation of production of aquatic animals. Academic Press, New York.
Whittaker, R. H. 1975. Communities and ecosystems. 2d ed. Macmillan Co., New York.
Zar, J. H. 1984. Biostatistical analysis. 2d ed. Prentice-Hall, Englewood Cliffs, N.J.

6b

aquatic productivity

1. Introduction

In aquatic ecosystems, producers and primary consumers are largely microscopic plankton suspended in the water. By measuring the rates of oxygen production and oxygen consumption (and/or carbon dioxide consumption and production) in a known volume of water, one can estimate the rate of photosynthetic carbon fixation (*primary productivity*) and the rate of respiration.

The measurement of primary productivity may be performed either by the "light and dark bottle" method or by more direct procedures that monitor diurnal and nocturnal concentrations of oxygen (O_2) and/or carbon dioxide (CO_2). In the first method, place samples of water in glass bottles and determine the initial oxygen content. Then completely cover one bottle to exclude light; expose the other bottle to the light of the habitat from which the samples came. After a known time, measure the O_2 content in both the light and the dark bottles. The difference between the initial O_2 measurement and the dark bottle O_2 measurement indicates the amount of oxygen consumed by respiration of the organisms in the bottle. The difference between the initial O_2 measurement and the light bottle O_2 measurement indicates the net primary production. In the second method, monitor the changes in O_2 or CO_2 concentration in a body of water over a twenty-four-hour period.

2. Light and dark bottle method

2.1. Procedure
Take water samples from a desired source; glass-stoppered bottles of identical known volumes are recommended. Each sample should contain the typical proportions of normal phytoplankton and zooplankton of the habitat. For a more thorough analysis of the body of water, take samples at various locations and depths in the water. Either of two methods can be used to determine oxygen concentration in the water: an electronic oxygen meter, or the Winkler chemical technique

(see section 2E.6). If you use the Winkler method, take three samples at each site. The first is used to determine the initial oxygen content, and the others are designated the light and dark bottles. If you use an oxygen meter, take only two samples, and record the initial O_2 content of both light and dark bottles.

Darken one bottle with black paint or cover it completely with plastic tape or aluminum foil. Stopper both bottles and return them to their original sampling locations in the pond or laboratory. If in the field, you may attach a series of bottles to a line so that each pair of bottles is suspended at a different depth. Allow respiration and photosynthesis to proceed for twenty-four hours. (Times much longer than this tend to yield invalid results.) Express the concentrations of O_2 in mg/l (i.e., ppm).

2.2. Data and calculations
Respiration rate (R) in terms of oxygen consumption is calculated as:

$$R = (C_o - C_D)/\Delta t, \qquad (1)$$

where C_o is the initial concentration of O_2 (in mg/1), C_D is the final O_2 concentration in the dark bottle (in mg/1), and Δt is the period of time over which respiration took place. If Δt is measured in days, as in the present discussion, then R will be mg O_2/liter/day; if Δt were in hours, then R would be mg O_2/liter/hr.[1]

The total, or gross, photosynthetic productivity of oxygen (P_G, in mg O_2/liter/day) is:

$$P_G = (C_L - C_D)/\Delta t, \qquad (2)$$

where C_L is the final concentration of oxygen in the light bottle.

The net productivity of oxygen (P_N) is:

$$P_N = (C_L - C_o)/\Delta t, \qquad (3)$$

or,

$$P_N = P_G - R, \qquad (4)$$

expressed in mg O_2/liter/day.

If water samples were obtained at different sites and depths in a water body, then average the various values of R, P_G, and P_N to express mean rates of respiration, gross production, and net production.

2.3. Assumptions
The light-dark bottle procedure assumes that the rates of respiration in the light and dark bottles are the same, but this is not always true. Further, one assumes that rates of production in a closed container are the same as in the natural environment. Also, this method measures only a portion of the respiration and productivity of the total community, and attached algae, as well as macrophytes, benthos, and nekton, are ignored.

---

1. Some express respiration and production in cubic meters of water, rather than liters. As one cubic meter is 1000 liters, any of the results in the present discussion may simply be multiplied by 1000 to obtain values referring to cubic meters.

3. Direct determination

Any sample of water enclosed in a bottle or other artificial container (as in section 6B.2) is not totally under natural conditions. For example, bottles absorb light at different wavelengths than water and thus alter the rate of photosynthesis. In addition, a bottled sample does not contain the nekton or decomposers that exist in an entire natural body of water. Hence the measurement of respiration is not representative of the water body. These and other deficiencies in the light-dark bottle method resulted in development of methods for determining aquatic productivity under natural conditions. One such procedure follows the daily cycles of oxygen and carbon dioxide concentration in the body of water instead of in immersed bottles. Determine the concentration of either (or both) O_2 and CO_2 at approximately four-hour intervals for twenty-four hours. If we assume that biologically caused changes in these gases in the water are large compared to gaseous exchanges with the atmosphere, then the O_2 and CO_2 changes represent rates of productivity or respiration.

3.1. Procedure
Sample a pond or small lake at approximately 4-hr intervals, including samples as close to dawn or dusk as possible. As the length of daylight is generally different from the length of night, adjust your sampling periods to make them approximately equal in length. For example, if dawn was at 0500 and dusk at 1700, then you might sample at 0500, 0830, 1300, 1630, 1900, 2230, 0130, and 0500.[2] A middle of the night reading may be omitted if the rate of respiration is not appreciably changing then. In fact, some ecologists have based their estimations on only measurements taken at dawn and dusk (Talling, 1961). However the more samples taken the less chance of error in your determinations. Though you can begin sampling at any time of day, choose a day when the water is relatively calm.

Take samples at various depths and from a number of locations. As productivity varies with depth and often displays local variation (e.g., the littoral vs. limnetic zones), an average productivity for the pond or lake may be desirable. Obtain water samples as described in section 2D and exercise care to avoid contaminating the samples with atmospheric oxygen or carbon dioxide (see sections 2E.2 and 2E.6 for proper handling of water samples for gas analysis). Make chemical determinations of oxygen or carbon dioxide as soon as possible on the samples, using the methods described in sections 2E.2 and 2E.6. If possible, perform the chemical determinations in the field. A method for rapid determination of carbon dioxide in the field is described by Beyers and Odum (1959) and Verduin (1964). This involves the construction of a standard curve for the concentration of CO_2 versus pH for a given

body of water. Then measure the pH of the water sample with a portable pH meter and read the equivalent concentration of CO_2 from the standard curve. This procedure is nor applicable to seawater, because of the $CaCO_3$ buffer system therein.

3.2. Data and Calculations
Determine the net oxygen productivity during each daylight period:

$$P_N = (C_{t+1} - C_t)/\Delta t, \qquad (5)$$

where C_t is the oxygen concentration at time t, C_{t+1} is the concentration at the next sampling time, and Δt is the time interval between measurements C_t and C_{t+1}. For example, if one finds 7.4 mg/1 O_2 per liter at 1500 hr and 8.1 mg O_2/liter at 1900 hr, then:

$$P_N = (8.1 - 7.4)/4 = 0.18 \text{ mg } O_2/1/hr.$$

The rate of respiration for each sampling period during the night is then estimated as:

$$R = (C_t - C_{t+1})/\Delta t; \qquad (6)$$

if the O_2 concentrations are in mg/1, and Δt is in hr, then R will be in mg/1/hr.

If we assume that the average rate of respiration during the night approximates the average daytime respiration rate, then we may estimate the gross primary productivity as:

$$P_G = P_N + R, \qquad (7)$$

where P_N and R represent the average values of mg O_2/1/hr determined from equations 5 and 6, respectively. For example, consider the following hypothetical data:

Daytime Measurements

start of time period	0500	0930	1400
end of time period	0930	1400	1900
Δt(hr)	4.5	4.5	5.0

Increase in oxygen concentration (net O_2 production):

(mg/1)	1.6	2.6	1.9
Net productivity:			
P_N (mg O_2/1/hr)	0.36	0.58	0.38

Then, the net oxygen productivity during the day is:

$$P_N = \frac{(1.6 + 2.6 + 1.9) \text{ mg/1}}{(4.5 + 4.5 + 5.0) \text{ hr.}}$$
$$= \frac{6.1 \text{ mg/1}}{14.0 \text{ hr}}$$
$$= 0.44 \text{ mg } O_2/1/hr.$$

Nighttime Measurements

start of time period	1900	2330	0200
end of time period	2330	0200	0500
Δt (hr)	3.5	3.5	3.0

Decrease in oxygen concentration (O_2 consumed by respiration):

(mg/1)	1.1	1.6	1.3
Respiration rate:			
R (mg O_2/1/hr)	0.31	0.46	0.43

Then, the rate of oxygen consumption during the night is:

$$R = \frac{(1.1 + 1.6 + 1.3) \text{ mg/1}}{(3.5 + 3.5 + 3.0) \text{ hr}} = \frac{4.0 \text{ mg/1}}{10.0 \text{ hr}}$$
$$= 0.40 \text{ mg } O_2/1/\text{hr.}$$

Therefore, the gross primary productivity of oxygen is $P_N + R = (0.44 + 0.40)$ mg/1/hr = 0.84 mg/1/hr during the 14.0 hours of daylight, or $(0.84)(14.0) = 11.8$ mg O_2/1 for the entire day. Since the molecular weight of O_2 is 32, 11.8 mg/1 is equivalent to $(11.8 \text{ mg/1})/(32 \text{ mg/mmole}) = 0.37$ mmole/1. (Some investigators prefer to report molar rather than weight units.)

Perhaps you want to measure carbon dioxide rather than oxygen. Then during the daylight, C_t is the CO_2 concentration at the beginning of the time period, C_{t+1} is the CO_2 concentration at the end of the time period, Δt is the length of the time period, and

$$P_N' = (C_t - C_{t+1})/\Delta t \tag{8}$$

is the primary productivity expressed as milligrams of carbon dioxide consumed per liter of water per hour.

The rate of respiration during the night is:

$$R' = (C_{t+1} - C_t)/\Delta t. \tag{9}$$

in milligrams of CO_2 produced per liter of water.

In a fashion analogous to equation 7, the gross primary productivity may be expressed in mg CO_2 consumed/liter/hr, as:

$$P_G' = P_N' + R'. \tag{10}$$

This is illustrated with the following data:

Daytime Measurements

start of time period	0500	0930	1400
end of time period	0930	1400	1900
Δt (hr)	4.5	4.5	5.0

Decrease in CO_2 concentration (net production):

(mg/1)	2.2	3.5	2.6
Net productivity:			
P_N' (mg CO_2/1/hr)	0.49	0.78	0.52

Then, the net productivity during the day is:

$$P_N' = \frac{(2.2 + 3.5 + 2.6) \text{ mg/1}}{(4.5 + 4.5 + 5.0) \text{ hr}}$$
$$= \frac{8.3 \text{ mg/1}}{14.0 \text{ hr.}}$$
$$= 0.59 \text{ mg } CO_2/1/\text{hr.}$$

Nighttime Measurements

start of time period	1900	2230	0200
end of time period	2230	0200	0500
Δt (hr)	3.5	3.5	3.0

Increase in CO_2 conc. (CO_2 produced by respiration):

(mg/1)	1.5	2.2	1.8
Respiration rate:			
R' (mg CO_2/1/hr)	0.43	0.63	0.60

Then, the rate of respiration during the night is:

$$R' = \frac{(1.5 + 2.2 + 1.8) \text{ mg/1}}{(3.5 + 3.5 + 3.0) \text{ hr}}$$
$$= \frac{5.5 \text{ mg/1}}{10.0 \text{ hr}}$$
$$= 0.55 \text{ mg } CO_2/1/\text{hr.}$$

Therefore, the gross productivity, as CO_2 consumption, is $P_N' + R' = (0.59 + 0.55)$ mg/1/hr = 1.14 mg/1/hr during the 14.0 hours of daylight, or $(1.14)(14.0) = 16.0$ mg CO_2/1 for the day. As the molecular weight of CO_2 is 44, 16.0 mg/1 is equivalent to $(16.0 \text{ mg/1})/(44 \text{ mg/mmole}) = 0.36$ mmole/1.

The rate of O_2 production or CO_2 fixation may be converted to milligrams of carbon or dry-weight biomass, or to kilocalories of energy, as described in sections 6B.4 and 6B.5.

3.3. Sources of Error   To estimate gross primary productivity, you must assume that the average rate of nocturnal respiration is equal to the average diurnal rate. During warmer months when waters are calm and photosynthetic rates are high, these assumptions are approximately correct and results should not be seriously biased. But the assumption of an insignificant bias from atmospheric gas exchange may not always be valid—there may be water turbulence, or the water may be supersaturated with O_2 or CO_2. Also, a lake with a large ratio of surface area to volume may have a greater rate of gaseous exchange with the atmosphere. And if the percent saturation of gas in the water is very low, the rate of exchange will be higher. If you need corrections for atmospheric gas exchange, consult Vollenweider (1969). Also, depth of the water has a large effect on productivity. Estimates may be biased unless you take an adequate sampling of depths. The assumption of equal rates of respiration during the night and day may not hold and ideally should be verified.

4. Biomass equivalents

Biomass productivity may be expressed as follows. Consider the photosynthesis equation:

$$6CO_2 + 6H_2O + \text{energy} \rightarrow C_6H_{12}O_6 + 6O_2, \tag{11}$$

from which we see that 6 moles of O_2 are produced for every 6 moles of carbon fixed (1 mole of O_2 is associated with 1 mole of C). The atomic weights of C and O are 12.0 and 16.0, respectively (appendix C). Therefore the production of 32.0 grams of oxygen results during the fixation of 12.0 grams of carbon. This can be stated as 0.375 g C per gram of O_2, or 0.375 mg C/mg O_2. However some photosynthesized glucose ($C_6H_{12}O_6$) is rapidly changed to other organic compounds, with the result that the *photosynthetic quotient*:

$$PQ = \frac{\text{moles (or volume) of } O_2 \text{ producted}}{\text{moles (or volume) of } CO_2 \text{ consumed}} \quad (12)$$

is usually not 1.0, but for phytoplankton is typically around 1.2 (Ryther, 1956; Strickland, 1960). Therefore, the *gross biomass productivity* may be estimated as:

$$GBP = 0.375P_G/1.2, \quad (13)$$

the units of *GBP* being mg C/liter/day.

The amount of biomass oxidized during respiration may be estimated by considering the respiration equation:

$$C_6H_{12}O_6 + 6O_2 \rightarrow 6CO_2 + 6H_2O + \text{energy}, \quad (14)$$

which is the reverse of equation 11. As in equation 13, we use the ratio of the molar weights of carbon and oxygen, namely 0.375 mg C/mgO₂. We shall use a value of 1.0 for the *respiratory quotient*:

$$RQ = \frac{\text{moles (or volume) of } CO_2 \text{ produced}}{\text{moles (or volume) of } O_2 \text{ consumed}}, \quad (15)$$

assuming that carbohydrate is the overwhelmingly predominant respiratory substrate for phytoplankton (Strickland, 1960), and that the amount of animal respiration is negligibly small compared to that of the algae. (Oxidation of fat gives an RQ of about 0.7, and the RQs for the catabolism of various proteins are in the neighborhood of 0.7–0.9.) Therefore the amount of biomass consumed during respiration is:

$$RB = 0.375R/1.0 = 0.375R \quad (16)$$

mg C/liter/day.

The *net biomass productivity* may be estimated as:

$$NBP = GBP - RB \quad (17)$$

mg C/liter/day.

If we measure carbon dioxide instead of oxygen, then note from equation 11 that photosynthetic fixation of 1 mole of carbon is associated with the consumption of 1 mole of CO_2. Using the atomic weights for C and O (appendix C), we see that the consumption of 44.0 grams of CO_2 is associated with the fixation of 12.0 grams of carbon. That is, 0.273 grams of carbon incorporated into glucose requires one gram of carbon dioxide.

Therefore the *gross biomass productivity* (in mg C/liter of CO_2 per day) is estimated as:

$$GBP' = (0.273P'_G)(1.2), \quad (18)$$

where 1.2 is the *photosynthetic quotient* explained in equation 12.

The amount of biomass consumed during respiration is estimated as:

$$RB' = RB' = 0.273R', \quad (19)$$

assuming a *respiratory quotient* of 1.0, as explained in equation 15.

The *net biomass productivity* may be estimated as:

$$NBP' = GBP' - RB' \quad (20)$$

mg carbon/liter CO_2/day.

The above equations assume a PQ of 1.2 or an RQ of 1.0, which may not always be valid. If you need more accurate values, determine CO_2 consumption or production (in mg/l/day) by using the techniques in section 2E.2 for total CO_2. Since equal volumes of gases represent equal numbers of moles,[3] we obtain the following from equations 12 and 15, respectively:

$$PQ = \left(\frac{\text{mg } O_2 \text{ produced}}{\text{mg } CO_2 \text{ consumed}}\right)\left(\frac{44.01}{32.00}\right), \quad (21)$$

which is equal to equation 12; and

$$RQ = \left(\frac{\text{mg } CO_2 \text{ consumed}}{\text{mg } O_2 \text{ produced}}\right)\left(\frac{32.00}{44.01}\right), \quad (22)$$

which is equal to equation 15.

The ash-free dry weight of phytoplankton is about twice the carbon biomass (Ryther, 1956; Vollenweider, 1969). Therefore, the biomass productivity expressed in mg C/liter/day can be converted to an estimate of mg ash-free dry weight/liter/day by multiplying the former by 2.

See the introduction to unit 6 for the calculation of important productivity ratios.

5. Energy equivalents

We can express the above amounts of oxygen or carbon dioxide consumption or production as energy:

Assume that phytoplankton respiration makes exclusive use of carbohydrate (Strickland, 1960), and that the combustion of carbohydrate with one liter of oxygen by respiration releases 5.05 kcal of energy (Brody, 1964). (If pure fat were catabolized, about 4.7 kcal would be released per liter consumed.) Since the density of oxygen is 1.429 g/liter (at standard temperature and pressure), carbohydrate catabolism involving one gram of O_2 releases 5.05/1.429 = 3.53 kcal of energy (and 1 mg O_2 is associated with 0.00353 kcal, or 3.53 cal). Therefore the amount of energy released by the respiration of organisms in the water sample (assumed to be overwhelmingly algae) is:

$$RE = 0.00353R \quad (23)$$

kcal/liter/day, where R is in mg O_2/l (i.e., ppm) per day.

3. This and other considerations of molar volumes of gases is based on the assumption that the gases behave "ideally." In fact, the behavior of carbon dioxide departs considerably from that of an "ideal gas," but, fortunately, the amounts of CO_2 of concern to us are small enough to make this departure negligible.

Rates of production have to take into account the PQ (assumed in section 6B.4 to be 1.2), because the total O_2 evolution does not reflect only carbohydrate synthesis. Metabolism involving one liter of oxygen and a PQ of 1.2 is associated with 4.84 kcal of energy (Brody, 1964); therefore, $(4.84 \text{ kcal/l})/(1.429 \text{ g/l}) = 3.39$ kcal/g, or 0.00339 kcal/mg of O_2. Thus gross primary productivity, in kilocalories/liter of water/day, is:

$$GPP = 0.00339 P_G, \qquad (24)$$

where P_G is in mg O_1/l (ppm) per day.

Net primary productivity is:

$$NPP = GPP - RE \qquad (25)$$

kcal/liter/day.

If we measure carbon dioxide instead of oxygen, we use the figure of 5.05 kcal of energy released for each liter of CO_2 produced during carbohydrate catabolism (Brody, 1964). (There would be about 6.7 kcal/liter of CO_2 if the respiration used only fat.) Since the density of carbon dioxide is 1.977/liter (at standard temperature and pressure), carbohydrate oxidation that releases 1 gram of CO_2 also releases $5.05/1.977 = 2.55$ kcal of energy (and, therefore, 1 mg CO_2 is associated with 0.00255 kcal).

If R' is in mg CO_2/l (i.e., ppm) per day, then:

$$RE' = 0.00255 R' \qquad (26)$$

is the amount of energy released during respiration (in kcal/liter of water/day). Metabolic conversions involving one liter of CO_2 and PQ of 1.2 are associated with 5.81 kcal (Brody, 1964); $(5.81 \text{ kcal/l})/(1.977 \text{ g/l}) = 2.94$ kcal/g $= 0.00294$ kcal/mg of CO_2. Therefore gross primary productivity, in kcal/liter of water/day, is:

$$GPP' = 0.00294 P_G', \qquad (27)$$

where P_G' is in mg CO_2/l (i.e., ppm) per day.

The net primary productivity is:

$$NPP' = GPP' - RE' \qquad (28)$$

kcal/liter of water per day.

6. Suggested exercises

1. Determine rates of respiration and gross and net productivity using pairs of light and dark bottles kept for twenty-four hours at different depths in a pond. What causes the differences in results? Measure the temperature (section 2D.3.1) and light intensity (section 2D.3.3) at each depth used.

2. Determine B (section 6A) and NPP for ponds of various ages or with different degrees of pollution. How do the turnover times and the productivity ratios (introduction to section 6) differ in these communities?

3. For two or more ponds or lakes, determine biomass and the GBP/B and GPP/R ratios. Estimate the turnover time for each body of water.

7. Selected references

Beyers, R. J., and H. T. Odum. 1959. The use of carbon dioxide to construct pH curves for the measurement of productivity. Limnol. Oceanogr. 4:499–502.

Brody, S. 1964. Bioenergetics and growth. Hafner Publishing Co., New York.

Brown, A. L. 1971. Ecology of fresh water. Harvard University Press, Cambridge, Mass.

Dýkyjová, D., and J. Květ (eds.). 1978. Pond littoral ecosystems. Structure and functioning. Springer-Verlag, New York.

Greenberg, A. E., J. J. Conners, and D. Jenkins (eds.). 1985. Standard methods for the examination of water and wastewater. 16th ed. American Public Health Association, Washington, D.C.

Kitchings, W. M., and B. J. Copeland. 1980. Succession in laboratory microecosystems subjected to thermal and nutrient addition stress, 536–561. In J. B. Giesy, Jr. (ed.), Microcosms in ecological research. U.S. Department of Energy. National Technical Information Service. Springfield, Va.

Likens, G. E. (ed.). 1985. An ecosystem approach to aquatic ecology. Mirror Lake and its environment. Springer-Verlag, New York.

Odum, H. T. 1956. Primary production in flowing waters. Limnol. Oceanogr. 1:102–117.

Parsons, T. R., M. Takahashi, and B. Hargrave. 1984. Biological oceanographic processes. 3d ed. Pergamon Press, New York.

Pratt, D. M., and H. Berkson. 1959. Two sources of error in the oxygen light and dark bottle method. Limnol. Oceanogr. 4:328–334.

Ryther, J. H. 1956. The measurement of primary production. Limnol. Oceanogr. 1:72–84.

Strickland, J. D. H. 1960. Measuring the production of marine phytoplankton. Fisheries Research Board of Canada Bull. 122.

Talling, J. F. 1957. Diurnal changes of stratification and photosynthesis in some tropical African waters. Proc. Royal Soc. B. 147:57–83.

Talling, J. F. 1961. Photosynthesis under natural conditions. Annu. Rev. Plant Physiol. 12:133–154.

Verduin, J. 1956. Primary production in lakes. Limnol. Oceanogr. 1:85–91.

Verduin, J. 1964. Principles of primary productivity: Photosynthesis under natural conditions, 221–238. In D. F. Jackson (ed.), Algae and man. Plenum Press, New York.

Vollenweider, R. A. (ed.). 1969. A manual on methods for measuring primary production in aquatic environments. IBP handbook no. 12. Blackwell Scientific Publications, Oxford, England.

Wetzel, R. G., and G. E. Likens. 1979. Limnological analysis. W. B. Saunders Co., Philadelphia.

1. Introduction

Field studies are often difficult because of the size and complexity of natural ecosystems. In contrast, laboratory investigations using protozoa or other microorganisms take only a short period of time, require only small containers, and occur under relatively controlled conditions. Unicellular organisms are relatively easy to study and to culture in the laboratory. In such a culture, one can study producers (algae), consumers (protozoa), and decomposers (bacteria and fungi). In these situations, ecological characteristics such as density, diversity, community similarity, succession, biomass, and productivity can be measured or manipulated in controlled experiments. Environmental variables such as temperature, pH, light, oxygen, food, and inorganic nutrients can also be controlled and manipulated.

A laboratory aquatic microecosystem, or microcosm, may be one of two types. In *autotrophic* systems, the energy source driving the system is light, directly used by the algae (producer trophic level). These microecosystems behave quite similarly to natural ecosystems (Odum, 1983; Heath, 1980). In *heterotrophic* systems, the ultimate source of energy is an organic food source. Controlled experiments can involve introducing known densities of various species of protozoans and bacteria into a sterile medium (Gause, 1936; Hairston et al., 1968; Hill and Wiegert, 1980). All heterotrophic systems and most laboratory autotrophic systems are not complete enough to be self-sustaining; therefore, they are not strictly analogous to natural ecosystems.

Aquatic microecosystems may be used in conjunction with other exercises to illustrate sampling methods (sections 2E and 3E), chemical analyses (section 2E), population distribution (section 4C), population dynamics (section 4B), ecological succession (section 5D), various community characteristics (unit 5), and considerations of productivity (sections 6A, 6B, and 6C). Recent reviews of microcosms in ecological research show the wide use of such systems (Giesy, 1980; Ringelberg and Kersting, 1978), but their limited applications to large-scale aquatic systems have also been noted (King, 1980; Heath, 1980).

2. Heterotrophic systems

These aquatic laboratory systems are commonly based on a detrital food chain (organic matter $\longrightarrow$ decomposers $\longrightarrow$ microphages $\longrightarrow$ tertiary consumers). Heterotrophic systems have been studied by many ecologists who have found them useful in demonstrating certain aspects of populations, communities, and ecosystems. As heterotrophic systems have a fixed initial supply of energy, the cultures will eventually die out unless organic material is supplied regularly. They may provide models for studying the ecology of energy-rich systems (e.g., polluted lakes and streams), detrital food chains (e.g., in benthic habitats), and energy-

6c

aquatic microecosystems

poor systems (e.g., in caves). The pattern of succession occurring under laboratory situations with a fixed energy source ("heterotrophic succession") is analogous to that naturally found in a rotting log, dung pile, or rotting carcass.

A good culture medium for a heterotrophic system is a boiled hay solution. Broth from boiling grains of wheat or brown rice (about 50 g in 500 ml of water) also makes a good stock medium, particularly good for culturing *Paramecium*. The culture medium is made with 10 ml of stock per 100 ml of sterile tap water. Fill a series of 200- to 300-ml bottles or flasks with the culture medium to just below the neck. Mark each bottle for identification; use some for a control and some for experimental manipulation. Add 1 ml of pond water, or a culture of a desired species of protozoan, to each bottle; plug the bottles with cotton, and set them aside for later examination. The experimental culture may be used for such studies as nutrient enrichment (e.g., with nitrate or phosphate), exposure to light or dark, changing of pH, variation of water temperature, or control of energy input to consumers by addition of known amounts of sugar.

An experimental situation may be designed to last from four to twelve weeks, with samples being taken once or twice weekly. Or, set up the experiment in advance so that cultures are begun at one- or two-week intervals, then analyze all cultures during one or two laboratory periods. Sample either a given depth (surface, middle, or bottom), or stir the culture and study a mixed sample. Place 0.01 to 0.5 ml of the culture in each of a series of 10 to 20 small drops on a microscope slide; count and identify all organisms in each drop. Use 40× to 100× magnification, depending on the size of the organisms to be tabulated. This procedure limits the movement of the organisms to the field of vision under the microscope and provides a sample size that can be counted easily without having to use a slowing agent or a coverslip. Take at least two samples from each container. The appropriate volume of each sample will depend on the concentration of organisms. If they are highly concentrated, use a 0.01-ml sample; if the density is very low, then use a 0.5-ml sample in a depression slide. Smaller sample volumes require the taking of more samples. Be careful that the number of small drops is not greater than can be analyzed before drying up. Express densities as the number of individuals per milliliter.

Confirming the identification of the taxa may require coverslips and a slowing agent such as a methylcellulose, cupric acetate, or polyvinyl alcohol solution, which will enable use of higher power. Become familiar with the taxa before sampling for smoother going. In these cultures most of the organisms will be small flagellates and ciliates (see Eddy and Hodson, 1961; Jahn, 1979; Needham and Needham, 1962).

Qualitative estimation of abundance is useful where time does not permit more detailed analysis. Place a random sample of 0.1 ml on a slide and cover it. Under 40 or 100×, examine twenty random microscopic fields and record the presence of each taxon. Abundance classes, based on frequency, may then be expressed as the following: *dense,* 80–100% of microscope fields (or drops) contain the species; *abundant,* the species is in 60–79% of the fields examined; *common,* present in 40–59% of the fields; *sparse,* present in 20–39% of the fields; and *rare,* present in only 1–19%. Examine a series of drops, as described above, but take care all drops are as nearly equal in size as possible.

3. Autotrophic microecosystems

Autotrophic systems permit the study of primary productivity, aquatic succession, and community structure under controlled laboratory conditions. In these systems, the major pathway of energy transfer is the grazing food chain (light $\longrightarrow$ producers $\longrightarrow$ primary consumers $\longrightarrow$ secondary consumers $\longrightarrow$ tertiary consumers). Therefore the energy supply may be considered unlimited. However, even with an abundant supply of energy, these systems rarely are indefinitely self-sustaining. They display senescence, largely due to inefficient recycling of nutrients. Despite their limitations, they have been beneficial in demonstrating concepts of ecological succession and productivity.

Either of two types of autotrophic system may be used. In one, defined chemical nutrients are used with selected species of algae (e.g., a cellular green, a flagellate, a cellular blue-green, and a diatom). You may study competitive relationships between similar species, or use various numbers of species of unrelated taxa to study the effect of diversity on the system. A second type of system contains producers, consumers, and decomposers and is made from sterile pond water inoculated with pond or aquarium water. This type of microcosm represents a closed system and is much simpler to maintain than complex regulated flow through ecosystems. The latter type, often called a *chemostat,* regulates concentrations of nutrients, dissolved gases, or other gases in the culture system (King, 1980).

It is best to include material from the sediments in your inoculum. Fill 250- or 500-ml flasks with culture medium or pond water, plug them with cotton and sterilize. After cooling, inoculate with either known species of algae or other suitable inoculum. Test the effect of one environmental variable on the system (pH of 6.5 vs. 8.5, temperature of 15° vs. 25°C, or added mineral nutrients giving an increased medium concentration of 1 ppm PO_4, 15 ppm NO_3, or 100 ppm HCO_3). Place cultures under a good fluorescent light source, preferably on a 16-hr light and 8-hr dark cycle.

Mix each culture prior to sampling using a mechanical shaker or a sterile glass rod. Sample each culture with a sterile pipette. Take 0.1-ml samples as described above and spot drop the sample on a slide. You may also choose to sample the periphyton (attached organisms), using a rubber scraper to remove them from a measured area on the side or bottom of the flask. Place the scraped sample on a microscope slide and examine the organisms. Most species of algae require higher magnification for identification and counting, so use the procedure presented in section 3E.4.5. Identify each taxon and determine the number of cells per milliliter.

For a rapid approximation of density, determine the frequency of occurrence for each taxon (the number of microscope fields containing that taxon). If you assume a random dispersion of cells, then approximate the density using table 4C.3. Thus for example, a frequency of 0.73 would represent a mean density of 1.31 organisms per field. By knowing the area of the microscope field and the depth of the sample under the coverslip, the density may be expressed as the mean number per milliliter.

4. Biomass

The size of microorganisms may be expressed as biovolume (section 6A.4.2). Or you can measure total biomass directly by determining the dry weight. Weigh a piece of dried filter paper to 0.1 mg. Filter a measured volume of the culture through the paper and dry at 70°–80°C. Weigh the dried paper plus filtered material and subtract the initial weight of the filter paper. Express your results in milligrams dry weight per milliliter of filtered water. If you want a more refined measurement of biomass use a millipore filter, because small microorganisms are not trapped by ordinary filter paper. Total organic biomass can then be determined as the ash-free dry weight (see section 6A.3). Bear in mind that much nonliving organic matter is trapped by filtration, so the dry weight determined is the total dry weight of *both* living and nonliving matter in the water. Another procedure for collecting samples for biomass involves spinning water samples in a centrifuge, decanting off the water, and drying and weighing the remaining material. You may also approximate biomass of the algae by using the method of section 6B.4, or express it as chlorophyll content, as described in section 6C, sampling the culture as described above for biomass.

5. Productivity and respiration

Autotrophic microecosystems are particularly useful for studying primary productivity under controlled conditions. One may stopper the flasks and perform a light-dark-bottle experiment (section 6B.2). You may also determine the primary productivity as described in section 6B.3 by following the changes in oxygen or carbon dioxide concentrations for 4-hr periods during the light cycle or the 8-hr dark cycle. Since there is a limited volume of culture, O_2 concentration is best determined with an oxygen meter or carbon dioxide determined using a pH meter (see section 6B.3.1).

Respiration in a heterotropic system is determined somewhat like that described for biochemical oxygen demand (BOD, section 2D.5.3), but by using an oxygen meter and measuring the oxygen consumption over a few hours instead of five days. Results are expressed as milliliters of O_2 consumed per milliliter of water per hour. These values may be converted to calories of energy consumed or grams of carbon consumed (see sections 6B.4 and 6B.5). Determination of CO_2 using a pH meter may not be reliable in heterotrophic systems, for organic acids are commonly formed during decomposition.

6. Suggested exercises

1. Follow a microecosystem over a period of three to four weeks and determine changes in any or all of the following as a function of time:
   a. species diversity (section 5B)
   b. total biomass (section 6A)
   c. primary productivity and respiration (section 6B)
   d. chlorophyll diversity and biomass (section 6C)
   What are the patterns of species diversity over time? What are the temporal patterns of the ratio of production to respiration and the ratio of production to biomass?
2. Culture systems differing in an experimental variable such as pH, temperature, or nutrient concentration. Allow the cultures to grow for three or four weeks. Compare them for productivity, biomass, species diversity (section 5B), densities of dominant species, community similarity (section 5C), and/or chlorophyll content (section 6C).
3. Follow the relative densities of the major species in a culture over time and plot the data as described in section 5D.2.3.
4. Examine your sampling procedure for density determination, according to the considerations in section 1A. Determine the number and volume of samples required using species-area curves, performance curves, and/or the desired magnitude of the standard error.

7. Selected references

Beyers, R. J. 1963. The metabolism of twelve aquatic laboratory microcosms. Ecol. Monogr. 33:281–306.

Bretthauer, R. 1980. Laboratory microcosms, 416–445. In J. B. Giesy, Jr. (ed.), Microcosms in ecological research. U.S. Department of Energy. National Technical Information Service. Springfield, Va.

Cooke, G. D. 1967. Pattern of autotrophic succession in laboratory microcosms. BioScience 17:717–721.

Cole, J. E. 1964. Preliminary investigation of succession of invertebrates in pond infusion cultures. School Sci. Math. 64:325–331.

Eddy, S. 1928. Succession of protozoa in cultures under controlled conditions. Trans. Amer. Microscop. Soc. 47:283–319.

Eddy, S., A. C. Hodson, J. C. Underhill, W. D. Schmid, and D. E. Gilbertson. 1982. Taxonomic keys to common animals of the north central states, exclusive of the parasitic worms, terrestrial insects, and birds. Burgess Publishing Co., Minneapolis.

Fine, M. S. 1912. Chemical properties of hay infusion with special reference to titratable acidity and its relationship to protozoan succession. J. Exper. Zool. 12:272–294.

Gause, G. F. 1936. The principles of biocoenology. Quart. Rev. Biol. 11:320–336.

George, E. A. 1976. A guide to algal keys (excluding seaweeds). Brit. Phycol. J. 11:49–55.

Giddings, J. M., and G. K. Eddleman. 1978. Some ecological and experimental properties of complex aquatic microcosms. Intern. J. Environ. Stud. 13:119–123.

Giesy, J. B., Jr. (ed.). 1980. Microcosms in ecological research. U.S. Department of Energy. National Technical Information Service. Springfield, Va.

Goldman, J. G. 1973. Carbon dioxide and pH: Effect on species succession of algae. Science 182:306–307. (A comment on Shapiro, 1973.)

Hairston, N. G., J. D. Allan, R. K. Colwell, D. J. Futuyma, J. Howell, M. D. Lubin, J. Mathias, and J. H. Vandermeer. 1968. The relationship between species diversity and stability: An experimental approach with protozoa and bacteria. Ecology 49:1091–1101.

Heath, R. T. 1980. Are microcosms useful for ecosystem analysis?, 333–347. In J. B. Giesy, Jr. (ed.), Microcosms in ecological research. U.S. Department of Energy. National Technical Information Service. Springfield, Va.

Hill, J., IV, and R. G. Wiegert. 1980. Microcosms in ecological modeling, 138–163. In J. B. Giesy, Jr. (ed.), Microcosms in ecological research. U.S. Department of Energy. National Technical Information Service. Springfield, Va.

Jahn, T. L. 1979. How to know the protozoa. Wm. C. Brown Co., Publishers, Dubuque, Iowa.

King, D. L. 1980. Some cautions when applying results from aquatic microcosms, 164–191. In J. B. Giesy, Jr. (ed.), Microcosms in ecological research. U.S. Department of Energy. National Technical Information Service. Springfield, Va.

Kitchings, W. M., and B. J. Copeland. 1980. Succession in laboratory microecosystems subjected to thermal and nutrient addition stress, 536–561. In J. B. Giesy, Jr. (ed.), Microcosms in ecological research. U.S. Department of Energy. National Technical Information Service. Springfield, Va.

Kudo, R. R. 1977. Protozoology. 5th ed. Charles E. Thomas Co., Springfield, Ill.

Mann, K. H. 1969. The dynamics of aquatic ecosystems. Adv. Ecol. Res. 6:1–81.

Margalef, R. 1963. On certain unifying principles in ecology. Amer. Natur. 97:357–374.

Margalef, R. 1975. Perspectives in ecological theory. University of Chicago Press, Chicago.

Needham, J. G., and P. R. Needham. 1962. A guide to the study of freshwater biology. 5th ed. Holden-Day, San Francisco.

Odum, E. P. 1983. Basic ecology. Saunders College Publishing, Philadelphia.

Prescott, G. W. 1978. How to know the fresh water algae. 3d ed. Wm. C. Brown Co. Publishers, Dubuque, Iowa.

Ringelberg, J., and K. Kersting. 1978. Properties of an aquatic microecosystem. I. General introduction to the prototypes. Arch. Hydrobiol. 83:46–68.

Shapiro, J. 1973. Blue-green algae: Why they become dominant. Science 179:382–384.

Taub, F. B. 1976. Demonstration of pollution effects in aquatic microcosms. Intern. J. Environ. Stud. 10:23–33.

Woodruff, L. L. 1912. Observations on the origin and sequence of the protozoa fauna of hay infusions. J. Exper. Zool. 12:205–264.

A.	acre (appendix B.2)
A	absorbance (2E.1.2, 2E.1.5)
A	index, or coefficient, of aggregation (4C.3.6)
A	habitat area (3A.5, 3G.6, 3H.6, 6A.1)
A	attack rate of predation (4E.2)
A'	index, or coefficient, of aggregation (4C.3.6)
$\overline{A}$	mean habitat area (3C.3)
Å	Angstrom (appendix B.1)
A_b	absorbance of a blank (2E.1.5)
A_p	absorbance of a prepared sample (2E.1.2, 2E.1.5)
A_s	absorbance of a standard (2E.1.2, 2E.1.5)
ASTM	American Society for Testing and Materials (2C.8)
a.	are (appendix B.2)
a	Y-intercept in regression (1B.6, 3G.3, 6A.4.1)
a	area covered (as measured by foliage area, basal area, or basal coverage) (3A.3)
a	lower limit of a scale (2F.3.3)
a_i	area covered by species i (3A.3, 3C.3)
antilog	antilogarithm of a common logarithm (logarithm to the base 10) (2D.6, 6A.4.1)
B	logarithmic base (5B.2.3)
B	biomass, or standing crop (3.3, unit 6, 6A.1)
B	logarithmic base (5B2.3)
B_i	biomass of species i (3A.3, 6A.7)
BOD	biochemical oxygen demand (2E.5, 2F.6)
BTU	British thermal unit (appendix B.7)
b	regression coefficient, or slope of regression line (1B.6, 2F.1.2, 3G.3, 6A.4.1)
b	upper limit of a scale (2F.3.3)
C	Celsius (appendix B.8)
C	concentration (2D.5.3, 6B.2.2, 6B.3.2)
C	coverage (3.4, 3A.3, 3C.3)
C_D	oxygen concentration in a dark bottle (6B.2.2)
C_L	oxygen concentration in a light bottle (6B.2.2)
C_i	concentration at time i (2D.5.3)
C_i	coverage of species i (3A.3, 3C.3)
C_j	coverage of a species in successional stage j (5D.2.3)
C_o	original concentration (2E.1.4, 6B.2.2)
C_p	concentration of a prepared sample (2E)
C_p	specific conductance of a prepared sample (2E.5.4)
C_s	concentration of a standard (2E.1.2, 2E.1.5)
C_s	specific conductance of a standard (2E.5.4)
C_t	concentration of titrant (2E.1.3, 2E.1.5)
C_t	concentration at time t (6B.3.2)
C_w	concentration per unit dry weight (2E.1.4)
C_{25}	specific conductance at 25°C (2E.5.4)
CBE	Council of Biological Editors (1C.2)
CC_J	Jaccard coefficient of community (5C.2.1)
CC_S	Sørensen coefficient of community (5C.2.1)
c	concentration or dilution factor (2D.5.3, 3D.4.5)
c	the number of species in common in two communities (5C.2.1)
c	constant, correction factor, or intermediate calculation (1B.5.5, 2B.4.5, 2C.8, 5B.2.4)
cal	calorie (appendix B.7)
cc	cubic centimeter (appendix B.3)
cd	candela (2B.4.1)
cm	centimeter (appendix B.1, B.6)
cm²	square centimeter (appendix B.2)
cm³	cubic centimeter (appendix B.3)
cu	cubic (appendix B.3)
D	death rate of predator population (4B.1)
D	density (number/area or weight/volume) (2C.8, 3.1, 3A.3, 3F.3, 3G.6, 3H.2, 4C.3.5, 4E.2, 6A.1)
D'	predicted density (4C.3.5)
D_a	Margalef's index of diversity (5B.2.1)
D_b	Menhinick's index of diversity (5B.2.1)
D_i	density of species i (3A.3, 3C.3)
D_i	density estimate from transect i (3H.2.1)
D_j	density of a species in successional stage j (5D.2.3)
D_{max}	maximum diversity, by Simpson's index (5B.2.4)
D_s	Simpson's index of diversity (5B.2.2)
DF	degrees of freedom (1B.2.2, 1B.4)
DF_i	degrees of freedom for sample i (1B.3.1)
DI	diversity index, using SCI (2D.5.2)
DO	dissolved oxygen (2E.6)
DO'	dissolved oxygen at saturation (2E.6.4)
d	depth (2B.4.1, 2D.2, 2D.3.3, 2D.4.5)
d	diameter (2C.8)
d, d'	distance (2A.4, 2D.3.3, 3C.3, 4C.3.6, 6C.3)
$\overline{d}$	mean distance (3C.3)
d_i	distance from observer to ith sighting (3H.2)
d_m	distance to the mth individual (3C.3)
d_{max}	maximum diversity, using index d_s (5B.2.4)
d_s	diversity index, the inverse of Simpson's l (5B.2.2)
d_x	number dying during age interval x (4A.4)
dN	instantaneous change in population size (4B.1, 4B.2)
dt	infinitesimal time interval (4B.1, 4B.2)
E	east
E	extinction coefficient (2D.3.3)
E	total energy gain from predation (4E.3)
E_1, E_2	energy gains from prey item 1 or 2 (4E.3)
E_s	evenness, using Simpson's diversity index (5B.2.4)
EDTA	ethylenediaminetetraacetic acid (or its salts) (2E.3, 2E.4)
EPA	Environmental Protection Agency (2E.1)
e	base of natural logarithms (= 2.71828 . . .) (4B.2, 4C.2)
e_s	evenness, using diversity index d_s (5B.2.4)
e_x	life expectancy for age class x (4A.4)
F	Fahrenheit (appendix B.8)
F	test statistic for "analysis of variance" (1B.3.2)
F	frequency hypothesized or expected (1B.4, 1B.5, 4D.2.1)
$F(X)$	expected frequency of occurrence of X (4C.3.3)
FHD	foliage height diversity (2A.8)
FW	formula weight (2E.1.2ff)

f	frequency of occurrence (1B.4, 1B.5, 3.2, 3A.3, 3B.3, 4C.3.5, 4D.2.1)
f_i	frequency of species i (3A.3, 3B.3, 3C.3)
f_j	frequency of a species in successional stage j (5D.2.3)
$f(X)$	frequency of occurrence of X (4C.2, 4C.3)
fc	footcandle (2B.4.1)
fl	fluid (appendix B.4)
ft	foot (appendix B.1, B.6, B.7)
ft^2	square foot (appendix B.2)
GBP, GBP'	gross biomass productivity (unit 6, 6B.4)
GPP, GPP'	gross primary productivity (6B.5)
g	gram (appendix B.5, B.7)
g	gravitational constant ($= 981$ cm/sec^2) (2D.3.2)
gal	gallon (appendix B.4)
gr	grain (appendix B.5)
H	Brillouin's diversity index (5B.2.3)
H'	Shannon's diversity index (2A.8, 5B.2.3)
H_A	alternate hypothesis (1B.3)
H_0	null hypothesis (1B.3)
H_{max}	maximum diversity for the Brillouin diversity index (5B.2.4)
H_{max}'	maximum diversity for the Shannon diversity index (5B.2.4)
H_1', H_2'	Shannon's H' for community 1 or 2 (5C.2.5)
H_3'	Shannon's H' for two communities combined (5C.2.5)
H_4'	Maximum H' for species abundances from two communities (5C.2.5)
H_5'	Minimum H' for species abundances from two communities (5C.2.5)
$[H^+]$	hydrogen ion concentration (2D.6)
h, h'	height (2A.4, 2D.3.2, 6A.4.2)
ha	hectare (appendix B.2)
hr	hour (appendix B.6)
I	light intensity (2B.4.1)
I_o	light intensity at zero depth (2C.4.1, 2D.3.3)
I_{BC}	Bray-Curtis index of community similarity (5C.2.3)
I_{CM}	"Canberra metric" index of community similarity (5C.2.3)
I_d	light intensity at depth d (2B.4.1, 2D.3.3)
I_d	Morisita's index of dispersion (4C.3.4)
I_M	Morisita's index of community similarity (5C.2.4)
I_S	Stander's index of community similarity (5C.2.4)
IC	index of coverage (3B.3)
IC_i	index of coverage of species i (3B.3)
ID	index of population density (3.1, 3B.3, 3H.6)
ID_i	index of density for species i (3B.3)
INM	international nautical mile (appendix B.1)
IR	rating index (2F.3.3)
IR_i	rating index for habitat i (2F.3.3)
IV	importance value (3A.3, 3B.3, 3C.3)
IV_i	importance value of species i (3A.3, 3B.3, 3C.3)
IV_j	importance value of a species in successional stage j (5D.2.3)
i	individual, species, or sample number (3A.3)
in.	inch (appendix B.1)
in.^{2}	square inch (appendix B.2)
J	joule (appendix B.7)
J	evenness, using the Brillouin diversity index (5B.2.4)
J'	evenness, using the Shannon diversity index (5B.2.4)
JTU	Jackson turbidity unit (2D.3.3)

j	number of samples in which a species occurs (3A.3, 3B.3, 3C.3)
j	successional stage (5D.2.3)
j_i	number of samples in which species i occurs (3A.3, 3B.3, 3C.3)
K	carrying capacity (4B.2.2, 4D.2.1)
k	number of samples or subsamples taken (3A.3, 3B.3, 3C.3, 3E.4.5, 3H.2.1)
kcal	kilocalorie (appendix B.7)
kg	kilogram (appendix B.5, B.7)
kJ	kilojoule (appendix B.7)
km	kilometer (appendix B.1, B.6)
km^2	square kilometer (appendix B.2)
L	length of an organism (6A.4.1, 6A.4.2)
L	total length of a transect (3B.3, 3E.4.5, 3H.2)
L_i	length of transect i (3H.2.1)
L_x	number living belonging to age class x (4A.4)
l	liter (appendix B.4)
l	intercept length (3B.3)
l	Simpson's index of dominance (5B.2.2)
l_1, l_2	Simpson's l for community 1 or 2 (5C.2.4)
l_i	intercept length for species i (3B.3)
l_x	number alive at start of age interval x (4A.4, 4A.5)
lb	pound (appendix B.5, B.7)
ln	natural logarithm (logarithm to the base e) (4B.2, 4C)
log	common logarithm (logarithm to the base 10) (2A.8, 2D.3.3, 2D.6, 4B.2, 4C, 5B.2, 6A.4.1)
M	molarity (said of solutions) (2E)
M	number marked individuals in a population (3F.2)
M_i	number of individuals marked prior to day i (3G.5)
m	meter (appendix B.1, B.6, B.7)
m	individual number (3C.3)
$\overset{*}{m}$	Lloyd's index of mean crowding (4C.3.4)
m^2	square meter (appendix B.2)
m^3	cubic meter (appendix B.3)
mb	millibar (2B.4.5)
meq	milliequivalent (2E.1.1, 2E.5.4)
mg	milligram (appendix B.5)
mi	statute mile (appendix B.1, B.6)
mi^2	square mile (appendix B.2)
min	minute (appendix B.6)
ml	milliliter (appendix B.4)
mm	millimeter (appendix B.1)
mm^2	square millimeter (appendix B.2)
mm^3	cubic millimeter (appendix B.3)
mmho	millimho (2D.6, 2F.5)
mmole	millimole (6B.3.2)
mμ	millimicron (appendix B.1)
N	north
N	number of individuals in a population (3.1, 3D.4.5, 3F.2, 3G, 4B.1, 4B.2, 4D.2.1)
N	number of individuals in an entire collection (4C.3.4, 5B.2)
N	number of individuals in two communities (5C.2.5)
N	normal (said of solutions) (2E)
N_o	population size at time zero (4B.2, 4D.2)
N_1, N_2	number of individuals in community 1 or 2 (5B.2)
N_p	number of prey eaten by predator (4E.2)
N_s	normality of a standard solution (2E.1.3)
N_t	normality of titrant (2E.1.3)
N_t	population size at time t (4B.2)
NBP, NBP'	net biomass productivity (unit 6, 6B.4)

NPP, NPP'	net primary productivity (6B.5)	*r*	radius (6A.4.2)
n	chemical valence (2F.1.3ff)	*r*	a remainder, in computation (5B.2.4)
n	number of individuals in a sample (1B.2.1ff, 2D.5.2, 3D.4.5, 3E.2, 3H.2, 3H.6, 6A.1)	*r*	intrinsic rate of population increase (per individual); the biotic potential (4B.2, 4D.2.1)
n	number of sample plots or points; number of replicates (4C, 4D.2.1)	*r*	sample correlation coefficient (1B.6.6)
n_i	number of individuals in sample *i* (1B.3.1, 3G)	S	south
n_i	number of individuals in species *i* (3A.3, 3B.3, 3C.3, 5B.2)	S	salinity (2E.6)
nm	nanometer (appendix B.1)	S	number of species in two communities (5C.2)
oz	ounce (appendix B.4, B.5)	S'	number of equally-abundant species yielding a given *H'* (5B.2.3)
P	pressure; barometric pressure (2B.4.5, 2D.2, 2E.6)	SCI	sequential comparison index (2D.5.2)
P	predicted probability, or proportion (4C.2, 4C.3)	SCS	Soil Conservation Service (2A.5)
P(X)	predicted probability, or proportion, of *X* (4C.2, 4C.3)	SD	sample standard deviation (1B.2.2)
PA	phenolphthalein alkalinity (2E.2)	SE	sample standard error (i.e., standard deviation of the mean) (1B.2.3, 3F.2, 3G.4, 3H.2.1)
P_G, P_G'	gross productivity (6B.2.2, 6B.3.2, 6C.4)	SP	sum of the cross products of the deviations of *X* and *Y* from their means (1B.6.1)
PIE	probability of interspecific encounter (5B.2.2)	SS	"sum of squares" (sum of the squares of deviations from the mean) (1B.2.2, 4C.3.2)
P_N, P_N'	net productivity (6B.2.2, 6B.3.2)	SS_i	"sum of squares" for sample *i* (1B.3.1)
PQ	photosynthetic quotient (6B.4, 6B.5)	SS_X	sum of squares of *X* (1B.6.1)
PS	percentage similarity between two communities (5C.2.2)	*s*	number of species (5A.7.1, 5B.2)
p	water vapor pressure (2E.6)	*s*	sample standard deviation (1B.2.2)
p	proportion or probability (2A.9, 2C.8, 4C.2, 4C.3)	s^2	sample variance (1B.2.2, 4C.3.2)
p_i	proportion in the *i*th category; relative abundance of category *i* (2A.8, 5B.2)	s^2	variance associated with sample data surrounding a regression line (1B.6.3)
p(X)	observed probability, or proportion, of *X* (4C.2, 4C.3)	s_b	standard error of the regression coefficient (or slope) (1B.5.3)
*p*H	common logarithm of the reciprocal of the hydrogen ion concentration (2D.6, 2F.7)	s_1, s_2	number of species in community 1 or 2 (5C.2)
ppm	parts per million (2D.5.3, 2D.6, 2F.1.1ff)	s_p^2	pooled sample variance (1B.3.1)
pt	pint (appendix B.4)	s_r	standard error of the correlation coefficient (1B.6.6)
q_x	probability of dying during age interval *x* (4A.4)	$s_{\bar{x}}$	sample standard error (i.e., standard deviation of the mean) (1B.2.3)
qt	quart (appendix B.4)	$s_{\bar{x}_1 - \bar{x}_2}$	sample standard error of the difference between means 1 and 2 (1B.3.1)
R, R'	respiration rate (6B.2.2, 6B.3.2, 6B.5)	s_x	probability of surviving age interval *x* (4A.4)
R	a hydrometer reading (2C.8)	s_Y	standard error of a predicted *Y* (1B.6.4)
R	number of recaptured animals (3E.2)	sec	second (appendix B.6)
R_b	a hydrometer reading of a blank suspension (2C.8)	sq	square (appendix B.2)
R_o	Horn's index of community overlap (5C.2.5)	T	turnover time for biomass (unit 6)
R_1	sum of the ranks in sample 1 (1B.3.4)	T	transmittance of light (2D.3.3)
RB, RB'	respiration-consumed biomass (unit 6, 6B.4)	T	total predation time (4E.2)
RB	relative biomass (6A.1)	T_s	searching time in predation (4E.2)
RB_i	relative biomass of species *i* (6A.7)	T_h	handling time, per prey item (4E.2)
RC	relative coverage (3.4, 3A.3, 3C.3)	T_{h_1}, T_{h_2}	handling time for prey item 1 or 2 (4E.3)
RC_i	relative coverage of species *i* (3A.3, 3B.3)	T_x	time units left to live from age class *x* onward (4A.4)
RC_j	relative coverage of a species in successional stage *j* (5D.2.3)	TA	total alkalinity (2F.2)
RD	relative density (3.1, 3A.3, 3B.3, 3C.3)	TC	total coverage for all species *(TC = ΣC)* (3A.3)
RD_i	relative density of species *i* (3A.3, 3B.3)	TD	total density for all species *(TD = ΣD)* (3A.3, 3C.3)
RD_j	relative density of a species in successional stage *j* (5D.2.3)	*t*	time (2C.8, 4B.1, 4B.2, 4D.2.1, 6B.2.2, 6B.3.2)
RE, RE'	respiration-released energy (6B.5)	*t*	Student's *t* (1B.2.3, 1B.3.1, 1B.6, 3F.2, 3F.3, 3G.4, 3H.2.1, 4C.3.2, 4C.3.6)
Rf	relative frequency (3.2, 3A.3, 3B.3, 3C.3)	*U, U'*	Mann-Whitney test statistic (1B.3.4)
Rf_i	relative frequency of species *i* (3A.3, 3B.3, 3C.3)	U.S.	United States
Rf_j	relative frequency of a species in successional stage *j* (5D.2.3)	USDA	U.S. Department of Agriculture (2C.8)
RH	relative humidity (2C.4.5)	USPHA	U.S. Public Health Association
RH_c	relative humidity, corrected for barometric pressure (2B.4.5)	*u*	number of units of measurement (3C.3)
RI	relative index (3F.3.4)	*V*	biovolume (6A.4.2)
RI_i	relative index for habitat *i* (3F.3.4)	*V*	total volume (3E.4.5)
RIV_j	relative importance value of a species in successional stage *j* (5D.2.3)	V_b	volume of titrant used for a blank (2E.1.5)
RQ	respiratory quotient (6A.4)	V_o	original volume (2F)
		V_p	volume of a prepared sample (2E)
		V_s	volume of a standard solution (2E.1.3)
		V_t	volume of titrant for standard (2E)

v	volume sampled (3E.4.5)	μ_i	mean of population i (1B.3.1, 1B.3.2)
v	velocity (2D.3.2)	μm	micron; micrometer (appendix B.1)
W	west	μmho	micromho (2D.6, 2E.5)
W	weight of an individual or of a chemical substance (2C.8, 2F.1.2, 6A.1, 6A.4.1)	v	degrees of freedom (1B.2.2, 1B.4ff)
W_a	weight of ash (6A.3)	II	taking the product (5B.2.3)
W_b	weight of container with ash (6A.3)	π	pi, the ratio of circumference to diameter ($= 3.14159 \ldots$) (6A.4.2)
W_c	weight of empty container (6A.2)		
W_d	dry weight (2C.8, 2F.1.4, 6A.2)	ρ	population correlation coefficient (1B.6.6)
W_f	ash-free dry weight (6A.3)	Σ	summation (1B.2ff)
W_o	weight of container with dried sample (6A.2, 6A.3)	Σa	sum of the areas covered for all individuals in a species (3C.3)
W_t	weight at time t (2C.8)	ΣB	sum of the biomasses for all species (6A.7)
$\bar{W}$	mean weight of individuals in a sample (6A.1)	ΣC	sum of the coverages for all species ($\Sigma C = TC$) (3A.3, 3C.3)
w	width (3E.4.5)	ΣC	sum of the coverages for a species in all successional stages (5D.2.3)
X	variable X (1B.2.1ff, 3G.3, 4C.2, 4C.3)		
X_i	ith value of variable X (3G.3)	ΣD	sum of the densities of all species ($\Sigma D = TD$) (3A.3, 3C.3)
X_{max}	maximum value of X (2F.3.3)		
X_{min}	minimum value of X (2F.3.3)	ΣD	sum of the densities for a species in all successional stages (5D.2.3)
X_s	standard value of X (2F.3.4)		
$\bar{X}$	sample mean of X (1B.2.1, 1B.6.1, 4C.2, 4C.3.2)	Σf	sum of the frequencies of all species (3A.3, 3B.3, 3C.3)
$\bar{X}_i$	mean of sample i (1B.3.1)	Σf	sum of the frequencies for a species in all successional stages (5D.2.3)
x	age interval, age class, or cohort (4A.4)		
x_i	number of individuals in species i in community 1 (5C.2)	ΣIC	sum of the linear coverage indices for all species (3B.3)
Y	variable Y (1B.6.1, 3G.3)	ΣID	sum of the linear density indices for all species (3B.3)
$\bar{Y}$	sample mean of Y (1B.5.1)		
Y_i	ith value of variable Y (3G.3)	ΣIV	sum of the importance values for a species in all successional stages (5D.2.3)
yd	yard (appendix B.1)		
yd²	square yard (appendix B.2)	Σl	sum of the intercept lengths for all species (3B.3)
y_i	number of individuals in species i in community 2 (5C.2)	Σn	total number of individuals of all species (3A.3, 3B.3, 3C.3)
Z	standard score (3F.3.4)		
Z_i	standard score for habitat i (3F.3.4)	σ	population standard deviation (1B.2.2)
α	statistical probability of falsely rejecting a true H_0 (1B.2.3, 1B.3, 1B.6)	σ^2	population variance (1B.2.2, 4C.3.2)
		χ^2	the chi-squared test statistic (1B.4, 1B.5, 4C.3.2, 4C.3.3, 4C.3.4)
α_{ij}	competition coefficient: the effect of population j on the growth of population i (4D.2.1)	$+$	plus
		$-$	minus
β	population regression coefficient (1B.5.3)	$\times$	times; multiplied by
Δ	difference, change, or interval (4B.1, 6B.2.2, 6B.3.2)	$/$	divided by; per
		$\pm$	plus and minus
ΔN	change in population size (4B.1)	$\sqrt{}$	square root
ΔT	difference between wet and dry bulb temperatures (2B.4.5)	$=$	equals
		$\neq$	does not equal
Δt	time interval (4B.1, 6B.2.2, 6B.3.2)	$\cong$	approximately equals
Δ_s	Simpson's index of diversity (5B.2.2)	$<$	less than
δ_s	Simpson's index of diversity (5B.2.2)	$>$	greater than
θ	angle (2A.4)	X^2	the square of X
λ	Simpson's index of dominance (5B.2.2)	$X!$	X factorial (4C.2, 5B.2.3, 5B.2.4)
λ_i	encounter rate for prey item i (4E.3)	$\lvert X \rvert$	absolute value of X (4B.2.2, 4C.3.2)
μ	micron (appendix B.1)	∞	infinity
μ	population mean (1B.2.1, 1B.3.1, 1B.3.2, 4C.2, 4C.3)	$\%$	percent (parts per hundred)
		$^o/_{oo}$	parts per thousand (2F.1.1)

equivalents for units of measurement

These measurement equivalents may be used as conversion factors. For example, to convert from kilometers to meters, multiply by 1000. To convert from kilometers to feet, multiply by 3281. At least four significant figures are used in these tables; five may be found for many of these units in Pennycuick (1988).

1. units of length

1 nanometer* (nm) = 10 Angstroms (Å)	1 inch (in.) = 0.08333 ft
1 micron (μ or μm) = 1000 nm	1 foot (ft) = 12 in.
1 millimeter (mm) = 1000 μ	= 0.3333 yd
1 centimeter (cm) = 10 mm	1 yard (yd) = 36 in.
1 meter (m) = 1000 mm	= 3 ft
= 100 cm	1 fathom† = 6 ft
1 kilometer (km) = 1000 m	1 statute mile (mi) = 5280 ft
	= 1760 yd
	= 0.8690 INM
	1 international nautical mile (INM)‡ = 6076 ft
	= 1.151 mi

1 mm = 0.03937 in.	1 in. = 25.40 mm
1 cm = 0.3937 in.	= 2.540 cm
= 0.03281 ft	1 ft = 304.8 mm
1 m = 39.37 in.	= 30.48 cm
= 3.281 ft	= 0.3048 m
= 1.094 yd	1 yd = 914.4 mm
1 km = 3281 ft	= 91.44 cm
= 0.6214 statute mile (mi)	= 0.9144 m
= 0.5400 nautical mile	1 fathom† = 1.829 m
	1 statute mile (mi) = 1609 m
	= 1.609 km
	1 nautical mile‡ = 1852 m
	= 1.852 km

*Maritime measure. The British Admiralty nautical mile is about 1.17 m longer than INM (Pennycuick, 1988).
†Maritime measure.
‡Also called a millimicron (mμ).

2. units of area

1 square centimeter (cm²)
  = 100 square millimeters (mm²)
1 square meter (m²)
  = 1,000,000 mm²
  = 10,000 cm²
1 are (a.) = 1,000,000 cm²
        = 100 m²
1 hectare (ha) = 10,000 m²
            = 100 a.
1 square kilometer (km²)
  = 1,000,000 m²
  = 10,000 a.
  = 100 ha

1 square inch (sq in. or in²) = 0.006944 sq ft
1 square foot (sq ft or ft²)
  = 144 square inches (sq in. or in.²)
  = 0.1111 sq yd
1 square yard (sq yd or yd²)
  = 1296 sq in.
  = 9 sq ft
1 acre (A.) = 43,560 sq ft
       = 4840 sq yd
       = 0.0015625 sq mi
       = 0.001180 sq INM
1 square mile (sq mi or mi²)
  = 27,878,400 sq ft
  = 3,097,600 sq yd
  = 640 A.
1 square international nautical mile (sq INM)
  = 0.7551 sq INM
  = 36,918,900 sq ft
  = 4.102,100 sq yd
  = 1.324 sq mi

1 mm² = 0.001550 sq in.
1 cm² = 0.1550 sq in.
 1 m² = 1550 sq in.
     = 10.76 sq ft
     = 1.196 sq yd
 1 a. = 1076 sq ft
     = 119.6 sq yd
     = 0.02471 A.
 1 ha = 107,639 sq ft
     = 11,960 sq yd
     = 2.471 A.
     = 0.003861 sq mi
     = 0.0029155 sq INM
1 km² = 247.1 A.
     = 0.3861 sq mi
     = 0.29155 sq INM

1 sq in. = 645.2 mm²
       = 6.452 cm²
1 sq ft = 929.0 cm²
     = 0.09290 m²
1 sq yd = 8361 cm²
     = 0.8361 m²
  1 A. = 4047 m²
     = 40.47 a.
     = 0.4047 ha
     = 0.004047 km²
1 sq mi = 259.0 ha
     = 2.590 km²
1 sq INM = 343.0 ha
     = 3.430 km²

3. units of volume

1 cubic centimeter (cm³ or cc)
  = 1000 cubic millimeters (mm³)
1 cubic meter (m³)
  = 1,000,000 cm³

1 cubic inch (cu in. or in.³) = 0.005787 cu ft
1 cubic foot (cu ft or ft³)
  = 1728 (cu in.)
  = 0.03704 cu yd
1 cubic yard (cu yd or yd³)
  = 46,656 cu in.
  = 27 cu ft

---

1 cm³ = 0.06102 cu in.
1 m³ = 61,024 cu in.
  = 35.31 cu ft
  = 1.308 cu yd

1 cu in. = 16,387 mm³
  = 16.39 cm³
1 cu ft = 28,317 cm³
  = 0.02832 m³
1 cu yd = 764,555 cm³
  = 0.7646 m³

4. units of liquid capacity

1 liter* (1) = 1000 milliliters (ml)

1 U.S. fluid ounce† (fl oz)
  = 1.805 cubic inches (in.³)
1 U.S. pint‡ (pt)
  = 28.875 in.³
  = 16 fl oz
1 U.S. quart‡ (qt)
  = 57.75 in.³
  = 32 fl oz
  = 2 pt
1 U.S. gallon‡§(gal)
  = 231 in.³
  = 128 fl oz
  = 8 pt
  = 4 qt
  = 0.1337 ft³

---

1 ml = 0.061025 in.³
1 liter‖ = 61.02 in.³
  = 33.81 U.S. fl oz
  = 2.113 pt
  = 1.057 U.S. qt
  = 0.2642 U.S. gal
  = 0.03531 ft³

1 in.³ = 16.39 ml
  = 0.01639 liter‖
1 ft³ = 28.32 liters
1 U.S. fl oz = 29.57 ml
  = 0.02957 liter
1 U.S. pt = 473.2 ml
  = 0.4732 liter
1 U.S. qtr = 946.4 ml
  = 0.9464 liter
1 U.S. gal = 3785 ml
  = 3.785 liters

---

*1 liter of water at 4°C weighs 1 kilogram (or 2.205 pounds); 1 milliliter of water at 4°C weighs 1 gram. 1 liter = 1000 cubic centimeters.
†1 British fluid ounce = 0.9608 U.S. fluid ounce; 1 U.S. fluid ounce = 1.041 British fluid ounce.
‡British gallon, quart, and pint = 1.201 U.S. gallon, quart, and pint, respectively; U.S. gallon, quart, and pint = 0.8327 gallon, quart, and pint, respectivley.
§1 U.S. gallon of water at 15°C weighs 3.782 kilograms (or 8.337 pounds).
‖Do not abbreviate *liter* when it stands alone.

5. units of mass

1 gram (g) = 1000 milligrams (mg)
1 kilogram (kg) = 1000 g
1 metric ton = 1000 kg
           = 1 tonne (t)

1 avoirdupois ounce (oz) = 437.5 grains (gr)
                                = 0.0625 lb
1 pound (lb) = 16 oz
1 ton (or short ton)* = 2000 lb

---

1 g = 15.43 gr
   = 0.03527 oz
   = 0.0020 lb
1 kg = 35.28 oz
    = 2.205 lb
1 metric ton = 2205 lb
              = 1.102 tons

1 gr = 64.80 mg
    = 0.06480 g
1 oz = 28.35 g
    = 0.02835 kg
1 lb = 453.6 g
    = 0.4536 kg
1 ton = 907.2 kg
     = 0.9072 metric ton

---

*long ton = 1.120 short ton = 2240 lb; 1 short ton = 0.8929 long ton.

6. units of speed

1 cm/sec = 0.6000 m/min
       = 0.0360 km/hr
1 m/sec = 3.600 km/hr
       = 0.0600 km/min
1 m/min = 1.667 cm/sec
       = 0.0600 km/hr
1 km/min = 1667 cm/sec
       = 60.00 km/hr
1 km/hr = 27.78 cm/sec
       = 16.67 m/min
       = 0.2778 m/sec
       = 0.01667 km/min

1 ft/sec = 0.6818 mi/hr
       = 0.5925 knot
       = 0.01136 mi/min
1 ft/min = 0.01667 ft/sec
       = 0.01136 mi/min
       = 0.009875 knot
1 mi/min = 5280 ft/min
       = 88.0 ft/sec
       = 60.00 mi/hr
       = 52.14 knots
1 mi/hr = 5280 ft/hr
       = 88.00 ft/min
       = 1.467 ft/sec
       = 0.8690 knot
       = 0.01667 mi/min
1 knot = 6076 ft/hr
       = 101.3 ft/min
       = 1.688 ft/sec
       = 1.151 mi/hr

1 cm/sec = 1.9685 ft/min
       = 0.03281 ft/sec
       = 0.02237 mi/hr
1 m/sec = 196.85 ft/min
       = 3.281 ft/sec
       = 2.237 mi/hr
       = 1.944 knots
       = 0.03728 mi/min
1 m/min = 3.281 ft/min
       = 0.05468 ft/sec
       = 0.03728 mi/hr
       = 0.03240 knots
1 km/min = 3281 ft/min
       = 54.68 ft/sec
       = 37.28 mi/hr
       = 32.40 knots
       = 0.6214 mi/min
1 km/hr = 54.68 ft/min
       = 0.9113 ft/sec
       = 0.6214 mi/hr
       = 0.5400 knot
       = 0.01036 mi/min

1 ft/sec = 30.48 cm/sec
       = 18.29 m/min
       = 1.0973 km/hr
       = 0.01829 km/min
1 ft/min = 0.5080 cm/sec
       = 0.3048 m/min
       = 0.01829 km/hr
1 mi/min = 26.28 m/sec
       = 1.609 km/min
1 mi/hr = 44.70 cm/sec
       = 26.82 m/min
       = 1.609 km/hr
       = 0.4470 m/sec
1 knot = 51.44 cm/sec
       = 30.87 m/min
       = 1.852 km/hr
       = 0.5144 m/sec

7. units of energy

1 calorie* (cal)
   = 4.184 J
   = 0.4266 kg-m
   = 0.001162 watt-hour

1 kilocalorie† (kcal)
   = 4184 J
   = 1000 cal
   = 426.6 kg-m
   = 1.162 watt-hours
   = 0.001162 kilowatt-hour

1 joule (J) = 10^7 ergs
       = 1 watt-second
       = 0.2390 cal
       = 0.1020 kg-m

1 kilojoule (kJ)
   = 10^4 ergs
   = 1000 watt-seconds
   = 1 kilowatt-second
   = 239.0 cal
   = 0.2390 kcal
   = 102.0 kg-m

1 British thermal unit (BTU) = 778.2 ft-lb

---

1 cal = 3.086 ft-lb
     = 0.003968 BTU

1 kcal = 3086 ft-lb
       = 3.968 BTU
       = 0.001559 horsepower-hour

1 BTU = 1055 J
      = 252.0 cal
      = 107.6 kg-m
      = 0.2931 watt-hour
      = 0.2520 kcal

---

*also called *gram-calorie* (g-cal)
†also called *kilogram-calorie* (kg-cal)
Equivalents given are those for the calorie as defined by the U.S. National Bureau of Standards.

8. units of temperature

1 degree Celsius (°C) = 9/5 (= 1.8) degrees Fahrenheit (°F)
1 degree Fahrenheit (°F) = 5/9 (= 0.56) degree Celsius (°C)

Water freezes at 0°C = 32°F
Water boils at 100°C = 212°F

To convert from a Fahrenheit temperature (F) to a Celsius temperature (C):
$$C = (F - 32)(5/9).$$
To convert from a Celsius temperature (C) to a Fahrenheit temperature (F):
$$F = (9/5)C + 32.$$

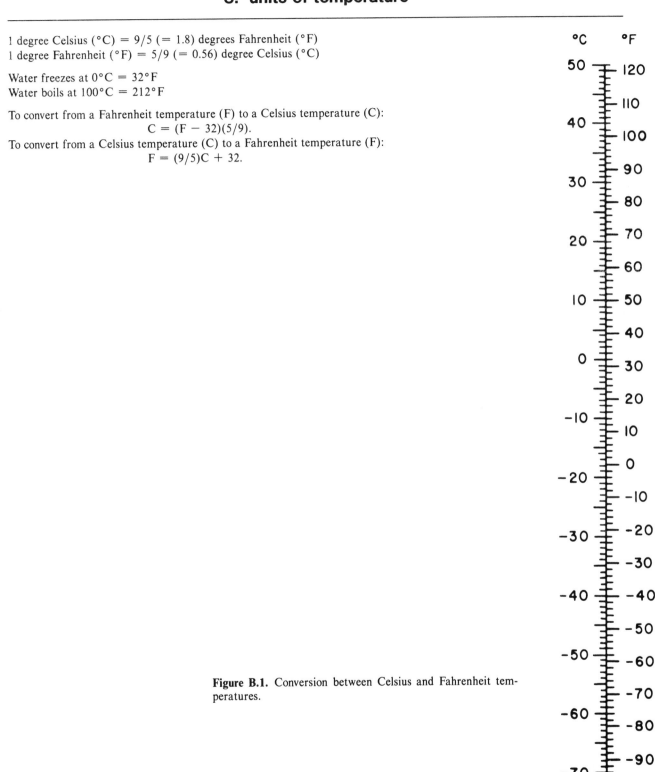

Figure B.1. Conversion between Celsius and Fahrenheit temperatures.

9. Reference

Pennycuick, C. J. 1988. Conversion factors. University of Chicago Press, Chicago.

appendix C

atomic weights of elements

Element	Symbol	Atomic number	Atomic weight
Actinium	Ac	89	227.0278
Aluminum	Al	13	26.9815
Antimony	Sb	51	121.75
Argon	Ar	18	39.948
Arsenic	As	33	74.9216
Astatine	At	85	210.9875
Barium	Ba	56	137.34
Beryllium	Be	4	9.01218
Bismuth	Bi	83	208.9806
Boron	B	5	10.81
Bromine	Br	35	79.904
Cadmium	Cd	48	112.40
Calcium	Ca	20	40.08
Carbon	C	6	12.011
Cerium	Ce	58	140.12
Cesium	Cs	55	132.9055
Chlorine	Cl	17	35.453
Chromium	Cr	24	51.996
Cobalt	Co	27	58.9332
Copper	Cu	29	63.546
Dysprosium	Dy	66	162.50
Erbium	Er	68	167.26
Europium	Eu	63	151.96
Fluorine	F	9	18.9984
Francium	Fr	87	223.0198
Gadolinium	Gd	64	157.25
Gallium	Ga	31	69.72
Germanium	Ge	32	72.59
Gold	Au	79	196.9665
Hafnium	Hf	72	178.49
Helium	He	2	4.00260
Holmium	Ho	67	164.9303
Hydrogen	H	1	1.0080
Indium	In	49	114.82
Iodine	I	53	126.9045
Iridium	Ir	77	192.22
Iron	Fe	26	55.847
Krypton	Kr	36	83.80
Lanthanum	La	57	138.9055
Lead	Pb	82	207.2
Lithium	Li	3	6.941
Lutetium	Lu	71	174.97
Magnesium	Mg	12	24.305
Manganese	Mn	25	54.9380
Mercury	Hg	80	200.59

Element	Symbol	Atomic number	Atomic weight
Molybdenum	Mo	42	95.94
Neodymium	Nd	60	144.24
Neon	Ne	10	20.179
Neptunium	Np	93	237.0482
Nickel	Ni	28	58.71
Niobium	Nb	41	92.9064
Nitrogen	N	7	14.0067
Osmium	Os	76	190.2
Oxygen	O	8	15.9994
Palladium	Pd	46	106.4
Phosphorus	P	15	30.9738
Platinum	Pt	78	195.09
Plutonium	Pu	94	242.0587
Polonium	Po	84	208.9825
Potassium	K	19	39.102
Praseodymium	Pr	59	140.9077
Prometheum	Pm	61	145
Protactinium	Pa	91	231.0359
Radium	Ra	88	226.0254
Radon	Rn	86	222.0175
Rhenium	Re	75	186.2
Rhodium	Rh	45	102.91055
Rubidium	Rb	37	85.4678
Ruthenium	Ru	44	101.07
Samarium	Sm	62	150.4
Scandium	Sc	21	44.9559
Selenium	Se	34	78.96
Silicon	Si	14	28.086
Silver	Ag	47	107.868
Sodium	Na	11	22.9898
Strontium	Sr	38	87.62
Sulfur	S	16	32.06
Tantalum	Ta	73	180.9479
Technicium	Tc	43	98.9062
Tellurium	Te	52	127.60
Terbium	Tb	65	158.9254
Thallium	Tl	81	204.37
Thorium	Th	90	232.0381
Thulium	Tm	69	168.9342
Tin	Sn	50	118.69
Titanium	Ti	22	47.90
Tungsten	W	74	183.85
Uranium	U	92	238.029
Vanadium	V	23	50.9414
Xenon	Xe	54	131.30
Ytterbium	Yb	70	173.04
Yttrium	Y	39	88.9059
Zinc	Zn	30	65.37
Zirconium	Zr	40	91.22

Data from Commission on Atomic Weights. Division of Inorganic Chemistry, International Union of Pure and Applied Chemistry. 1969. Atomic weights of the elements. Pure Appl. Chem. 21:91–108; and Gordon, A. J., and R. A. Ford. 1972. The Chemist's Companion: A Handbook of Practical Data, Techniques, and References. John Wiley & Sons, Inc., New York, 82–87.
These atomic weights, adopted in 1969, are based on an atomic weight of 12 for carbon-12.

common logarithms

Table D.1. *Common logarithms.*

	.00	.01	.02	.03	.04	.05	.06	.07	.08	.09
1.0	.0000	.0043	.0086	.0128	.0170	.0212	.0253	.0294	.0334	.0374
1.1	.0414	.0453	.0492	.0531	.0569	.0607	.0645	.0682	.0719	.0755
1.2	.0792	.0828	.0864	.0899	.0934	.0969	.1004	.1038	.1072	.1106
1.3	.1139	.1173	.1206	.1239	.1271	.1303	.1335	.1367	.1399	.1430
1.4	.1461	.1492	.1523	.1553	.1584	.1614	.1644	.1673	.1703	.1732
1.5	.1761	.1790	.1818	.1847	.1875	.1903	.1931	.1959	.1987	.2014
1.6	.2041	.2068	.2095	.2122	.2148	.2175	.2201	.2227	.2253	.2279
1.7	.2304	.2330	.2355	.2380	.2405	.2430	.2455	.2480	.2504	.2529
1.8	.2553	.2577	.2601	.2625	.2648	.2672	.2695	.2718	.2742	.2765
1.9	.2788	.2810	.2833	.2856	.2878	.2900	.2923	.2945	.2967	.2989
2.0	.3010	.3032	.3054	.3075	.3096	.3118	.3139	.3160	.3181	.3201
2.1	.3222	.3243	.3263	.3284	.3304	.3324	.3345	.3365	.3385	.3404
2.2	.3424	.3444	.3464	.3483	.3502	.3522	.3541	.3560	.3579	.3598
2.3	.3617	.3636	.3655	.3674	.3692	.3711	.3729	.3747	.3766	.3784
2.4	.3802	.3820	.3838	.3856	.3874	.3892	.3909	.3927	.3945	.3962
2.5	.3979	.3997	.4014	.4031	.4048	.4065	.4082	.4099	.4116	.4133
2.6	.4150	.4166	.4183	.4200	.4216	.4232	.4249	.4265	.4281	.4298
2.7	.4314	.4330	.4346	.4362	.4378	.4393	.4409	.4425	.4440	.4456
2.8	.4472	.4487	.4502	.4518	.4533	.4548	.4564	.4579	.4594	.4609
2.9	.4624	.4639	.4654	.4669	.4683	.4698	.4713	.4728	.4742	.4757
3.0	.4771	.4786	.4800	.4814	.4829	.4843	.4857	.4871	.4886	.4900
3.1	.4914	.4928	.4942	.4955	.4969	.4983	.4997	.5011	.5024	.5038
3.2	.5051	.5065	.5079	.5092	.5105	.5119	.5132	.5145	.5159	.5172
3.3	.5185	.5198	.5211	.5224	.5237	.5250	.5263	.5276	.5289	.5302
3.4	.5315	.5328	.5340	.5353	.5366	.5378	.5391	.5403	.5416	.5428
3.5	.5441	.5453	.5465	.5478	.5490	.5502	.5514	.5527	.5539	.5551
3.6	.5563	.5575	.5587	.5599	.5611	.5623	.5635	.5647	.5658	.5670
3.7	.5682	.5694	.5705	.5717	.5729	.5740	.5752	.5763	.5775	.5786
3.8	.5798	.5809	.5821	.5832	.5843	.5855	.5866	.5877	.5888	.5899
3.9	.5911	.5922	.5933	.5944	.5955	.5966	.5977	.5988	.5999	.6010
4.0	.6021	.6031	.6042	.6053	.6064	.6075	.6085	.6096	.6107	.6117
4.1	.6128	.6138	.6149	.6160	.6170	.6180	.6191	.6201	.6212	.6222
4.2	.6232	.6243	.6253	.6263	.6274	.6284	.6294	.6304	.6314	.6325
4.3	.6335	.6345	.6355	.6365	.6375	.6385	.6395	.6405	.6415	.6425
4.4	.6435	.6444	.6454	.6464	.6474	.6484	.6493	.6503	.6513	.6522
4.5	.6532	.6542	.6551	.6561	.6571	.6580	.6590	.6599	.6609	.6618
4.6	.6628	.6637	.6646	.6656	.6665	.6675	.6684	.6693	.6702	.6712
4.7	.6721	.6730	.6739	.6749	.6758	.6767	.6776	.6785	.6794	.6803
4.8	.6812	.6821	.6830	.6839	.6848	.6857	.6866	.6875	.6884	.6893
4.9	.6902	.6911	.6920	.6928	.6937	.6946	.6955	.6964	.6972	.6981

Table D.1. *Common logarithms (continued)*

	.00	.01	.02	.03	.04	.05	.06	.07	.08	.09
5.0	.6990	.6998	.7007	.7016	.7024	.7033	.7042	.7050	.7059	.7067
5.1	.7076	.7084	.7093	.7101	.7110	.7118	.7126	.7135	.7143	.7152
5.2	.7160	.7168	.7177	.7185	.7193	.7202	.7210	.7218	.7226	.7235
5.3	.7243	.7251	.7259	.7267	.7275	.7284	.7292	.7300	.7308	.7316
5.4	.7324	.7332	.7340	.7348	.7356	.7364	.7372	.7380	.7388	.7396
5.5	.7404	.7412	.7419	.7427	.7435	.7443	.7451	.7459	.7466	.7474
5.6	.7482	.7490	.7497	.7505	.7513	.7520	.7528	.7536	.7543	.7551
5.7	.7559	.7566	.7574	.7582	.7589	.7597	.7604	.7612	.7619	.7627
5.8	.7634	.7642	.7649	.7657	.7664	.7672	.7679	.7686	.7694	.7701
5.9	.7709	.7716	.7723	.7731	.7738	.7745	.7752	.7760	.7767	.7774
6.0	.7782	.7789	.7796	.7803	.7810	.7818	.7825	.7832	.7839	.7846
6.1	.7853	.7860	.7868	.7875	.7882	.7889	.7896	.7903	.7910	.7917
6.2	.7924	.7931	.7938	.7945	.7952	.7959	.7966	.7973	.7980	.7987
6.3	.7993	.8000	.8007	.8014	.8021	.8028	.8035	.8041	.8048	.8055
6.4	.8062	.8069	.8075	.8082	.8089	.8096	.8102	.8109	.8116	.8122
6.5	.8129	.8136	.8142	.8149	.8156	.8162	.8169	.8176	.8182	.8189
6.6	.8195	.8202	.8209	.8215	.8222	.8228	.8235	.8241	.8248	.8254
6.7	.8261	.8267	.8274	.8280	.8287	.8293	.8299	.8306	.8312	.8319
6.8	.8325	.8331	.8338	.8344	.8351	.8357	.8363	.8370	.8376	.8382
6.9	.8388	.8395	.8401	.8407	.8414	.8420	.8426	.8432	.8439	.8445
7.0	.8451	.8457	.8463	.8470	.8476	.8482	.8488	.8494	.8500	.8506
7.1	.8513	.8519	.8525	.8531	.8537	.8543	.8549	.8555	.8561	.8567
7.2	.8573	.8579	.8585	.8591	.8597	.8603	.8609	.8615	.8621	.8627
7.3	.8633	.8639	.8645	.8651	.8657	.8663	.8669	.8675	.8681	.8686
7.4	.8692	.8698	.8704	.8710	.8716	.8722	.8727	.8733	.8739	.8745
7.5	.8751	.8756	.8762	.8768	.8774	.8779	.8785	.8791	.8797	.8802
7.6	.8808	.8814	.8820	.8825	.8831	.8837	.8842	.8848	.8854	.8859
7.7	.8865	.8871	.8876	.8882	.8887	.8893	.8899	.8904	.8910	.8915
7.8	.8921	.8927	.8932	.8938	.8943	.8949	.8954	.8960	.8965	.8971
7.9	.8976	.8982	.8987	.8993	.8998	.9004	.9009	.9015	.9020	.9025
8.0	.9031	.9036	.9042	.9047	.9053	.9058	.9063	.9069	.9074	.9079
8.1	.9085	.9090	.9096	.9101	.9106	.9112	.9117	.9122	.9128	.9133
8.2	.9138	.9143	.9149	.9154	.9159	.9165	.9170	.9175	.9180	.9186
8.3	.9191	.9196	.9201	.9206	.9212	.9217	.9222	.9227	.9232	.9238
8.4	.9243	.9248	.9253	.9258	.9263	.9269	.9274	.9279	.9284	.9289
8.5	.9294	.9299	.9304	.9309	.9315	.9320	.9325	.9330	.9335	.9340
8.6	.9345	.9350	.9355	.9360	.9365	.9370	.9375	.9380	.9385	.9390
8.7	.9395	.9400	.9405	.9410	.9415	.9420	.9425	.9430	.9435	.9440
8.8	.9445	.9450	.9455	.9460	.9465	.9469	.9474	.9479	.9484	.9489
8.9	.9494	.9499	.9504	.9509	.9513	.9518	.9523	.9528	.9533	.9538
9.0	.9542	.9547	.9552	.9557	.9562	.9566	.9571	.9576	.9581	.9586
9.1	.9590	.9595	.9600	.9605	.9609	.9614	.9619	.9624	.9628	.9633
9.2	.9638	.9643	.9647	.9652	.9657	.9661	.9666	.9671	.9675	.9680
9.3	.9685	.9689	.9694	.9699	.9703	.9708	.9713	.9717	.9722	.9727
9.4	.9731	.9736	.9741	.9745	.9750	.9754	.9759	.9763	.9768	.9773
9.5	.9777	.9782	.9786	.9791	.9795	.9800	.9805	.9809	.9814	.9818
9.6	.9823	.9827	.9832	.9836	.9841	.9845	.9850	.9854	.9859	.9863
9.7	.9868	.9872	.9877	.9881	.9886	.9890	.9894	.9899	.9903	.9908
9.8	.9912	.9917	.9921	.9926	.9930	.9934	.9939	.9943	.9948	.9952
9.9	.9956	.9961	.9965	.9969	.9974	.9978	.9983	.9987	.9991	.9996

To determine the logarithm of a number larger or smaller than those in the left-hand margin of this table, multiply that number by a power of 10 such that the result does fall within the range in the table. This power of 10 is then added to the logarithm in the table. For example, the logarithm of 2.71 can be read from the table as 0.4330; the logarithm of 271 (i.e., 2.71×10^2) is $0.4330 + 2$, or 2.4330; and the logarithm of 0.271 (i.e., 2.71×10^{-1}) is $0.4330 - 1$, or -0.5670. The antilogarithm of a number is the value whose logarithm is that number. For example, the antilogarithm of 0.4330 is 2.71; the antilogarithm of 2.4330 is 271; and the antilogarithm of -0.5670 is 0.271.

Table D.2. *Common logarithms of proportions.*

	0.000	0.001	0.002	0.003	0.004	0.005	0.006	0.007	0.008	0.009
0.00		−3.000	−2.699	−2.523	−2.398	−2.301	−2.222	−2.155	−2.097	−2.046
0.01	−2.000	−1.959	−1.921	−1.886	−1.854	−1.824	−1.796	−1.770	−1.745	−1.721
0.02	−1.699	−1.678	−1.658	−1.658	−1.638	−1.602	−1.585	−1.569	−1.553	−1.538
0.03	−1.523	−1.509	−1.495	−1.481	−1.469	−1.456	−1.444	−1.432	−1.420	−1.409
0.04	−1.398	−1.387	−1.377	−1.367	−1.357	−1.347	−1.337	−1.328	−1.319	−1.310
0.05	−1.301	−1.292	−1.284	−1.276	−1.268	−1.260	−1.252	−1.244	−1.237	−1.229
0.06	−1.222	−1.215	−1.208	−1.201	−1.194	−1.187	−1.180	−1.174	−1.167	−1.161
0.07	−1.155	−1.149	−1.143	−1.137	−1.131	−1.125	−1.119	−1.114	−1.108	−1.102
0.08	−1.097	−1.092	−1.086	−1.081	−1.076	−1.071	−1.066	−1.060	−1.056	−1.051
0.09	−1.046	−1.041	−1.036	−1.032	−1.027	−1.022	−1.018	−1.013	−1.009	−1.004
0.10	−1.000	−0.996	−0.991	−0.987	−0.983	−0.979	−0.975	−0.971	−0.967	−0.963
0.11	−0.959	−0.955	−0.951	−0.947	−0.943	−0.939	−0.936	−0.932	−0.928	−0.924
0.12	−0.921	−0.917	−0.914	−0.910	−0.907	−0.903	−0.900	−0.896	−0.893	−0.889
0.13	−0.886	−0.883	−0.879	−0.876	−0.873	−0.870	−0.866	−0.863	−0.860	−0.857
0.14	−0.854	−0.851	−0.848	−0.845	−0.842	−0.839	−0.836	−0.833	−0.830	−0.827
0.15	−0.824	−0.821	−0.818	−0.815	−0.812	−0.810	−0.807	−0.804	−0.801	−0.799
0.16	−0.796	−0.793	−0.790	−0.788	−0.785	−0.783	−0.780	−0.777	−0.775	−0.772
0.17	−0.770	−0.767	−0.764	−0.762	−0.759	−0.757	−0.754	−0.752	−0.750	−0.747
0.18	−0.745	−0.742	−0.740	−0.738	−0.735	−0.733	−0.730	−0.728	−0.726	−0.724
0.19	−0.721	−0.719	−0.717	−0.714	−0.712	−0.710	−0.708	−0.706	−0.703	−0.701
0.20	−0.699	−0.697	−0.695	−0.693	−0.690	−0.688	−0.686	−0.684	−0.682	−0.680
0.21	−0.678	−0.676	−0.674	−0.672	−0.670	−0.668	−0.666	−0.664	−0.662	−0.660
0.22	−0.658	−0.656	−0.654	−0.652	−0.650	−0.648	−0.646	−0.644	−0.642	−0.640
0.23	−0.638	−0.636	−0.635	−0.633	−0.631	−0.629	−0.627	−0.625	−0.623	−0.622
0.24	−0.620	−0.618	−0.616	−0.614	−0.613	−0.611	−0.609	−0.607	−0.606	−0.604
0.25	−0.602	−0.600	−0.599	−0.597	−0.595	−0.593	−0.592	−0.590	−0.588	−0.587
0.26	−0.585	−0.583	−0.582	−0.580	−0.578	−0.577	−0.575	−0.573	−0.572	−0.570
0.27	−0.569	−0.567	−0.565	−0.564	−0.562	−0.561	−0.559	−0.558	−0.556	−0.554
0.28	−0.553	−0.551	−0.550	−0.548	−0.547	−0.545	−0.544	−0.542	−0.541	−0.539
0.29	−0.538	−0.536	−0.535	−0.533	−0.532	−0.530	−0.529	−0.527	−0.526	−0.524
0.30	−0.523	−0.521	−0.520	−0.519	−0.517	−0.516	−0.514	−0.513	−0.511	−0.510
0.31	−0.509	−0.507	−0.506	−0.504	−0.503	−0.502	−0.500	−0.499	−0.498	−0.496
0.32	−0.495	−0.493	−0.492	−0.491	−0.489	−0.488	−0.487	−0.485	−0.484	−0.483
0.33	−0.481	−0.480	−0.479	−0.478	−0.476	−0.475	−0.474	−0.472	−0.471	−0.470
0.34	−0.469	−0.467	−0.466	−0.465	−0.463	−0.462	−0.461	−0.460	−0.458	−0.457
0.35	−0.456	−0.455	−0.453	−0.452	−0.451	−0.450	−0.449	−0.447	−0.446	−0.445
0.36	−0.444	−0.442	−0.441	−0.440	−0.439	−0.438	−0.437	−0.435	−0.434	−0.433
0.37	−0.432	−0.431	−0.429	−0.428	−0.427	−0.426	−0.425	−0.424	−0.423	−0.421
0.38	−0.420	−0.419	−0.418	−0.417	−0.416	−0.415	−0.413	−0.412	−0.411	−0.410
0.39	−0.409	−0.408	−0.407	−0.406	−0.405	−0.403	−0.402	−0.401	−0.400	−0.399
0.40	−0.398	−0.397	−0.396	−0.395	−0.394	−0.393	−0.391	−0.390	−0.389	−0.388
0.41	−0.387	−0.386	−0.385	−0.384	−0.383	−0.382	−0.381	−0.380	−0.379	−0.378
0.42	−0.377	−0.376	−0.375	−0.374	−0.373	−0.372	−0.371	−0.370	−0.369	−0.368
0.43	−0.367	−0.366	−0.365	−0.364	−0.363	−0.362	−0.361	−0.360	−0.359	−0.358
0.44	−0.357	−0.356	−0.355	−0.354	−0.353	−0.352	−0.351	−0.350	−0.349	−0.348
0.45	−0.347	−0.346	−0.345	−0.344	−0.343	−0.342	−0.341	−0.340	−0.339	−0.338
0.46	−0.337	−0.336	−0.335	−0.334	−0.333	−0.333	−0.332	−0.331	−0.330	−0.329
0.47	−0.328	−0.327	−0.326	−0.325	−0.324	−0.323	−0.322	−0.321	−0.321	−0.320
0.48	−0.319	−0.318	−0.317	−0.316	−0.315	−0.314	−0.313	−0.312	−0.312	−0.311
0.49	−0.310	−0.309	−0.308	−0.307	−0.306	−0.305	−0.305	−0.304	−0.303	−0.302
0.50	−0.301	−0.300	−0.299	−0.298	−0.298	−0.297	−0.296	−0.295	−0.294	−0.293
0.51	−0.292	−0.292	−0.291	−0.290	−0.289	−0.288	−0.287	−0.287	−0.286	−0.285

Table D.2. *Common logarithms of proportions (continued)*

	0.000	0.001	0.002	0.003	0.004	0.005	0.006	0.007	0.008	0.009
0.52	−0.284	−0.283	−0.282	−0.281	−0.281	−0.280	−0.279	−0.278	−0.277	−0.277
0.53	−0.276	−0.275	−0.274	−0.273	−0.272	−0.272	−0.271	−0.270	−0.269	−0.268
0.54	−0.268	−0.267	−0.266	−0.265	−0.264	−0.264	−0.263	−0.262	−0.261	−0.260
0.55	−0.260	−0.259	−0.258	−0.257	−0.256	−0.256	−0.255	−0.254	−0.253	−0.253
0.56	−0.252	−0.251	−0.250	−0.249	−0.249	−0.248	−0.247	−0.246	−0.246	−0.245
0.57	−0.244	−0.243	−0.243	−0.242	−0.241	−0.240	−0.240	−0.239	−0.238	−0.237
0.58	−0.237	−0.236	−0.235	−0.234	−0.234	−0.233	−0.232	−0.231	−0.231	−0.230
0.59	−0.229	−0.228	−0.228	−0.227	−0.226	−0.225	−0.225	−0.224	−0.223	−0.223
0.60	−0.222	−0.221	−0.220	−0.220	−0.219	−0.218	−0.218	−0.217	−0.216	−0.215
0.61	−0.215	−0.214	−0.213	−0.213	−0.212	−0.211	−0.210	−0.210	−0.209	−0.208
0.62	−0.208	−0.207	−0.206	−0.206	−0.205	−0.204	−0.203	−0.203	−0.202	−0.201
0.63	−0.201	−0.200	−0.199	−0.199	−0.198	−0.197	−0.197	−0.196	−0.195	−0.194
0.64	−0.194	−0.193	−0.192	−0.192	−0.191	−0.190	−0.190	−0.189	−0.188	−0.188
0.65	−0.187	−0.186	−0.186	−0.185	−0.184	−0.184	−0.183	−0.182	−0.182	−0.181
0.66	−0.180	−0.180	−0.179	−0.178	−0.178	−0.177	−0.177	−0.176	−0.175	−0.175
0.67	−0.174	−0.173	−0.173	−0.172	−0.171	−0.171	−0.170	−0.169	−0.169	−0.168
0.68	−0.167	−0.167	−0.166	−0.166	−0.165	−0.164	−0.164	−0.163	−0.162	−0.162
0.69	−0.161	−0.161	−0.160	−0.159	−0.159	−0.158	−0.157	−0.157	−0.156	−0.156
0.70	−0.155	−0.154	−0.154	−0.153	−0.152	−0.152	−0.151	−0.151	−0.150	−0.149
0.71	−0.149	−0.148	−0.148	−0.147	−0.146	−0.146	−0.145	−0.144	−0.144	−0.143
0.72	−0.143	−0.142	−0.141	−0.141	−0.140	−0.140	−0.139	−0.138	−0.138	−0.137
0.73	−0.137	−0.136	−0.135	−0.135	−0.134	−0.134	−0.133	−0.133	−0.132	−0.131
0.74	−0.131	−0.130	−0.130	−0.129	−0.128	−0.128	−0.127	−0.127	−0.126	−0.126
0.75	−0.125	−0.124	−0.124	−0.123	−0.123	−0.122	−0.121	−0.121	−0.120	−0.120
0.76	−0.119	−0.119	−0.118	−0.117	−0.117	−0.116	−0.116	−0.115	−0.115	−0.114
0.77	−0.114	−0.113	−0.112	−0.112	−0.111	−0.111	−0.110	−0.110	−0.109	−0.108
0.78	−0.108	−0.107	−0.107	−0.106	−0.106	−0.105	−0.105	−0.104	−0.103	−0.103
0.79	−0.102	−0.102	−0.101	−0.101	−0.100	−0.100	−0.099	−0.099	−0.098	−0.097
0.80	−0.097	−0.096	−0.096	−0.095	−0.095	−0.094	−0.094	−0.093	−0.093	−0.092
0.81	−0.092	−0.091	−0.090	−0.090	−0.089	−0.089	−0.088	−0.088	−0.087	−0.087
0.82	−0.086	−0.086	−0.085	−0.085	−0.084	−0.084	−0.083	−0.082	−0.082	−0.081
0.83	−0.081	−0.080	−0.080	−0.079	−0.079	−0.078	−0.078	−0.077	−0.077	−0.076
0.84	−0.076	−0.075	−0.075	−0.074	−0.074	−0.073	−0.073	−0.072	−0.072	−0.071
0.85	−0.071	−0.070	−0.070	−0.069	−0.069	−0.068	−0.068	−0.067	−0.067	−0.066
0.86	−0.066	−0.065	−0.064	−0.064	−0.063	−0.063	−0.062	−0.062	−0.061	−0.061
0.87	−0.060	−0.060	−0.059	−0.059	−0.058	−0.058	−0.057	−0.057	−0.057	−0.056
0.88	−0.056	−0.055	−0.055	−0.054	−0.054	−0.053	−0.053	−0.052	−0.052	−0.051
0.89	−0.051	−0.050	−0.050	−0.049	−0.049	−0.048	−0.048	−0.047	−0.047	−0.046
0.90	−0.046	−0.045	−0.045	−0.044	−0.044	−0.043	−0.043	−0.042	−0.042	−0.041
0.91	−0.041	−0.040	−0.040	−0.040	−0.039	−0.039	−0.038	−0.038	−0.037	−0.037
0.92	−0.036	−0.036	−0.035	−0.035	−0.034	−0.034	−0.033	−0.033	−0.032	−0.032
0.93	−0.032	−0.031	−0.031	−0.030	−0.030	−0.029	−0.029	−0.028	−0.028	−0.027
0.94	−0.027	−0.026	−0.026	−0.025	−0.025	−0.025	−0.024	−0.024	−0.023	−0.023
0.95	−0.022	−0.022	−0.021	−0.021	−0.020	−0.020	−0.020	−0.019	−0.019	−0.018
0.96	−0.018	−0.017	−0.017	−0.016	−0.016	−0.015	−0.015	−0.015	−0.014	−0.014
0.97	−0.013	−0.013	−0.012	−0.012	−0.011	−0.011	−0.011	−0.010	−0.010	−0.009
0.98	−0.009	−0.008	−0.008	−0.007	−0.007	−0.007	−0.006	−0.006	−0.005	−0.005
0.99	−0.004	−0.004	−0.003	−0.003	−0.003	−0.002	−0.002	−0.001	−0.001	−0.000

microcomputer programming

1. Introduction

Recent years have witnessed a dramatic increase in the availability of microcomputers, some of which are referred to as personal computers, or desk-top computers, in schools, laboratories, businesses, and homes. These devices are not only immensely useful in the collection, storage, and processing of information, but they also are very easy to program to perform one's particular needs. This appendix will introduce you to the most common programming language for microcomputers; it is named BASIC (for Beginner's All-purpose Symbolic Instruction Code) and is the standard language on most microcomputers. Although there are many different versions of BASIC, the following discussion contains the most elementary and most needed parts of the language, which are to be found in nearly all versions. It will become apparent that very few different commands are necessary to perform even complicated mathematical manipulations. Instructions on the use of additional kinds of commands will be found in the manuals that accompany particular microcomputers and in the many books now available on the BASIC language.

As BASIC statements (i.e., instructions to the computer) are introduced below, examples will be given. All the major mathematical computations in this book will be presented in the form of BASIC computer programs so that this appendix will give the reader both an introduction to the art of programming and a set of workable programs to accompany the quantitative calculations called for in this book. The programs presented are not necessarily the best possible, but they work. To guard against errors that are likely in entering the program statements into a computer, it is advisable to test the program using data for which the computational results are known (as from examples from this book); it should be noted, however, that the results may differ slightly, depending upon how many significant figures are employed.

2. Anatomy of a computer program

A computer program is a series of statements, written in a programming language, that issue instructions to the computer. In order to issue these instructions one must know the vocabulary of the programming language used; we shall see how a very few phrases of such vocabulary can elicit a great deal of work from a microcomputer.

Arithmetic statements in BASIC are written very simply. If one wishes to define a variable, X, as being equal to 1.23, the following statement will accomplish the task:

LET X = 1.23

To assign another variable, Y, the sum of 4.56 and 7.89, we simply write

LET Y = 4.56 + 7.89

If we wish Z to be set equal to the sum of X and Y minus 2.5, then we can state

LET Z = X + Y − 2.5

Incidentally, the computer does not consider spaces between elements of these statements to be important, so the preceding could just as well have been written as

LETZ=X+Y−2.5

If we now wish Z to be set equal to its previous value divided by 4, we can write

LET Z = Z/4

Addition is indicated by "+", subtraction by "−", multiplication by "*", and division by "/". Also, exponentiation is a typical feature of the BASIC language and may be indicated by "**", by "↑", or by " ^ ", depending upon the computer. The computer used in this discussion employs " ^ ", so we could set another variable (call it S) equal to the square of Z by writing

LET S = Z ^ 2

or we could have obtained the fourth power of Z by writing the following:

LET S = Z ^ 4

In order to keep a set of instructions in a desired sequence, a BASIC program places a statement number (an integer) in front of each instruction. Therefore, the above statements could be numbered as follows:

```
1 LET X = 1.23
2 LET Y = 4.56 + 7.89
3 LET Z = X + Y − 2.5
4 LET Z = Z/4
5 LET S = Z ^ 2
6 END
```

where the END statement (which is not required by all computers) is placed at the end of the series of statements that comprise our program. It should be noted that the relative positions of statement numbers rather than the actual values of the numbers are what are important in defining the sequence of operations; that is, we could just as well have written the above program as

```
1 LET X = 1.23
2 LET Y = 4.56 + 7.89
5 LET Z = X + Y − 2.5
12 LET Z = Z/4
33 LET S = Z ^ 2
156 END
```

or as

```
10 LET X = 1.23
20 LET Y = 4.56 + 7.89
30 LET Z = X + Y − 2.5
40 LET Z = Z/4
50 LET S = Z ^ 2
60 END
```

Computer programmers typically like to number by tens, as shown immediately above, for this readily allows one to insert statements in between two existing program statements if modifications are desired after the program has been written. (Some computers have the capability of renumbering all program statements on command, so statement numbers of 1, 2, 5, 12, 33, 156 could be changed automatically to 1, 2, 3, 4, 5, 6, or to 10, 20, 30, 40, 50, 60, automatically.) Also, it is vitally important to distinguish the letter "O" from the numeral zero, so zero is often displayed and printed as "$\emptyset$."

Variables must be named according to the convention of the computer being used. It is typical to allow for a variable name to be composed of one or two characters, where the first must be a letter and the second may be a letter or a numeral. Therefore, the following are acceptable variable names: X, X2, X9, Y, Y2, YA, YY, ME, HE. (As some computers only allow a numeral as the second character in a variable name, we shall here name variables in that fashion.)

On most computers the command LET is optional when assigning a value to a variable. Therefore, the above computer program could be written as follows, which is the way programs will be presented below:

```
1Ø X = 1.23
2Ø Y = 4.56 + 7.89
3Ø Z = X + Y - 2.5
4Ø Z = Z/4
5Ø S = Z ^ 2
6Ø END
```

Lastly, most computers allow for "comments" to be written into programs; these are messages that are not acted on by the computer but serve only as little reminders to the human programmer. Such a comment may be indicated in one of several ways, depending on the computer; very commonly it is designated by REM (standing for "remark"), and the above program would be executed in exactly the same fashion if it appeared as follows:

```
5 REM FEBRUARY CLASS PROJECT
1Ø X = 1.23
2Ø Y = 4.56 + 7.89
29 REM Z IS MEASURED IN METERS
3Ø Z = X + Y - 2.5
4Ø Z = Z/4
5Ø S = Z ^ 2
6Ø END
```

Once a program is in the computer it is executed by entering the command RUN. It is also important to have the ability to save a program on a magnetic disk or tape (e.g., with a command such as SAVE) and to be able to cause a program to be brought into the computer from a disk or tape (e.g., with a command such as LOAD), so that a program does not have to be reentered by hand into the computer each time it is to be used. Procedures and commands for storing and retrieving programs and data using disks or tapes vary with the computer, so users' manuals should be consulted for details.

3. Input and output

The labors of a computer program must have provision for some result to be presented to the user. This is most often in the form of a display on the screen of the computer monitor or output on a printer attached to the computer, but disk or tape output is not uncommon. In BASIC, the word PRINT causes the items that follow it to appear on the screen: The current value of a variable is displayed if called for, and an item enclosed in quotation marks is displayed verbatim. For example, the statement

15 PRINT X

in the above program would result in

1.23

appearing on the screen, and the statement

25 PRINT X, Y

would cause this display:

1.23   12.45

The statement

15 PRINT "X"

would display

X

and this statement:

15 PRINT "THE VALUE OF X IS"; X

would result in this display:

THE VALUE OF X IS 1.23

Note that a comma between two items in a PRINT statement causes two pieces of output to be displayed widely separated, whereas a semicolon displays items close to each other. Also, notice that any characters enclosed within quotation marks in a PRINT statement will be displayed verbatim in the output. Printing a blank line is demonstrated in Section 4. There are many additional printing capabilities in BASIC, whereby, for example, one can specify the numbers of digits to be displayed or use tabulator-type settings to have output appear in a columnar format. These capabilities will not be discussed here.

The most common procedure for providing input to a BASIC program is by the INPUT statement. The statement

1Ø INPUT X

in a program causes the computer to display a question mark; the computer then waits for the user to insert a value into the computer keyboard and to then press the ENTER or RETURN key. This quantity is thereby entered into the program as the value of variable X. Many computers will allow material enclosed in quotation marks to precede the question mark, so, for example,

1Ø INPUT "ENTER A VALUE FOR X"; X

would result in the following appearing on the screen

ENTER A VALUE FOR X ?

The above input and output statements (INPUT and PRINT) will be demonstrated further in each program that follows.

4. Computing a mean

Here is simple BASIC program that allows the user to enter any number of data and then calculates and displays the mean of the data (in accordance with equation 1 in section 1B.2.1):

```
10 REM CALCULATING A MEAN
20 S=0
30 PRINT
40 INPUT "NUMBER OF DATA ";N
50 PRINT
60 PRINT "ENTER DATA ONE AT A TIME"
70 FOR I=1 TO N
80 INPUT X
90 S=S+X
100 NEXT I
110 M=S/N
120 PRINT
130 PRINT "THE MEAN =";M
140 END
```

This program first asks the user how many data are to be entered, after which we enter that many data one at a time. There is in this program what is known as a program *loop;* a loop is a portion of a program that is repeated. The BASIC statements from 70 through 100 comprise this loop, and this portion of the program obtains the sum of all *N* values of *X*. The loop begins with a FOR-TO statement which simply acts as a counter. In the above example we use a variable we call *I* to count how many times the loop is executed and we state that we wish to execute it a total of *"N"* times. The NEXT statement defines the end of the program loop. The computer screen finally looks like this for the data in section 1B.2:

```
RUN

NUMBER OF DATA ? 5

ENTER DATA ONE AT A TIME
? 10.1
? 11.4
? 11.7
? 12.1
? 13.3

THE MEAN = 11.72
```

Note that when a PRINT statement occurs with no indication of what to print (e.g., statements 30 and 50 above), then a blank line appears as output.

The IF statement in BASIC may be introduced here to demonstrate another way to write a program for the computation of the mean. Here, we do not require that the number of data be given to the computer program; instead, we inform the program that all the data have been entered by entering the number "9999." Statement 70 below says to the computer, "*If* the number 9999 has been entered through the keyboard, *then go to* program statement 110."

```
10 REM CALCULATING A MEAN
20 REM "X" IS A MEASUREMENT
30 REM "S" IS THE SUM OF THE X'S
40 W=0
50 N=0
60 INPUT "MEASUREMENT ";X
70 IF X=9999 THEN 110
80 S=S+X
90 N=N+1
100 GOTO 60
110 M=S/N
120 PRINT
130 PRINT "THE MEAN =";M
140 END
```

For the data in section 1B.2, the computer screen would look like this after running the above program:

```
RUN
MEASUREMENT ? 10.1
MEASUREMENT ? 11.4
MEASUREMENT ? 11.7
MEASUREMENT ? 12.1
MEASUREMENT ? 13.3
MEASUREMENT ? 9999

THE MEAN = 11.72
```

On some computers statement 70 would be written in the form

70 IF X=9999 GOTO 110

Note that the number "9999" is not considered to be a datum by the program. This number does not enter into any computations; it serves only as an indicator that the user has finished entering all the data. (Of course, if 9999 were a datum the program would not consider it as such so this indicator number should be chosen as one that we are certain will not appear among the data.)

Another approach to the computation of a mean will be shown here to introduce another important programming capability: that of defining arrays and subscripts. An array is a set of variables denoted by the same variable name, with each variable distinguished by a subscript. For example, the set of five data above could be referred to as $X_1 = 10.1$, $X_2 = 11.4$, $X_3 = 11.7$, $X_4 = 12.1$, and $X_5 = 13.3$. In BASIC, X_i is written as X(I), and the following program reads all five data into the array X, using a loop; then, using a second loop, it sums the five values of X_i in the array. A dimension (DIM) statement must appear in a BASIC program prior to the use of a subscripted variable (except that many computers assume an array of ten if a dimension statement is not provided). The dimension statement, DIMX (N), tells the computer to set aside an array of *n* data storage locations for the variable X. The DIM statement also sets the initial value of each member of the array to zero.

```
10 REM CALCULATING A MEAN
20 S=0
30 PRINT
40 INPUT "NUMBER OF DATA ";N
50 DIM X(N)
60 PRINT
70 PRINT "ENTER DATA ONE AT A TIME"
80 FOR I=1 TO N
90 INPUT X(I)
100 NEXT I
110 FOR I=1 TO N
120 S=S+X(I)
130 NEXT I
140 M=S/N
150 PRINT
160 PRINT "THE MEAN =";M
180 END
```

5. Computing the range

Computation of the range (see section 1B.2.2) will demonstrate another kind of IF statement, one that causes a value to be assigned to a variable (rather than causing a jump to another program statement, as is the case in the example immediately above).

We refer to the IF-THEN statement appearing in statements 120 and 130. On some computers "greater than" is denoted by "GT" instead of by ">," and "less than" is written as "LT" instead of as "<."

The strategy of this program is as follows: The first datum is read in, at which point the maximum and minimum is each set equal to that datum. Then, each other datum is read in one at a time. As a datum is entered into the program it is compared to the maximum; if it is larger than the maximum, then it becomes the new maximum (program statement 120). Then (in statement 130) each datum is compared to the previous minimum; if it is smaller than the minimum, then it replaces the minimum.

```
10  REM CALCULATING A RANGE
20  PRINT
30  INPUT "NUMBER OF DATA "; N
40  PRINT
50  PRINT "ENTER DATA ONE AT A TIME:"
60  INPUT X
70  REM M1=MINIMUM, M2=MAXIMUM
80  M1=X
90  M2=X
100 FOR I=2 TO N
110 INPUT X
120 IF X>M2 THEN M2=X
130 IF X<M1 THEN M1=X
140 NEXT I
150 PRINT
160 PRINT "MINIMUM VALUE = ";M1
170 PRINT
180 PRINT "MAXIMUM VALUE = ";M2
190 PRINT
200 PRINT "RANGE OF VALUES = ";M2-M1
210 PRINT
220 END
```

6. Computing the standard deviation

In calculating the standard deviation we need to take a square root (see section 1B.2.2), and we shall pause here to consider that the BASIC language typically has several built-in functions, the number and identity of which depend on the computer used. The most commonly needed of such functions are the following:

ABS(X)	to take the absolute value of X
EXP(X)	to take e (the base of natural logarithms) to the power X
INT(X)	to take the integer part of X
LOG(X)	to compute the natural logarithm of X [on some computers this is written as LN(X) or LOGE(X)]
SQR(X)	to take the square root of X [on some computers this is written as SQRT(X)]

Trigonometric functions (e.g., COS, SIN, TAN) are also commonly available. Some computers provide for obtaining common logarithms (i.e., logarithms in base 10), using a built-in function called LOG10, LGT, CLG, or CLOG, instead of LOG. The computer programs in this book employ natural logarithms (i.e., base e, via the LOG function); to convert them to common logarithms (i.e., base 10), see table 5B.3. The alternative designations given above for the square root and logarithmic functions are from the very informative book by Lien (1986) that describes variations among the BASIC languages used by over 250

different computers. (Lien's book also describes programming methods that can be used if a particular computer does not have the above functions built in.)

The program below has the user enter the number of data and then uses that number to set up a loop to obtain X and X^2 (see section 1B.2.2 for the equations needed). After all the data are read in, the program computes and displays the mean, the sum of squares, the degrees of freedom, the variance, the standard deviation, the coefficient of variation, and the standard error. Then the program asks the user to provide as input the value of Student's t with which to calculate confidence limits (see section 1B.2.4) and the confidence limits are computed and displayed.

```
10  REM MEAN AND VARIANCE
11  REM X = DATUM; W = SUM OF X
12  REM W2 = SUM OF S^2
13  REM N = NO. OF DATA; M = MEAN
14  REM Q = SUM OF SQUARES
15  REM S2 = VARIANCE
16  REM S = STANDARD DEVIATION
17  REM S3 = STANDARD ERROR
20  W=0
30  W2=0
40  PRINT
50  INPUT "NUMBER OF DATA ";N
60  PRINT
70  PRINT "ENTER DATA ONE AT A TIME"
80  FOR I=1 TO N
90  INPUT X
100 W=W+X
110 W2=W2+X^2
120 NEXT I
130 M=W/N
140 Q=W2-W^2/N
150 D=N-1
160 S2=Q/D
170 S=SQR(S2)
180 PRINT
190 PRINT "MEAN =";M
200 PRINT "SUM OF SQUARES =";Q
210 PRINT"DEGREES OF FREEDOM = DF =";D
220 PRINT "VARIANCE =";S2
230 PRINT "STANDARD DEVIATION =";S
240 PRINT "COEFFICIENT OF"
250 PRINT "  VARIATION =";S/M
260 S3=SQR(S2/N)
270 PRINT "STANDARD ERROR =";S3
280 PRINT
290 PRINT "ENTER STUDENT'S T FOR";
300 PRINT D; "DF"
310 INPUT T
320 PRINT
330 PRINT "CONFIDENCE LIMITS:"
340 PRINT "  LOWER =";M-T*S3
350 PRINT "  UPPER =";M+T*S3
360 PRINT
370 END
```

7. The two-sample t test

In the following program, two means are compared by the procedure of section 1B.3.1, using nothing more than some of the programming techniques already presented. The data for one of the groups are entered, and a mean and sum of squares are computed for that sample; then the second group of data is entered and a mean and sum of squares determined for it. Finally, the computation of t takes place as described in section 1B.3.1.

```
10  REM TWO-SAMPLE T TEST
20  W=0
30  W2=0
40  PRINT
50  INPUT "NO. OF DATA IN GROUP 1";N1
60  PRINT
70  PRINT "ENTER GROUP 1 DATA"
80  FOR I=1 TO N1
90  INPUT X
100 W=W+X
110 W2=W2+X^2
120 NEXT I
130 M1=W/N1
140 Q=W2-W^2/N1
150 D1=N1-1
160 PRINT
170 W=0
180 W2=0
190 PRINT
200 INPUT "NO. OF DATA IN GROUP 2";N2
210 PRINT
220 PRINT "ENTER GROUP 2 DATA"
230 FOR I=1 TO N2
240 INPUT X
250 W=W+X
260 W2=W2+X^2
270 NEXT I
280 M2=W/N2
290 D2=N2-1
300 S2=(Q+W2-W^2/N2)/(D1+D2)
310 S3=SQR(S2/N1+S2/N2)
320 PRINT
330 PRINT "MEAN FOR GROUP 1 =";M1
340 PRINT "MEAN FOR GROUP 2 =";M2
350 PRINT "DF FOR GROUP 1 =";D1
360 PRINT "DF FOR GROUP 2 =";D2
370 PRINT "POOLED VARIANCE =";S2
380 PRINT "STD. ERROR OF DIFFERENCE"
390 PRINT "  BETWEEN MEANS =";S3
400 PRINT
410 T=(M1-M2)/S3
420 PRINT "T =";T
430 PRINT "WITH DF =";D1+D2
440 PRINT
450 END
```

8. The Mann-Whitney test

In this statistical test (see section 1B.3.4), an array called X is established (via dimension statement 80) with $N = n_1 + n_2$ storage positions. The n_1 data of sample 1 are stored in the first n_1 positions in the array, and the n_2 data of sample 2 are stored in the remaining n_2 positions. The n_1 data of sample 1 are then assigned ranks. In the following program, this is done by the loop composed of statements 210 through 280, in which each of the data in sample one is examined one at a time. Within this loop is another loop, statements 240 through 270, by which each of the N data in both samples combined is compared to each datum in sample 1. Using this loop nested within a loop, the program counts how many data are less than or equal to each datum in sample 1, and that number is the rank of that datum.

In this program (statement 250) we see the use of "$<=$" to indicate "less than or equal to" (i.e., "$\leq$," which on some computers is indicated by "LE"). Similarly, "$>=$" in a BASIC

statement would indicate "greater than or equal to" (i.e., "$\geq$," which on some computers is indicated as "GE"), and "$<>$" means "not equal to" (i.e., $\neq$, which on some computers is written as "NE" or "$><$" or "#"). And statement 250 is another example of an IF-THEN statement: If the comparison following the word "IF" is true, the program executes the statement following the word "THEN" (and such a comparison may contain "=" or "<" or ">" or "<=" or ">=" or "<>").

Program statements 430 through 480 below compute the value of t needed if the size of either sample is greater than 20 (by equation 15 in section 1B.3.4). In statement 430 we see the use of the word "AND" when performing comparisons. This statement reads, "If both $n_1 \leq 20$ and $n_2 \leq 20$, then go to statement 490, which is the end of the program." If either sample size is greater than 20, then the program execution does not proceed to statement 490; rather, it continues with statement 440 onward.

```
10  REM MANN-WHITNEY TEST
20  PRINT
30  INPUT "NO. OF DATA IN SAMPLE 1 ";N1
40  PRINT
50  INPUT "NO. OF DATA IN SAMPLE 2 ";N2
60  PRINT
70  N=N1+N2
80  DIM X(N)
90  PRINT "ENTER DATA FOR SAMPLE 1"
100 FOR I = 1 TO N1
110 INPUT X(I)
120 NEXT I
130 PRINT
140 PRINT "ENTER DATA FOR SAMPLE 2"
150 FOR I = 1 TO N2
160 INPUT X(I+N1)
170 NEXT I
180 REM RANK ALL DATA IN SAMPLE 1
190 REM R(I) = RANK OF DATUM I
200 DIM R(N1)
210 FOR I = 1 TO N1
220 REM S = HOW MANY DATA ARE <= X(I)
230 S=0
240 FOR J = 1 TO N
250 IF X(J) <= X(I) THEN S = S+1
260 R(I) = S
270 NEXT J
280 NEXT I
290 W=0
300 REM SUM RANKS IN SAMPLE 1
310 FOR I = 1 TO N1
320 W = W+R(I)
330 NEXT I
340 U = N1*N2+N1*(N1+1)/2-W
350 U1 = N1*N2-U
360 PRINT
370 PRINT "MANN-WHITNEY U  = "; U
380 PRINT
390 PRINT "          AND U' = "; U1
400 PRINT
410 PRINT"N(1) = ";N1;"AND N(2) = ";N2
420 PRINT
430 IF N1<=20 AND N2<=20 THEN 490
440 A=ABS(U-N1*N2/2)
450 B=SQR(N1*N2*(N+1)/12)
460 T=A/B
470 PRINT "T = ";T;"WITH INFINITY DF"
480 PRINT
490 END
```

The above program for the Mann-Whitney test will only work properly if there are no tied data. If some data are tied (see the example in section 1B.3.4), then a modified procedure is needed to assign ranks correctly. This is embodied in the following program. The loop from statements 210 through 410, which encompasses two other loops, assigns ranks whether or not data are tied. (The reader is not expected to grasp exactly what is done here, but it would be instructive to decipher it.) Lines 290 and 360 of this program demonstrate the GOTO statement, which instructs the computer to proceed directly to the statement number indicated.

```
10  REM MANN-WHITNEY TEST
20  PRINT
30  INPUT "NO. OF DATA IN SAMPLE 1 ";N1
40  PRINT
50  INPUT "NO. OF DATA IN SAMPLE 2 ";N2
60  PRINT
70  N=N1+N2
80  DIM X(N)
90  PRINT "ENTER DATA FOR SAMPLE 1"
100 FOR I = 1 TO N1
110 INPUT X(I)
120 NEXT I
130 PRINT
140 PRINT "ENTER DATA FOR SAMPLE 2"
150 FOR I = 1 TO N2
160 INPUT X(I+N1)
170 NEXT I
180 REM RANK ALL DATA IN SAMPLE 1
190 REM R(I) = RANK OF DATUM I
200 DIM R(N)
210 FOR I = 1 TO N1
220 IF R(I) > 0 THEN 410
230 S=0
240 T=0
250 Y=X(I)
260 FOR J = 1 TO N
270 IF X(J) >= Y THEN 300
280 S = S+1
290 GOTO 330
300 IF X(J) > Y THEN 330
310 T=T+1
320 R(J)=-1
330 NEXT J
340 IF T > 1 THEN 370
350 R(I) = S+1
360 GOTO 410
370 R0 = S+T*(T+1)/(2*T)
380 FOR J = 1 TO N
390 IF R(J)=-1 THEN R(J)=R0
400 NEXT J
410 NEXT I
420 W=0
430 REM SUM RANKS IN SAMPLE 1
440 FOR I = 1 TO N1
450 W = W+R(I)
460 NEXT I
470 U = N1*N2+N1*(N1+1)/2-W
480 U1 = N1*N2-U
490 PRINT
500 PRINT "MANN-WHITNEY U  = "; U
510 PRINT
520 PRINT "              AND U' = "; U1
530 PRINT
540 PRINT"N(1) = ";N1;"AND N(2) = ";N2
```

```
550 PRINT
560 IF N1<=20 AND N2<=20 THEN 620
570 A=ABS(U-N1*N2/2)
580 B=SQR(N1*N2*(N+1)/12)
590 T=A/B
600 PRINT "T = ";T;"WITH INFINITY DF"
610 PRINT
620 END
```

9. Goodness of fit

The following BASIC program performs a goodness of fit test (see section 1B.4) for any number of categories. Note that program statement 390 checks to see if the number of categories is greater than 2, in which case the program computes chi-square according to equation 1B.16; if there are only two categories, then the program segment from statements 400 through 430 calculates chi-square using the Yates correction for continuity (by equation 1B.17).

In program statements 420 and 470 a quantity is squared (i.e., taken to the second power) by using "^{2}." The computer does not mistakenly take that quantity to the "$2/F0(I)$" power, because exponentiation is always performed before any other arithmetic operation (and multiplication and division are always performed before addition and subtraction), with clarification possible through the use of parentheses. Thus, if one really wanted to have a quantity raised to the $2/F0(I)$ power, the appropriate BASIC command would be "$^{} (2/F0(I))$," for expressions within parentheses are always operated on first.

```
10  REM CHI-SQUARE GOODNESS OF FIT
20  PRINT
30  INPUT "HOW MANY CATEGORIES "; K
40  PRINT
50  PRINT "ENTER RELATIVE FREQUENCY"
60  PRINT   "HYPOTHESIZED FOR EACH"
70  PRINT   "CATEGORY"
80  DIM F0(K)
90  REM M0 = SUM OF RELATIVE FREQS
100 M0=0
110 FOR I=1 TO K
120 INPUT F0(I)
130 M0=M0+F0(I)
140 NEXT I
150 REM CHANGE REL FREQS TO PROP'NS
160 FOR I=1 TO K
170 F0(I)=F0(I)/M0
180 NEXT I
190 PRINT "ENTER OBSERVED FREQUENCY"
200 PRINT   "IN EACH CATEGORY"
210 REM M=SUM OF OBSERVED FREQS
220 M=0
230 DIM F(K)
240 FOR I=1 TO K
250 INPUT F(I)
260 M=M+F(I)
270 NEXT I
280 REM CALCULATE CHI-SQUARE
290 REM CHANGE F0 TO EXPECTED FREQS
300 FOR I=1 TO K
310 F0(I)=F0(I)*M
320 NEXT I
330 REM CALCULATE CHI-SQUARE
340 D=K-1
350 PRINT
360 PRINT "N = "; M
```

```
370 PRINT
380 X2=0
390 IF K>2 THEN 460
400 FOR I=1 TO K
420 X2=X2+(ABS(F(I)-F0(I))-.5)^2/F0(I)
430 NEXT I
440 PRINT "WITH YATES CORRECTION,"
450 GOTO 490
460 FOR I=1 TO K
470 X2=X2+(F(I)-F0(I))^2/F0(I)
480 NEXT I
490 PRINT "  CHI-SQUARE = ";X2
500 PRINT "  WITH DF = ";D
510 PRINT
520 END
```

10. Contingency tables

Chi-square analysis of contingency tables (see section 1B.5) may be performed with the following program, for which the table may have any number of rows and columns. This program employs two two-dimensional arrays, which are defined by the BASIC declarations DIM F(R,C) and DIM F0(R,C) found in program statements 60 and 360; here, F is the observed frequency, F0 is the expected frequency, R is the number of rows, and C is the number of columns. Also note that a dimension statement may define the size of more than one array at a time by separating the arrays with commas (as does statement 60 in this example). If the contingency table has only two rows and two columns, then this program automatically applies the Yates correction for continuity (using equation 6.8 of Zar, 1984: 62).

```
10 REM CHI-SQUARE CONTINGENCY TABLES
20 PRINT
30 INPUT "HOW MANY ROWS ";R
40 PRINT
50 INPUT "HOW MANY COLUMNS ";C
60 DIM F(R,C), R9(R), C9(C)
80 REM N = TOTAL FREQ
90 N=0
100 FOR I = 1 TO R
110 PRINT
120 PRINT "ENTER, ONE AT A TIME, THE"
130 PRINT "  OBSERVED FREQS IN ROW ";I
140 REM R9 IS ARRAY OF ROW TOTALS
150 FOR J = 1 TO C
160 REM C9 IS ARRAY OF COLUMN TOTALS
170 INPUT F(I,J)
180 C9(J)=C9(J)+F(I,J)
190 R9(I)=R9(I)+F(I,J)
200 NEXT J
210 N=N+R9(I)
220 NEXT I
230 PRINT
240 PRINT "N = "; N
250 PRINT
260 IF R>2 OR C>2 THEN 350
270 REM CALCULATE CHI-SQUARE
280 A=F(1,1)*F(2,2)-F(1,2)*F(2,1)
290 A=ABS(A)-N/2
300 X2=N*A*A/R9(1)/R9(2)/C9(1)/C9(2)
310 D=1
320 PRINT
330 PRINT "WITH YATES CORRECTION,"
340 GOTO 500
```

```
350 REM CALCULATE EXPECTED FREQS
360 DIM F0(R,C)
370 FOR I = 1 TO R
380 FOR J = 1 TO C
390 F0(I,J)=R9(I)*C9(J)/N
400 NEXT J
410 NEXT I
420 REM CALCULATE CHI-SQUARE
430 X2=0
440 FOR I = 1 TO R
450 FOR J = 1 TO C
460 X2=X2+(F(I,J)-F0(I,J))^2/F0(I,J)
470 D=(R-1)*(C-1)
480 NEXT J
490 NEXT I
500 PRINT "  CHI-SQUARE ";X2
510 PRINT "  WITH DF = "; D
520 PRINT
530 END
```

11. Regression and correlation

The regression and correlation computations of sections 1B.6.1, 1B.6.3, and 1B.6.6 are performed by the following BASIC program. The data are entered into the program in a pairwise fashion; that is, a value of X is entered, followed by the Y associated with it, followed by another X, and so on.

```
10 REM REGRESSION AND CORRELATION
20 PRINT
30 PRINT "NUMBERS OF PAIRS OF DATA"
40 INPUT "  (NO. OF DATA POINTS)";N
50 PRINT
60 REM X5=SUM OF X'S, Y5=SUM OF Y'S
70 REM X6=SUM OF X SQ, Y6=SUM OF Y SQ
80 REM S=SUM OF X*Y
90 X5=0
100 Y5=0
110 X6=0
120 Y6=0
130 P=0
140 FOR I = 1 TO N
150 INPUT "ENTER AN X"; X
160 X5=X5+X
170 X6=X6+X^2
180 INPUT "ENTER THE ASSOCIATED Y ";Y
190 Y5=Y5+Y
200 Y6=Y6+Y^2
210 P=P+X*Y
220 NEXT I
230 REM X3=SS(X), Y3=SS(Y), P3=SP
240 X3=X6-X5^2/N
250 Y3=Y6-Y5^2/N
260 P3=P-X5*Y5/N
270 REM M1=MEAN OF X, M2=MEAN OF Y
280 M1=X5/N
290 M2=Y5/N
300 B=P3/X3
310 A=M2-B*M1
320 D=N-2
330 S2=(Y3-P3^2/X3)/D
340 REM S3 = SE OF B
350 S3=SQR(S2/X3)
360 T=ABS(B)/S3
370 R=P3/SQR(X3*Y3)
```

```
380 PRINT
390 PRINT "N = ";N
400 PRINT
410 PRINT "A = ";A;"    B = ";B
420 PRINT
430 PRINT "SE OF B = ";S3
440 PRINT
450 PRINT "R = ";R
460 PRINT
470 PRINT "T = ";T;"    WITH DF =";D
480 PRINT
490 END
```

There are instances (e.g., section 6A.4.1) where one wishes to consider a regression using the logarithms of X and/or of Y. Slight modification of the above program will allow this: simply insert statement

155 X = LOG(X)

if logarithms of X are desired, and statement

185 Y = LOG(Y)

to employ logarithms of Y.

Once one has the regression statistics, a and b, this program may be used to obtain predictions and confidence intervals for predictions:

```
10 REM PREDICTING WITH REGRESSION
20 PRINT
30 INPUT "ENTER B "; B
40 PRINT
50 INPUT "ENTER A "; A
60 PRINT
70 INPUT "ENTER N "; N
80 PRINT
90 PRINT "ENTER STUDENT'S T"
100 PRINT "  FOR DF = "; N-2
110 INPUT T
120 PRINT
130 INPUT "ENTER SE OF B "; S9
140 PRINT
150 INPUT "ENTER SS OF X "; X3
160 PRINT
170 INPUT "ENTER MEAN OF X "; M1
180 PRINT
190 S=S9*SQR(X3)
200 PRINT "TO PREDICT Y FROM X,"
210 PRINT "  ENTER A 1;"
220 PRINT "TO PREDICT X FROM Y,"
230 PRINT "  ENTER A 2;"
240 PRINT "TO TERMINATE, ENTER A ZERO"
250 INPUT E
260 PRINT
270 IF E = 0 THEN 620
280 IF E = 1 THEN 310
290 IF E = 2 THEN 450
300 GOTO 180
310 INPUT "ENTER X FOR PREDICTION ";X
320 PRINT
330 Y=A+B*X
340 PRINT "Y = "; Y
350 PRINT
360 S0=S*SQR(1/N+(X-M1)^2/X3)
370 PRINT "SE OF Y = "; S0
380 PRINT
```

```
390 D = T*S0
400 L1=Y-D
410 L2=Y+D
420 PRINT "CONFIDENCE LIMITS ARE"
430 PRINT L1; "AND ";L2
440 GOTO 180
450 INPUT "ENTER MEAN OF Y "; M2
460 PRINT
470 INPUT "ENTER Y FOR PREDICTION ";Y
480 PRINT
490 X=(Y-A)/B
500 PRINT "X = ";X
510 PRINT
520 C=B^2-T^2*S9^2
530 F1=(Y-M2)^2/X3
540 F2=C*(1+1/N)
550 D=T*S*SQR(F1+F2)/C
560 F=M1+B*(Y-M2)/C
570 L1=F-D
580 L2=F+D
590 PRINT "CONFIDENCE LIMITS ARE"
600 PRINT L1; "AND "; L2
610 GOTO 180
620 PRINT
630 END
```

12. Population size estimation

Estimation of population size using capture-recapture methods (section 3F) may be done using the following BASIC program:

```
10 REM CAPTURE-RECAPTURE ESTIMATION
20 PRINT
30 INPUT "ENTER NUMBER MARKED ";M
40 PRINT
50 INPUT "ENTER SIZE OF 2ND SAMPLE ";N
60 PRINT
70 INPUT "ENTER NUMBER RECAPTURED ";R
80 PRINT
90 PRINT "POPULATION SIZE ESTIMATES:"
100 PRINT
110 P=M*N/R
120 PRINT "LINCOLN-PETERSON EST. = ";P
130 P=M*(N+1)/(R+1)
140 PRINT " BAILY MODIFIED EST. = ";P
150 A=M^2*(N+1)*(N-R)
160 B=(R+1)^2*(R+2)
170 S=SQR(A/B)
180 PRINT "  STANDARD ERROR = ";S
190 C=1.96*S
200 PRINT "  95% CONFIDENCE LIMITS:"
210 PRINT P-C; "AND "; P+C
220 PRINT
230 END
```

Population size may be estimated by the regression method of section 3G.3 by employing the regression computer program of section 11 followed by the prediction program for estimating X from Y. The Moran-Zippin method of estimating population size (section 3G.4) is handled by the following computer program:

```
10  REM ZIPPIN'S POPULATION ESTIMATION
20 PRINT
30 PRINT "ENTER NUMBER CAUGHT:"
40 INPUT "  IN 1ST SAMPLING ";N1
50 INPUT "  IN 2ND SAMPLING ";N2
```

```
60 PRINT
70 N=N1^2/(N1-N2)
80 S=N1*N2*SQR(N1+N2)/(N1-N2)^2
90 D=1.96*S
100 PRINT "POPULATION ESTIMATE = ";N
110 PRINT
120 PRINT "STANDARD ERROR = ";S
130 PRINT
140 PRINT "95% CONFIDENCE LIMITS:"
150 PRINT N-D; "AND "; N+D
160 PRINT
170 END
```

The calculation of animal density from line transect data is discussed in section 3H.2 and is achieved by this program:

```
10 REM DENSITY BY LINE TRANSECT
20 PRINT
30 PRINT "LENGTH OF TRANSECT"
40 INPUT "   (IN METERS) ";L
50 PRINT
60 INPUT "HOW MANY SIGHTINGS ";N
70 PRINT
80 S=0
90 S0=0
100 PRINT "ENTER, ONE AT A TIME,"
110 PRINT "  THE SIGHTING DISTANCES"
120 PRINT "   (IN METERS):"
130 FOR I=1 TO N
140 INPUT D
150 S=S+D
160 S0=S0+1/D
170 NEXT I
180 P1=10000*N^2/(2*L*S)
190 P2=10000*S0/(2*L)
200 PRINT
210 PRINT "KING ESTIMATE = ";P1
220 PRINT
230 PRINT "HAYNE ESTIMATE = ";P2
240 PRINT
250 PRINT "THE ABOVE ARE NO./HECTARE"
260 PRINT
270 END
```

13. Population structure and growth

The following computer program uses the number of individuals alive in each of several age cohorts and provides life table information: the number alive at the start of each cohort, the number dying during each cohort, the probability of dying during each cohort, the probability of surviving each cohort, the animal-years left to live for each cohort, and the life expectancy for each cohort. The program assumes that data for age cohorts are available up through the cohort in which no animals are found. (See section 4A.4.)

```
10 REM LIFE TABLE
20 PRINT
30 INPUT "ENTER NO. OF COHORTS ";N
40 PRINT
50 M=N-1
60 DIM L(N),K(N),D(M),S(M),T(M)
70 PRINT "ENTER, ONE AT ATIME,"
80 PRINT "  NO. ALIVE IN EACH COHORT"
90 FOR X=1 TO N
100 INPUT L(X)
110 NEXT X
120 K(N)=0
130 FOR I=1 TO M
140 X=N-I
150 K(X)=2*L(X)-K(X+1)
160 NEXT I
170 PRINT"NO. ALIVE, START OF COHORT:"
180 FOR X=1 TO N
190 PRINT K(X)
200 NEXT X
210 INPUT "ENTER ZERO TO CONTINUE ";A
220 IF A<>0 THEN 600
230 PRINT "NO. DYING IN COHORT:"
240 FOR X=1 TO M
250 D(X)=K(X)-K(X+1)
260 PRINT D(X)
270 NEXT X
280 INPUT "ENTER ZERO TO CONTINUE ";A
290 IF A<>0 THEN 600
300 PRINT "PROB. OF DYING IN COHORT"
310 FOR X=1 TO M
320 Q=D(X)/K(X)
330 S(X)=1-Q
340 PRINT Q
350 NEXT X
360 INPUT "ENTER ZERO TO CONTINUE ";A
370 IF A<>0 THEN 600
380 PRINT "PROB. OF SURVIVING COHORT"
390 FOR X=1 TO M
400 PRINT S(X)
410 NEXT X
420 INPUT "ENTER ZERO TO CONTINUE ";A
430 IF A<>0 THEN 600
440 PRINT "ANIMAL-YEARS LEFT TO LIVE"
450 T(M)=L(M)
460 FOR I=2 TO M
470 X=N-I
480 T(X)=T(X+1)+L(X)
490 NEXT I
500 FOR X=1 TO M
510 PRINT T(X)
520 NEXT X
530 INPUT "ENTER ZERO TO CONTINUE ";A
540 IF A<>0 THEN 600
550 PRINT "LIFE EXPECTANCY BY COHORT"
560 FOR X=1 TO M
570 E=T(X)/K(X)
580 PRINT E
590 NEXT X
600 END
```

For a population growing exponentially (see section 4B.2), the population size at any time may be calculated by the following program, and it is a program like this one that was linked to a computer plotter to produce figure 4B.1. Statement 90 in this program employs the exponentiation function, which is in the form EXP(X) to indicate the operation e^x. See Lien (1986:123–124) if your computer does not have this built-in function, or simply use "2.71828 ^ X."

```
10 REM EXPONENTIAL POPULATION GROWTH
15 PRINT
20 INPUT "ENTER NO. AT TIME ZERO ";N0
25 PRINT
30 PRINT "ENTER INTRINSIC RATE OF"
```

```
40 INPUT "   INCREASE ";R
45 PRINT
50 PRINT "ENTER ZERO TO TERMINATE"
60 PRINT
70 INPUT "ENTER TIME, T "; T
80 IF T=0 THEN 160
90 N=N0*EXP(R*T)
100 PRINT "N = ";N
140 PRINT
150 GOTO 50
160 END
```

Population size during logistic growth (section 4B.2.2) may be computed with the following program, which is the type of program used to produce figure 4B.3.

```
10 REM LOGISTIC POPULATION GROWTH
20 PRINT
30 INPUT "ENTER NO. AT TIME ZERO ";N0
40 PRINT
50 PRINT "ENTER INTRINSIC RATE OF"
60 INPUT "   INCREASE ";R
70 PRINT
80 INPUT "ENTER CARRYING CAPACITY ";K
90 A=LOG((K-N0)/N0)
100 PRINT
110 PRINT "ENTER ZERO TO TERMINATE"
120 INPUT "ENTER TIME, T ";T
130 IF T=0 THEN 170
140 N=K/(1+EXP(A-R*T))
150 PRINT "N = ";N
160 GOTO 100
170 END
```

14. Spatial distribution

Poisson probabilities (see section 4C.2) may be computed as follows. These probabilities may then be used for graphical analysis or as input into the goodness of fit program of section 9.

```
10 REM POISSON PROBABILITIES
20 PRINT
30 INPUT "ENTER THE MEAN ";M
40 PRINT
50 PRINT "ENTER -1 TO TERMINATE"
60 INPUT "ENTER X ";X
70 P=EXP(-M)
80 IF X=-1 THEN 170
90 IF X=0 THEN 150
100 P=P*M^X
110 IF X=1 THEN 150
120 FOR I=2 TO X
130 P=P/I
140 NEXT I
150 PRINT "PROBABILITY ";P
160 GOTO 40
170 END
```

The following program provides for the assessment of spatial distribution by most of the methods of section 4C. Included are the variance-to-mean ratio and its t test, the chi-square test for dispersion, Morisita's index of dispersion and its chi-square, the index of mean crowding, and the plotless methods of Holgate, Hopkins, and Johnson and Zimmer.

```
10 REM ASSESSING SPATIAL DISTRIBUTION
20 PRINT
30 PRINT "IF DATA ARE FOR PLOTS,"
40 PRINT "   ENTER 1;"
50 PRINT "IF DATA ARE FOR PLOTLESS"
60 PRINT "   SAMPLING, ENTER ZERO"
70 INPUT Z
80 IF Z=0 THEN 490
90 IF Z<>1 THEN 30
100 N=0
110 W=0
120 W2=0
130 PRINT
140 PRINT "ENTER DATA FOR ONE PLOT"
150 PRINT "   AT A TIME:"
160 PRINT
170 PRINT "(ENTER -1 TO TERMINATE)"
180 PRINT
190 INPUT "ENTER NUMBER IN A PLOT ";X
200 IF X=-1 THEN 270
210 PRINT "ENTER HOW MANY PLOTS HAVE"
220 INPUT "   THAT NUMBER ";F
230 N=N+F
240 W=W+F*X
250 W2=W2+F*X^2
260 GOTO 160
270 M=W/N
280 Q=W2-W^2/N
290 D=N-1
300 S2=Q/D
310 R=S2/M
320 T=ABS(R-1)/SQR(2/(N-1))
330 PRINT
340 PRINT "N = ";N
350 PRINT "MEAN = ";M
360 PRINT "SUM OF SQUARES = ";Q
370 PRINT "DEGREES OF FREEDOM = ";D
380 PRINT "VARIANCE = ";S2
390 PRINT "VARIANCE TO MEAN RATIO = ";R
400 PRINT "T FOR THIS RATIO = ";T
410 PRINT "CHI-SQUARE = ";Q/M
420 I0=N*(W2-W)/W/(W-1)
430 PRINT "MORISITA'S INDEX = ";I0
440 X2=N*W2/W-W
450 PRINT "CHI-SQUARE = ";X2
460 M0=W2/W-1
470 PRINT "MEAN CROWDING INDEX = ";M0
480 GOTO 1100
490 PRINT
500 PRINT"ENTER THE NUMBER OF POINTS,"
510 INPUT" N: "; N
520 PRINT
530 PRINT "ENTER 1 FOR HOLGATE METHOD"
540 PRINT "ENTER 2 FOR HOPKINS METHOD"
550 PRINT "ENTER 3 FOR JOHNSON AND"
560 PRINT "   AND ZIMMER METHOD"
570 INPUT Z
580 PRINT
590 IF Z = 3 THEN 940
600 IF Z=2 THEN 740
```

```
610 IF Z<>1 THEN 530
620 S=0
630 FOR I=1 TO N
640 INPUT "ENTER A D ";D
650 INPUT"ENTER THE ASSOCIATED D' ";D0
660 S=S+D^2/D0^2
670 NEXT I
680 A=S/N-.5
690 T=ABS(A)/SQR(N/12)
700 PRINT
710 PRINT "HOLGATE'S INDEX OF"
720 PRINT "  AGGREGATION = ";A
730 GOTO 1070
740 PRINT "ENTER VALUES OF D,"
750 PRINT "  ONE AT A TIME:"
760 S=0
770 FOR I=1 TO N
780 INPUT D
790 S=S+D^2
800 NEXT I
810 PRINT
820 S0=0
830 PRINT "ENTER VALUES OF D',"
840 PRINT "  ONE AT A TIME:"
850 FOR I=1 TO N
860 INPUT D
870 S0=S0+D^2
880 NEXT I
890 A=S/S0-1
900 T=2*ABS((A+1)/(A+2)-.5)*SQR(2*N+1)
910 PRINT "HOPKINS' INDEX OF"
920 PRINT "  AGGREGATION =";A
930 GOTO 1070
940 PRINT "ENTER VALUES OF D,"
950 PRINT "  ONE AT A TIME:"
960 S2 = 0
970 S4 = 0
980 FOR I = 1 TO N
990 INPUT D
1000 S2 = S2 + D^2
1010 S4 = S4 + D^4
1020 NEXT I
1030 A = (N+1)*S4/S2^2-2
1040 T = A/SQR(4*(N-1)/(N+2)/(N+3))
1050 PRINT "JOHNSON & ZIMMER'S"
1060 PRINT "  INDEX OF AGGRESSION =";A
1070 PRINT
1080 PRINT "T =";T
1090 PRINT "  WITH INFINITY DF"
1100 PRINT
1110 END
```

15. Species diversity measures

The various measures of species diversity in section 5B may be
computed with the following program; and, optionally, measures
of evenness may be obtained as well.

```
10 REM SPECIES DIVERSITY INDICES
20 PRINT
30 N=0
40 N8=0
50 N7=0
60 N6=0
70 N5=0
80 INPUT "ENTER NO. OF SPECIES ";S
90 PRINT
100 PRINT "ENTER NO. IN EACH SPECIES,"
110 PRINT "  ONE SPECIES AT A TIME:"
120 FOR I=1 TO S
130 INPUT X
140 N=N+X
150 N8=N8+X*(X-1)
160 N7=N7+X^2
170 N6=N6+X*LOG(X)
180 IF X=1 THEN 220
190 FOR J=2 TO X
200 N5=N5+LOG(J)
210 NEXT J
220 NEXT I
230 PRINT
240 PRINT "NUMBER OF SPECIES = ";S
250 PRINT "NO. OF INDIVIDUALS = ";N
260 A=1/LOG(10)
270 B=1/LOG(2)
280 D=(S-1)/(A*LOG(N))
290 PRINT "MARGALEF DIVERSITY: ";D
300 D=S/SQR(N)
310 PRINT "MENHINICK DIVERSITY: ";D
320 D1=N8/N/(N-1)
330 PRINT "SIMPSON DOMINANCE: ";D1
340 D2=1-D1
350 PRINT "SIMPSON DIVERSITY: ";D2
360 D3=1/D1
370 PRINT "INVERSE, SIMPSON DIV.: ";D3
380 PRINT "IF NOT A RANDOM SAMPLE:"
390 D4=N7/N^2
400 PRINT "  SIMPSON DOMINANCE: ";D4
410 D5=1-D4
420 PRINT "  SIMPSON DIV.: ";D5
430 D6=1/D4
440 PRINT"  INVERSE OF DOMINANCE: ";D6
450 A=1/LOG(10)
460 B=1/LOG(2)
470 H1=(N*LOG(N)-N6)/N
480 PRINT "SHANNON DIVERSITY*:"
490 PRINT "  "; A*H1; H1; B*H1
500 PRINT "NO. OF EQUALLY ABUNDANT"
510 PRINT "  SPECIES:"; EXP(H1)
520 L=0
530 FOR I=2 TO N
540 L=L+LOG(I)
550 NEXT I
560 H2=(L-N5)/N
570 PRINT "BRILLOUIN DIVERSITY*:"
580 PRINT "  "; A*H2; H2; B*H2
590 PRINT "  *BASE 10, BASE E, BASE 2"
600 PRINT
610 PRINT "IF EVENNESS MEASURES ARE"
620 PRINT "  NOT DESIRED, ENTER ZERO;"
630 PRINT "IF EVENNESS MEASURES ARE"
640 PRINT "  DESIRED, ENTER 1:"
650 INPUT Z
660 IF Z=0 THEN 990
670 IF Z<>1 THEN 600
680 PRINT
690 PRINT "EACH EVENNESS MEASURE IS"
700 PRINT "  FOLLOWED BY MAX. DIV.:"
710 PRINT
720 M=(S-1)*N/S/(N-1)
730 E=D2/M
740 PRINT "SIMPSON DIVERSITY: ";E;M
750 M=S*(N-1)/(N-S)
760 E=D3/M
```

```
770 PRINT"INVERSE, SIMPSON DOM.: ";E;M
780 M=LOG(S)
790 E=H1/M
800 PRINT "SHANNON DIVERSITY: ";E;M
810 PRINT "  (MAX. IS IN BASE E)"
820 F=N/S
830 C=INT(F)
840 R=F-C
850 C0=0
860 FOR I=1 TO C
870 C0=C0+LOG(I)
880 NEXT I
890 C1=C0+LOG(C+1)
900 M=(L-(S-R)*C0-R*C1)/N
910 E=H2/M
920 PRINT "BRILLOUIN DIVERSITY: ";E;M
930 PRINT "  (MAX. IS IN BASE E)"
940 E=EXP(H1)
950 PRINT "SHELDON EVENNESS: "; E/S
960 D=(E-1)/(S-1)
970 PRINT "HEIP EVENNESS: "; D
980 PRINT "  (ABOVE TWO USE BASE E)"
990 PRINT
1000 END
```

16. Community similarity measures

The community similarity measures of section 5C may be calculated with the following program:

```
10 REM COMMUNITY SIMILARITY MEASURES
20 PRINT
30 INPUT "HOW MANY SPECIES ";S
40 PRINT
50 PRINT "FOR EACH SPECIES, ONE AT A"
60 PRINT "  TIME, ENTER THE NUMBER OF"
70 PRINT "  INDIVIDUALS IN COMMUNITY"
80 PRINT "  1 (X) AND THE NUMBER IN "
90 PRINT "  COMMUNITY 2 (Y):"
100 S1=0
110 S2=0
120 C=0
130 N1=0
140 N2=0
150 DIM X0(S),Y0(S)
160 D2=0
170 D3=0
180 D4 = 0
190 X2 = 0
200 Y2 = 0
210 X9=0
220 Y9=0
230 X2=0
240 P=0
250 G1=0
260 G2=0
270 G3=0
280 FOR I=1 TO S
290 INPUT "X ";X
300 X0(I)=X
310 N1=N1+X
320 X2 = X2 + X*X
330 IF X=0 THEN 370
340 S1=S1+1
350 X9=X9+X*(X-1)
360 G1=G1+X*LOG(X)
370 INPUT "Y ";Y
380 Y0(I)=Y
390 IF X>0 AND Y> 0 THEN C=C+1
400 N2=N2+Y
```

```
410 Y2 = Y2 + Y*Y
420 E1 = ABS(X-Y)
430 E2 = X+Y
440 IF Y=0 THEN 480
450 S2=S2+1
460 Y9=Y9+Y*(Y-1)
470 G2=G2+Y*LOG(Y)
480 D2=D2+E1
490 D3=D3+E2
500 D4 = D4 + E1/E2
510 P=P+X*Y
520 G3=G3+E2*LOG(E2)
530 NEXT I
540 N=N1+N2
550 F=0
560 FOR I=1 TO S
570 X=X0(I)/N1
580 Y=Y0(I)/N2
590 M=X
600 IF Y<X THEN M=Y
610 F=F+M
620 NEXT I
630 PRINT
640 K=C/S
650 PRINT "JACCARD COEFFICIENT: ";K
660 PRINT
670 K=2*C/(S1+S2)
680 PRINT "SORENSEN COEFFICIENT: ";K
690 PRINT
700 PRINT "PERCENT SIMILARITY: ";F
710 PRINT
720 K=1-D2/D3
730 PRINT"BRAY-CURTIS INDEX:";K
740 PRINT
750 K=1-D4/S
760 PRINT"CANBERRA METRIC:";K
770 PRINT
780 K=P/SQR(X2*Y2)
790 PRINT"STANDER'S INDEX:";K
800 PRINT
810 L1=X9/N1/(N1-1)
820 L2=Y9/N2/(N2-1)
830 K=2*P/((L1+L2)*N1*N2)
840 PRINT "MORISITA INDEX: ";K
850 G4=N*LOG(N)
860 H1=(N1*LOG(N1)-G1)/N1
870 H2=(N2*LOG(N2)-G2)/N2
880 H3=(G4-G3)/N
890 H4=(G4-G1-G2)/N
900 H5=(N1*H1+N2*H2)/N
910 K=(H4-H3)/(H4-H5)
920 PRINT
930 PRINT "HORN INDEX: ";K
940 PRINT
950 END
```

17. Selected references

Lien, D. A. 1986. The BASIC handbook. 3d ed. CompuSoft Publishing, San Diego, Calif.

Zar, J. H. 1984. Biostatistical analysis. 2d ed. Prentice-Hall, Englewood Cliffs, N.J.

Instruction books on the BASIC language are not listed here. There are approximately 250 such books published today in the U.S. Some are oriented to specific computers, some address certain programming needs (e.g., business, engineering, teaching, or games), and some are relatively general in their approach and coverage. Most of them assume no previous computer programming experience and would be instructive to the novice.

For an index to symbols (Latin letters, Greek letters, and mathematical symbols), see appendix A.

index